Table 7.2 Important Fourier transform pairs. The left-hand column includes only *energy signals* $f(t)$ (see Section 7.2), while the right-hand column includes *power signals* and distributions (covered in Chapter 9).

	$f(t) \leftrightarrow F(\omega)$		
1	$e^{-at}u(t) \leftrightarrow \frac{1}{a+j\omega},\ a>0$	14	$\delta(t) \leftrightarrow 1$
2	$e^{at}u(-t) \leftrightarrow \frac{1}{a-j\omega},\ a>0$	15	$1 \leftrightarrow 2\pi\delta(\omega)$
3	$e^{-a\lvert t\rvert} \leftrightarrow \frac{2a}{a^2+\omega^2},\ a>0$	16	$\delta(t-t_o) \leftrightarrow e^{-j\omega t_o}$
4	$\frac{a^2}{a^2+t^2} \leftrightarrow \pi a e^{-a\lvert\omega\rvert},\ a>0$	17	$e^{j\omega_o t} \leftrightarrow 2\pi\delta(\omega-\omega_o)$
5	$te^{-at}u(t) \leftrightarrow \frac{1}{(a+j\omega)^2},\ a>0$	18	$\cos(\omega_o t) \leftrightarrow \pi[\delta(\omega-\omega_o)+\delta(\omega+\omega_o)]$
6	$t^n e^{-at}u(t) \leftrightarrow \frac{n!}{(a+j\omega)^{n+1}},\ a>0$	19	$\sin(\omega_o t) \leftrightarrow j\pi[\delta(\omega+\omega_o)-\delta(\omega-\omega_o)]$
7	$\mathrm{rect}(\frac{t}{\tau}) \leftrightarrow \tau\,\mathrm{sinc}(\frac{\omega\tau}{2})$	20	$\cos(\omega_o t)u(t) \leftrightarrow$ $\frac{\pi}{2}[\delta(\omega-\omega_o)+\delta(\omega+\omega_o)]+\frac{j\omega}{\omega_o^2-\omega^2}$
8	$\mathrm{sinc}(Wt) \leftrightarrow \frac{\pi}{W}\mathrm{rect}(\frac{\omega}{2W})$	21	$\sin(\omega_o t)u(t) \leftrightarrow$ $j\frac{\pi}{2}[\delta(\omega+\omega_o)-\delta(\omega-\omega_o)]+\frac{\omega_o}{\omega_o^2-\omega^2}$
9	$\triangle(\frac{t}{\tau}) \leftrightarrow \frac{\tau}{2}\mathrm{sinc}^2(\frac{\omega\tau}{4})$	22	$\mathrm{sgn}(t) \leftrightarrow \frac{2}{j\omega}$
10	$\mathrm{sinc}^2(\frac{Wt}{2}) \leftrightarrow \frac{2\pi}{W}\triangle(\frac{\omega}{2W})$	23	$u(t) \leftrightarrow \pi\delta(\omega)+\frac{1}{j\omega}$
11	$e^{-at}\sin(\omega_o t)u(t) \leftrightarrow$ $\frac{\omega_o}{(a+j\omega)^2+\omega_o^2},\ a>0$	24	$\sum_{n=-\infty}^{\infty}\delta(t-nT) \leftrightarrow \frac{2\pi}{T}\sum_{n=-\infty}^{\infty}\delta(\omega-n\frac{2\pi}{T})$
12	$e^{-at}\cos(\omega_o t)u(t) \leftrightarrow$ $\frac{a+j\omega}{(a+j\omega)^2+\omega_o^2},\ a>0$	25	$\sum_{n=-\infty}^{\infty}f(t)\delta(t-nT) \leftrightarrow$ $\sum_{n=-\infty}^{\infty}\frac{1}{T}F(\omega-n\frac{2\pi}{T})$
13	$e^{-\frac{t^2}{2\sigma^2}} \leftrightarrow \sigma\sqrt{2\pi}e^{-\frac{\sigma^2\omega^2}{2}}$		

ANALOG SIGNALS AND SYSTEMS

ANALOG SIGNALS AND SYSTEMS

ERHAN KUDEKI

University of Illinois at Urbana-Champaign

DAVID C. MUNSON JR.

University of Michigan

PEARSON

Prentice
Hall

Upper Saddle River, New Jersey 07458

Library of Congress Cataloging-in-Publication Data

CIP data on file

Editorial Director, Computer Science, Engineering: *Marcia J. Horton*
Associate Editor: *Alice Dworkin*
Editorial Assistant: *William Opulach*
Senior Managing Editor: *Scott Disanno*
Production Editor: *James Buckley*
Art Director: *Jayne Conte*
Cover Designer: *Bruce Kenselaar*
Art Editor: *Greg Dulles*
Media Editor: *Dave Alick*
Manufacturing Manager: *Alan Fischer*
Manufacturing Buyer: *Lisa McDowell*

© 2009 by Pearson Education Inc.
Pearson Prentice Hall
Pearson Education, Inc.
Upper Saddle River, NJ 07458

Pearson Prentice Hall™ is a trademark of Pearson Education, Inc.

The author and publisher of this book have used their best efforts in preparing this book. These efforts include the development, research, and testing of the theories and programs to determine their effectiveness. The author and publisher make no warranty of any kind, expressed or implied, with regard to these programs or the documentation contained in this book. The author and publisher shall not be liable in any event for incidental or consequential damages in connection with, or arising out of, the furnishing, performance, or use of these programs.

Printed in the United States of America
5 2023

ISBN 0-13-143506-X
      978-0-13-143506-3

Pearson Education Ltd., *London*
Pearson Education Australia Pty. Ltd., *Sydney*
Pearson Education Singapore, Pte. Ltd.
Pearson Education North Asia Ltd., *Hong Kong*
Pearson Education Canada, Inc., *Toronto*
Pearson Educación de Mexico, S.A. de C.V.
Pearson Education—Japan, *Tokyo*
Pearson Education Malaysia, Pte. Ltd.
Pearson Education, Inc., *Upper Saddle River, New Jersey*

to
Beverly, Melisa, and Deren
and to
Nancy, David, Ryan, Mark, and Jamie

Contents

Appendix B Labs   471

Appendix C Further Reading   507

INDEX   509

Preface

Dear student: This textbook will introduce you to the exciting world of analog signals and systems, explaining in detail some of the basic principles that underlie the operation of radio receivers, cell phones, and other devices that we depend on to exchange and process information (and that we enjoy in our day-to-day lives). This subject matter constitutes a fundamental core of the modern discipline of electrical and computer engineering (ECE). The overall scope of our book and our pedagogical approach are discussed in an introductory chapter numbered "0" before we get into the nitty-gritty of electrical circuits and analog systems, beginning in Chapter 1. Here, in the Preface, we tell you and our other readers – mainly your instructors – about the reasons underlying our choice of topics and the organization of material presented in this book. We hope that you will enjoy using this text and then perhaps return to this Preface after having completed your study, for then you will be able to better appreciate what we are about to describe.

This textbook traces its origins to a major curriculum revision undertaken in the middle 1990s at the University of Illinois. Among the many different elements of the revision, it was decided to completely restructure the required curriculum in the area of circuits and systems. In particular, both the traditional sophomore-level circuit analysis course and the junior-level signals and systems course were phased out, with the material in these courses redistributed within the curriculum in a new way. Some of the circuits material, and the analog part of the signals and systems material, were integrated into a new sophomore-level course on analog signals and systems. This course, for which this book was written, occupies the same slot in the curriculum as the old circuit analysis course. Other material in the circuit analysis course was moved to a junior-level electronics elective and to an introductory course on power systems. The discrete-time topics from the old junior-level signals and systems course were moved into a popular course on digital signal processing. This restructuring consolidated the curriculum (saved credit hours); but, more importantly, it offered pedagogical benefits that are described below.

Similar to the trend initiated in *DSP First: A Multimedia Approach,* and its successor *Signal Processing First* (both by McClellan, Schafer, and Yoder, Prentice-Hall), our approach takes some of the focus in the early curriculum off circuit analysis, which no longer is *the* central topic of ECE. And, it permits the introduction of signal processing concepts earlier in the curriculum, for immediate use in subsequent courses. However, unlike *DSP First* and *Signal Processing First*, we prefer "analog first" as the portal into ECE curricula, for three reasons. First, this treatment follows more naturally onto required courses on calculus, differential equations, and physics, which model primarily analog phenomena. Second, this approach better serves as a cornerstone of the broader ECE curricula, preparing students for follow-on courses in electronics (requiring circuit analysis and frequency response), electromagnetics (needing phasors, capacitance, and inductance), solid state electronics (using differential equations), and power systems (requiring circuit analysis, complex numbers, and phasors). Third, the concept of digital frequency is entirely foreign to students familiar with only trigonometry and physics. Humans perceive most of the physical world to be analog. Indeed, it is not possible to fully understand digital signal processing without considering a complete system composed of an analog-to-digital converter, a digital filter (or other digital processing), and a digital-to-analog converter. An analysis of this system requires substantial knowledge of Fourier transforms and analog frequency response, which are the main emphases of this book.

Beginning with simple course notes, versions of this textbook have been used successfully at the University of Illinois for more than a decade. As we continued to work on this project, it became clear to us that the book would have broader appeal, beyond those schools following the Illinois approach to the early ECE curriculum. Indeed, this text is equally useful as either "analog first" (prior to a course on discrete-time signal processing) or "analog second" (after a course on discrete-time signal processing). In that sense, the text is universal. And, the book would work well for a course that follows directly onto a standard sophomore-level course on circuit analysis.

We invite instructors to try the approach in this text that has succeeded so well for us. And, we urge you to look beyond the topical headings, which may sound standard. We believe that instructors who follow the path of this book will encounter new and better ways of teaching this foundational material. We have observed that integration of circuit analysis with signals and systems allows students to see how circuits are used for signal processing, and not just as mathematical puzzles where the goal is to solve for node voltages and loop currents. Even more important, we feel that our students develop an unusually thorough understanding of Fourier analysis and complex numbers (see Appendix A), which is the core of our book. Students who complete this text can *design* simple filters and can explain in the Fourier domain the workings of a superheterodyne AM radio receiver. Finally, through the introduction of a small number of labs that are intimately tied to the theory covered in lecture (see Appendix B), we have constructed a well-rounded learning environment by integrating theory with applications, design, and implementation.

We extend sincere thanks to all of our faculty colleagues at the University of Illinois and elsewhere who participated at one time or another in the "ECE 210 project," and who contributed in so many ways to the course notes that evolved into our book. We especially thank Tangul Basar, Douglas Jones, George Papen, Dilip Sarwate, and Timothy Trick, who have used, critiqued, and helped improve many versions of the notes. We also thank countless students and graduate TAs – Andrea

Mitofsky in particular – for their helpful comments and catching our mistakes. Finally, we acknowledge the influence of our prior education and reading on what we have put to paper – this influence is so far-reaching and untraceable that we have avoided the task of compiling a comprehensive reference list. Instead, we have included a short list of further reading (see Appendix C). This list should be useful to readers who wish to explore the world of signals and systems beyond what can be reached with this book. Our very best wishes to all who are about to begin their learning journey!

ERHAN KUDEKI AND DAVID C. MUNSON, JR.

ANALOG SIGNALS
AND SYSTEMS

0

Analog Signals and Systems—The Scope and Study Plan

THE SCOPE

The world around us is teeming with *signals* from natural and man-made sources—stars and galaxies, radio and TV stations, computers and WiFi cards, cell phones, video cameras, MP3 players, musical instruments, temperature and pressure sensors, and countless other devices and systems. In their natural form many of these signals are *analog*, or continuous in time. For example, the electrical signal received by a radio antenna may be represented as an analog voltage waveform $v(t)$, a function of a continuous time variable t. Similarly, sound traveling through the air can be thought of as a pressure waveform having a specific numerical value at *each instant* in time and each position in space.

Not all signals are analog. Nowadays, many signals are *digital*. Digital signals are sequences of numbers. Most often, we acquire digital signals by sampling analog signals at uniformly spaced points in time and rounding off (quantizing) the samples to values that can be stored in a computer memory. This process of producing digital signals is called *analog-to-digital (A/D) conversion*, or *digitization*. The acquired sequence of numbers can be stored or processed (manipulated) by a computer. Then it often is desired to return the values from the digital realm back to the analog world. This is accomplished by a process called *digital-to-analog (D/A) conversion*, whereby a smooth, continuous (analog) waveform, say, some $v(t)$, is constructed that passes through the numerical values of the digital signal.

Modern-day signal processing systems commonly involve both analog and digital signals. For example, a so-called *digital cell phone* has many analog components. Your vocal chords create an acoustic signal (analog), which is captured by a microphone in the cell phone and converted into an electrical signal (analog). The analog electrical signal is then digitized—that is, sampled and quantized—to produce a (digital) signal that is further manipulated by a computer in the cell phone to create a new digital signal that requires fewer bits for storage and transmission and that is more resistant to errors encountered during transmission.

Next, this digital signal is used as the input to a modulator that creates a high-frequency analog waveform that carries the information in the coded digital signal away from your cell phone's antenna. At the receiving cell phone these processes are reversed. The receiving antenna captures an analog signal (voltage versus time), which then is passed through the demodulator to retrieve the digitally coded speech. A computer in the receiving cell phone processes, or decodes, this sequence to recreate the set of samples of the original speech waveform. This set of samples is passed through a D/A converter to create the analog speech waveform. This signal is amplified and passed to the speaker in the cell phone, which in turn creates the analog acoustic waveform that is heard by your ear.

Figure 0.1 uses a block diagram language to summarize what we have just described. From a high-level point of view—that is, ignoring what is happening in individual blocks or subsystems shown in the figure—we see that the overall task of the transmitting phone depicted at the top is to convert the analog voice input $p_i(t)$ into a propagating analog radio wave $E_i(t)$. The receiving phone's task, on the bottom, is to extract from many analog radio waves hitting its antenna ($E_1(t)$, $E_2(t)$, etc.) a

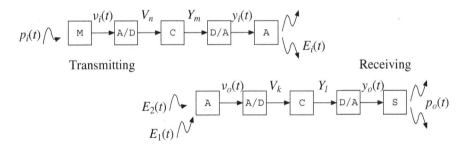

Figure 0.1 Simplified cell phone models, transmitting (at the top) and receiving (on the bottom). Blocks A denote a pair of antennas, A/D and D/A denote analog-to-digital and digital-to-analog converters, C stands for a computer or digital processing unit, and M and S represent a microphone and a loudspeaker, respectively. The transmitting phone at the top converts the sound input $p_i(t)$ (an analog pressure waveform) into a propagating radio wave $E_i(t)$, whereas the receiving phone on the bottom is designed to reconstruct a sound wave $p_o(t)$, which is a delayed and scaled copy of $p_i(t)$.

delayed and scaled copy of $p_i(t)$ and render it as sound. These tasks are carried out through a combination of analog and digital signal processing steps represented by the individual blocks of Figure 0.1. Such combinations of analog and digital processing are common in many devices that we encounter everyday: MP3 players, digital cable or satellite TV, digital cameras, anti-lock brakes, and household appliance controls.

In this book, we focus on the mathematical analysis and design of *analog signal processing,* which is carried out in the analog parts of the previously mentioned systems. Analog processing frequently is employed in amplification, filtering (to remove noise or interference) or equalization (to shape the frequency content of a signal), and modulation (to piggyback a low-frequency signal onto a higher frequency carrier signal for wireless transmission), as well as in parts of A/D and D/A converters. Processing of analog signals typically is accomplished by electrical circuits composed of elements such as resistors, capacitors, inductors, and operational amplifiers. Thus, it will be necessary for us to learn more about circuit analysis than you may have seen in preceding courses.

We wrote this book under the assumption that the reader may not be familiar with electrical circuits much beyond the most fundamental level, which is reviewed in Chapter 1. Chapters 2 through 4 introduce DC circuit analysis, time-varying circuit response, and sinusoidal steady-state AC circuits, respectively. The emphasis is on *linear* circuit analysis, because the discussions of *linear time-invariant* (LTI) systems in later chapters develop naturally as generalizations of the concepts from Chapter 4 on sinusoidal steady-state response in linear circuits.

To utilize the cell phone models shown in Figure 0.1, we need unambiguous descriptions of the input–output relations of their interconnected subsystems or processors. In Chapters 5 through 7 we will learn how to formulate such relations for linear, electrical circuits as well as for other types of analog LTI systems. Chapters 5 through 7 focus on how LTI circuits and systems respond to arbitrary periodic and non-periodic inputs, and how such responses can be represented by Fourier series and transform techniques, respectively.

As you will see in this book, there are two different approaches to understanding signal processing systems: Both *frequency-domain* and *time-domain* techniques are widely employed. Our initial approach in Chapters 5 through 7 takes the frequency-domain path, because that happens to be the easier and more natural approach to follow if our starting point is sinusoidal steady-state circuits (i.e., in Chapter 4). This path quickly takes us to a sufficiently advanced stage to enable a detailed discussion of AM radio receivers in Chapter 8. However, fluency in *both* time- and frequency-domain methods is necessary because, depending on the problem, one approach generally will be easier to use or will offer more insight than the other. We therefore will turn our attention in Chapters 9 and 10 to time-domain methods. We will translate the frequency-domain results of Chapter 7 to their time-domain counterparts and illustrate the use of the time-domain convolution method, as well as impulse and impulse response concepts. We also will learn in Chapter 9 about sampling and reconstruction of bandlimited analog signals, and get a glimpse of digital signal processing techniques.

An analog system can produce an output even in the absence of an input signal if there is initial energy stored in the system (like the free oscillations of a stretched spring after its release). In Chapters 11 and 12 we will investigate the full response of LTI systems and circuits, including their energy-driven outputs, and learn the Laplace transform technique for solving LTI system initial-value problems. The emphasis throughout most of the course is on circuit and system *analysis*—that is, determining how a circuit or system functions and processes its input signals. However, in Chapter 12 we will discuss system *design* and learn how to build stable analog filter circuits.

The book ends with three appendices. Appendix A is a review of complex numbers. Both at the circuit level and at the block diagram level, much of the analysis and design of signal processing systems relies on mathematics. You will be familiar with the required mathematical subjects: algebra, trigonometry, calculus, differential equations, and complex numbers. However, most students using this text will find that their background in complex numbers is entirely insufficient. For example, can you explain the meaning of $\sqrt{-1}$? If not, then Appendix A is for you. Most students will want to study this appendix carefully, parallel to Chapters 1 and 2, and then refer to it later as needed in Chapter 3 and beyond.

Appendix B includes five laboratory worksheets that are used at the University of Illinois in a required lab that accompanies the sophomore-level course ECE 210, Analog Signal Processing, based on this text. The lab component of ECE 210 starts approximately five weeks into the semester, and the biweekly labs involve simple measurement and/or design projects related to circuit and systems concepts covered in class. In the fourth lab session, an AM radio receiver—the topic of Chapter 8—is assembled with components built in the earlier labs. In the fifth session, the receiver is modified to include a PC sound card (and software), replacing the back-end hardware. The labs provide a taste of how signal and system theory applies in practice and illustrate how real-life signals and circuit behavior may differ from the idealized versions described in class.

Appendix C provides a list of further reading for students who may wish to learn more about a topic or who seek an alternative explanation.

STUDY PLAN

This book was written with a "just-in-time" approach. This means that the book tells a story (so to speak), and new ideas and topics relevant to the story are introduced only when needed to help the story advance. You will not find here "encyclopedic" chapters that are *stand-alone* and complete treatments of distinct topics. (Chapter 1, which is a review of circuit fundamentals, may be an exception.) Instead, topics are developed throughout the narrative, and individual ideas make multiple appearances just when needed as the story unfolds, much like the dynamics of individual characters in a novel or a play.

For example, although the title of Chapter 7 contains the words "Fourier transform," the concept of the Fourier transform is foreshadowed as early as in Chapter 3, and discussions of the Fourier transform continue into the final chapter of the book. Thus, to learn about the Fourier transform, a full reading of the text is necessary—reading just Chapter 7 will provide only a partial understanding. And that is true with many of the main ideas treated in the book. We hope that students will enjoy the story line enough to stick with it.

In ECE 210 at the University of Illinois, the full text is covered from Chapter 1 through Chapter 12 in one semester (approximately 15 weeks) in four lecture hours per week, including a first-week lecture on complex numbers as treated in Appendix A. Chapter 1 and much of Chapter 2 are of a review nature for most students; consequently, they are treated rapidly to devote the bulk of classroom time to Chapters 3 through 12. Exposure to circuit analysis in Chapters 1 through 4 prepares students for junior-level courses in electronics and electromagnetics, while signal processing and system analysis tools covered throughout the entire text provide the background for advanced courses in digital signal processing, communications, control, remote sensing, and other areas where linear systems notions are essential. Exposure of sophomores to the tools of linear system theory opens up many creative options for later courses in their junior and senior years.

The story line of our book is, of course, open ended, in the sense that student learning of the Fourier transform and its applications, and other important ideas introduced here, will continue beyond Chapter 12. Because of that, we trust our students to question "what happens next" and to pursue the plot in subsequent courses.

1

Circuit Fundamentals

Review
of voltage,
current,
and power;
KVL and KCL;
two-terminal
elements

Electrostatic attraction and repulsion between charged particles are fundamental to all electrical phenomena observed in nature. Free charge carriers (e.g., electrons and protons) transported against electrostatic forces gain potential energy just like a pebble lifted up from the ground against gravitational pull. Conversely, charge carriers release or lose their potential energy when they move in the direction of an electrostatic pull.

In *circuit models* of electrical systems and devices the movement of charge carriers is quantified in terms of *current* variables such as i_R, i_L, and i_C marked on the circuit diagram shown in Figure 1.1. *Voltage* variables such as v_s, v_R, and v_o are used to keep track of energy gains and losses of carriers moving against or with electrostatic forces. Flow channels of the carriers are represented by two-terminal circuit elements such as R, L, and C, which are distinguished from one another by unique *voltage–current*, or *v–i*, *relation*s. The *v–i* relations plus Kirchhoff's voltage and current laws representing energy and charge conservation are sufficient to determine quantitatively how a circuit functions and at what rates energy is generated

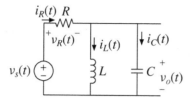

Figure 1.1   An electrical circuit.

and lost in the circuit.[1] These fundamental circuit concepts will be reviewed in this chapter.

1.1   Voltage, Current, and Power

Most elementary circuit models are constructed in terms of *two-terminal* elements representing the possible flow paths of electrical charge carriers. The energetics and flow rate of charge carriers transiting each circuit element are described in terms of the *voltage*, *current*, and *power* variables defined in Table 1.1.

Voltage

The definition of *element voltage v* given in Table 1.1 can be interpreted as energy loss per unit charge transported from an element terminal marked by a + sign to the second terminal marked by −; equivalently, as the energy gain per Coulomb moved from the − to the + terminal:

Example 1.1
In Figure 1.2a, $v_b = 4$ V stands for energy loss per unit charge transported from the left terminal of element b marked by the + sign to the right terminal marked by −. Equivalently, $v_b = 4$ V can be interpreted as energy gain per unit charge transported from the − to + terminals, or from right to left. Thus, electrical potential energy per unit charge, or the electrical potential, is higher at the + terminal of element b compared with its − terminal by an amount 4 V.

---

[1]Charge and energy conservation are fundamental to nature: Net electrical charge can neither be generated nor destroyed; if a room contains 1 C of net charge, the only way to change this amount is to move some charged particles in or out of the room; likewise, for energy. When we talk about electrical energy generation, use, or loss, what we really mean is *conversion* between electrical potential energy and some other form of energy, e.g., mechanical, chemical, thermal, etc.

Circuit element and variables	Units
$\xrightarrow{i}$ $+ \quad \boxed{} \quad -$ v	
Element *voltage* $$v \equiv \lim_{\Delta q \to 0} \frac{\Delta w}{\Delta q} = \frac{dw}{dq},$$ where Δw denotes the potential energy loss of Δq amount of charge transported from $+$ to $-$ terminal.	$v[=]\dfrac{\text{Joule (J)}}{\text{Coulomb (C)}} = \text{Volt (V)}$
Element *current* $$i \equiv \lim_{\Delta t \to 0} \frac{\Delta q}{\Delta t} = \frac{dq}{dt},$$ where Δq denotes the net amount of electrical charge transported in direction $\to$ during the time interval Δt.	$i[=]\dfrac{\text{Coulomb (C)}}{\text{second (s)}} = \text{Ampere (A)}$
Absorbed *power* $$p \equiv vi = \frac{dw}{dq}\frac{dq}{dt} = \frac{dw}{dt} = \lim_{\Delta t \to 0} \frac{\Delta w}{\Delta t},$$ where Δw denotes the net energy loss of charge carriers moving through the element during the time interval Δt.	$p[=]\dfrac{\text{Joule (J)}}{\text{second (s)}} = \text{Watt (W)}$

Table 1.1 Definitions of element voltage, current, and absorbed power, and the associated units in *Systeme International* (SI). The symbol [=] stands for "has the unit."

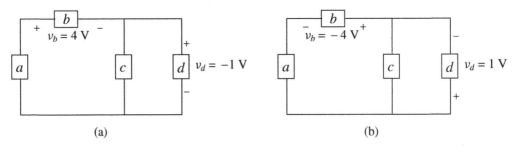

                        (a)   (b)

Figure 1.2 Circuits with several two-terminal elements.

Example 1.2

In Figure 1.2a, $v_d = -1$ V indicates that energy per unit charge is lower at the + terminal of element d on the top relative to the − terminal on the bottom, since v_d is negative. Or, equivalently, energy per unit charge is higher at the bottom terminal relative to the top terminal.

The plus and minus signs assigned to element terminals are essential for describing what the voltage variable stands for. These signs are said to indicate the *polarity* of the element voltage, but the polarity does not indicate whether the voltage is positive or negative. Figure 1.2b shows the same circuit as Figure 1.2a, but with the voltage polarities assigned in the opposite way. As the next two examples show, these revised voltage definitions describe the same physical situation as in Figure 1.2a, because the algebraic signs of the voltage values also have been reversed in Figure 1.2b.

Polarity

Example 1.3

In Figure 1.2b, $v_b = -4$ V stands for energy loss per unit charge transported from the right terminal of element b marked by a + sign to the left terminal marked by −. Therefore, energy gain per unit charge transported from right to left is 4 V, consistent with what we found in Example 1.1

Example 1.4

In Figure 1.2b, $v_d = 1$ V indicates that energy per unit charge is higher at the + terminal on the bottom relative to the − terminal on the top, because $v_d > 0$. This is consistent with what we found in Example 1.2.

We call an element voltage such as v_b a *voltage rise* from the − terminal to the + terminal, or, equivalently, a *voltage drop* from + to −. So, $v_b = 4$ V in Figure 1.2a is a 4 V rise from the right to left terminals (− to +) *and* a 4 V drop from left to right. Likewise, $v_d = 1$ V in Figure 1.2b is a 1 V drop from bottom to top and a 1 V rise from top to bottom.

Voltage drop and rise

Example 1.5

What is the voltage drop associated with v_b in Figure 1.2b?

Solution   In Figure 1.2b, v_b is a −4 V drop from right (+) to left (−) across element b. Remember, by definition, a drop is always from + to − (and rise, always from − to +).

Notions of voltage drop and rise will play a useful role in formulating Kirchhoff's voltage law in the next section.

Closely associated with the notion of element voltage is the concept of *node voltage*, or *electrical potential*. The connection points of element terminals in a circuit are called the *nodes* of the circuit. In Figure 1.3, two of the three nodes are marked by dots and the third node is marked by a ground symbol. The node marked by the ground symbol is called the *reference node*. Distinct node voltage variables are assigned to

Circuit nodes, reference, node voltage

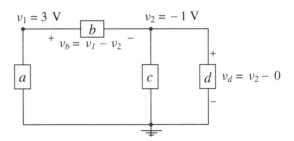

Figure 1.3   A circuit with four two-terminal elements and three nodes.

all the remaining nodes.[2] In Figure 1.3, we have labeled node voltages v_1 and v_2. By definition, the *node voltage*, or *electrical potential*, v_n stands for energy gain per unit charge transported from the reference node to node n.[3] The electrical potential v_0 of the reference node is, of course, zero.

Example 1.6
Because $v_1 = 3$ V in Figure 1.3, the energy gain per unit charge transported from the reference node to node 1 is 3 V, or 3 J/C. Thus, the electrical potential at node 1 is 3 V higher than at the reference node. Equivalently, charges transported from node 1 to the reference node lose energy at a 3 J/C rate.

Example 1.7
In Figure 1.3, $v_2 = -1$ V indicates that 1 C of charge transported from the reference node to node 2 gains -1 J of energy (same as losing 1 J). Thus, the electrical potential is higher at the reference node than at node 2.

The voltage across each element in a circuit is the difference of electrical potentials of the nodes at the element terminals. To express an element voltage as a potential difference, we subtract from the electrical potential of the + terminal the electrical potential of the − terminal. For instance, in Figure 1.3,

$$v_b = v_1 - v_2 = 3 \text{ V} - (-1 \text{ V}) = 4 \text{ V}.$$

This expression reflects the fact that energy loss per unit charge transferred from node 1 to node 2 is 4 V, because electrical potential at node 1 is 4 V higher than at node 2. Similarly, again referring to Figure 1.3,

$$v_d = v_2 - v_0 = (-1 \text{ V}) - 0 = -1 \text{ V}.$$

---

[2]The reference node generally is not electrically connected to the earth. Instead, it is an arbitrarily chosen node that is used as the baseline for defining the other node voltages, much as sea level is arbitrarily chosen as zero elevation.

[3]An analogy is the gravitational potential energy gain of a rock lifted from the ground and placed on a windowsill.

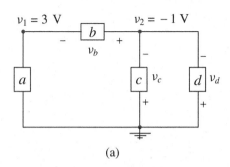

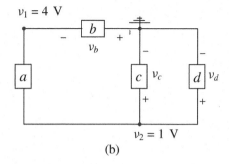

Figure 1.4 (a) The same as the circuit in Figure 1.3, but with reversed polarities assigned to element voltages v_b and v_d. (b) The same circuit, but with a different reference node.

Example 1.8

Express the voltage v_b in Figure 1.4a in terms of node voltages $v_1 = 3$ V and $v_2 = -1$ V.

Solution   Since the + terminal of element b in Figure 1.4a has an electrical potential of $v_2 = -1$ V and the − terminal has potential $v_1 = 3$ V, it follows that

$$v_b = v_2 - v_1 = (-1 \text{ V}) - 3 \text{ V} = -4 \text{ V}.$$

In Figure 1.4a, elements c and d are placed *in parallel*, making terminal contacts with the same pair of nodes. Thus, both elements have the same terminal potentials. Since the assigned polarities of v_c and v_d are in agreement, it follows that the potential differences, or element voltages, are $v_c = v_d = 0 - (-1 \text{ V}) = 1$ V. **Elements in parallel**

Finally, we note that any node in a circuit can be chosen as the reference. The choice impacts the values of the node voltages of the circuit, but not the element voltages, since the latter are potential differences. Changing the reference node causes equal changes in all node voltage values, so that element voltages (potential differences) remain unchanged. See Figure 1.4b for an illustration.

Current

Figure 1.5 shows a circuit with four elements, each representing a possible path for electrical charge flow. Current variables i_a, i_b, etc., shown in the figure have been defined to quantify these flows. (See Table 1.1 for the definition of current.) A current variable such as i_b indicates the amount of net electrical charge that transits an element per unit time in the direction indicated by the arrow $\rightarrow$ shown next to the element.

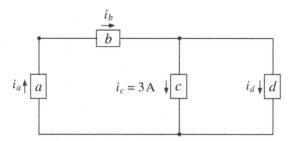

Figure 1.5   The same as Figure 1.3, but showing only the element currents.

Example 1.9

Suppose that, every second, 2 C of charge move through element b in Figure 1.5 from left to right. Then $i_b = 2$ A, because the arrow assigned to i_b points from left to right, the direction of 2 C/s net charge transport. If, on the other hand, the flow from left to right is -2 C/s (same as 2 C/s from right to left), then $i_b = -2$ A.

Example 1.10

In Figure 1.5, $i_c = 3$ A indicates that the amount of net charge transported through element c from top to bottom (the direction indicated by the arrow accompanying i_c) is 3 C/s. If the arrow direction were reversed, the same 3 C/s net flow from top to bottom would be described by $i_c = -3$ A.

Electrical current is a measure of net charge flow. Let's see what this really means: If, for instance, equal numbers of positive and negative charge carriers were to move in the same direction every second,[4] then there would be no net charge transport and the electrical current would be zero. Net charge transport requires unequal numbers of positive and negative charge carriers moving in the same direction (per unit time) or carriers with opposite signs moving in different directions. For instance, $i_c = 3$ A in Figure 1.5 could be due to top-to-bottom movement of only positive charge carriers, bottom-to-top transport of only negative charge carriers, or a combination of these two possibilities, such as negative carriers moving from bottom to top at a -2 C/s rate *simultaneously with* a 1 C/s transport of positive carriers from top to bottom. An element current indicates not how charge transport occurs through the element, but what the net charge flow is.

In Figure 1.5 elements a and b are positioned *in series* on the same circuit branch and therefore constitute parts of the same charge flow path. Since the assigned flow directions of i_a and i_b agree, it follows that $i_a = i_b$.

**Elements
in series**

---

[4]This movement is as in the interior of the sun, where free electrons and protons of the solar plasma move at equal rates in the direction perpendicular to solar magnetic field lines in response to electric fields.

Example 1.11
Given that $i_b = 2$ A in Figure 1.5, determine i_a.

Solution   Since the flow directions of i_a and i_b are the same and since elements a and b are in series, it follows that $i_a = i_b = 2$ A.

Example 1.12
Describe $i_c + i_d$ in Figure 1.5.

Solution   In Figure 1.5, $i_c + i_d = 3$ A $+ i_d$ is the net amount of charge transported from the top of the circuit to the bottom per unit time, since the direction arrows of both i_c and i_d point from top to bottom.

Absorbed power

In Figure 1.6, $v_b = 4$ V is a voltage drop in the direction of element current $i_b = 2$ A from left to right. Therefore, each coulomb of net charge moving through element b loses 4 J of energy, and since 2 C move every second, the product 4 V $\times$ 2 A $= 8$ W stands for the total energy loss of charge carriers passing through the element per unit time. We will refer to this product as the *absorbed power* for element b, since the energy loss of charge carriers is the energy gain of the element, according to the principle of energy conservation.

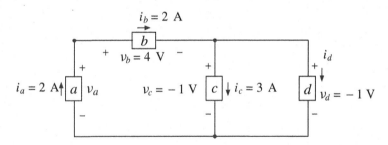

Figure 1.6   The same as Figure 1.2a, but showing the voltage and current variables defined for each circuit element.

In general, for any two-terminal element, *absorbed power* is defined as (see Table 1.1)

$$p \equiv vi,$$

where v denotes the voltage drop across the element in the direction of element current i. Depending on the algebraic signs of v and i, the absorbed power $p = vi$ for an element may have a positive or negative numerical value. For instance, with $v_b = 4$ V and $i_b = 2$ A, the absorbed power $p_b = v_b i_b = 8$ W is positive, indicating that element b "burns," or dissipates, 8 J of charge carrier energy every second. However, for element c, with $v_c = -1$ V and $i_c = 3$ A, the absorbed power $p_c = v_c i_c = -3$ W is negative. The reason for this is that charge carriers gain energy (1 J/C) as they pass through element c. Although it would be correct to say that the element dissipates -3 J every second, a better description would be to say the element injects 3 J of energy into the circuit (via moving the carriers against electrostatic forces) per second.

Circuit elements that are capable of injecting energy into circuits will be referred to as *sources* (e.g., independent voltage source, controlled current source, etc.). Such elements model physical devices that can convert nonelectrical forms of energy into electrical potential energy of charge carriers (a battery, for instance). It is also possible that an element stores, rather than dissipates, the absorbed energy in some form and then re-injects the stored energy to the circuit at a later time. This process can, of course, occur only in circuits with time-varying voltages and currents; in such circuits capacitors and inductors act as energy storage elements, as we will see later on. Resistors, however, literally burn (i.e., turn the absorbed energy into heat) under all possible operation conditions.

Energy conservation requires, at each instant, that the sum of all the energy that is burned or put into storage within a circuit must be exactly compensated by all the energy that is injected into the circuit. Since the absorbed power variable $p = vi$ for each element can quantify either the absorption rate or the injection rate (with positive and negative numerical values, respectively), we find that all the absorbed powers (taking into account the powers for each and every element in the circuit) must sum to zero. For instance, for the circuit shown in Figure 1.6, it is required that

Powers absorbed in a circuit sum to zero

$$p_a + p_b + p_c + p_d = p_a + 8\,\text{W} + (-3\,\text{W}) + p_d = 0,$$

or

$$p_a + p_d = -5\,\text{W},$$

where

$$p_a = -v_a i_a = -v_a 2,$$

and

$$p_d = v_d i_d = (-1) i_d.$$

In the next section we will determine the numerical values of v_a and i_d, and confirm that $p_a + p_d = -5$ W. Notice that the absorbed power p_a for element a is $-v_a i_a$ rather than $v_a i_a$, because v_a is a voltage rise (rather than a drop) in the direction of current i_a. (See Figure 1.6.)

1.2   Kirchhoff's Voltage and Current Laws: KVL and KCL

The values of v_a and i_d in Figure 1.6 can be determined with the aid of Kirchhoff's voltage and current laws (KVL and KCL), reviewed next. These laws, which are the basic *axioms of circuit theory*, correspond to principles of energy and charge conservation expressed in terms of element voltages and currents.

Kirchhoff's voltage law: Around any closed loop in a circuit, $\sum v_{rise} = \sum v_{drop}$

Translating this axiom into words, KVL demands that the sum of all voltage rises encountered around any closed loop of elements in a circuit equals the sum of all voltage drops encountered around the same loop.

In applying this rule, you should remember that each element voltage can be interpreted as a rise or a drop, depending on the transit direction across the element; in constructing a KVL equation, the voltage of each element should be added to only one side of the equation, depending on whether the loop traverses the element from minus to plus (called a voltage rise) or from plus to minus (called a voltage drop), as illustrated next:

Example 1.13
We traverse Loop 1 of Figure 1.7(a) in the clockwise direction and obtain the KVL equation

$$v_a = v_b + v_c,$$

since in the clockwise direction v_a appears as a voltage rise (we rise from the $-$ to the $+$ terminal as we traverse element a) and v_b and v_c appear as voltage drops (we drop from the $+$ to $-$ terminals in each case). Substituting the values for v_b and v_c gives

$$v_a = 4 \text{ V} + (-1) \text{ V} = 3 \text{ V}.$$

Notice that this result does not depend on the direction of travel around the loop. Traversing Loop 1 in the counterclockwise direction yields

$$v_c + v_b = v_a,$$

which is simply a rewritten version of the earlier equation.

Finally, we can apply KVL to Loop 2 in the clockwise direction, yielding

$$v_c = v_d.$$

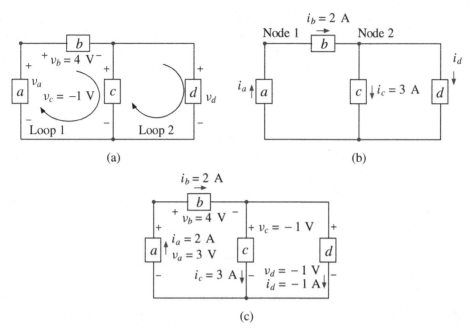

Figure 1.7   Figure 1.6 redrawn (a) without the element currents, (b) without the element voltages, and (c) after application of KVL and KCL.

Substituting for v_c in the prior equation gives

$$v_b + v_d = v_a,$$

which is a KVL equation for the outer loop of the circuit.

Kirchhoff's current law: At any node in a circuit, $\sum i_{in} = \sum i_{out}$

In plain words, KCL demands that the sum of all the currents flowing into a node equals the sum of all the currents flowing out.

In applying this rule, we just need to pay attention to the arrows indicating the flow directions of the element currents present at each node.

Example 1.14
KCL applied to Node 2 of Figure 1.7(b) states that

$$i_b = i_c + i_d,$$

since i_b flows into the node, while both i_c and i_d flow out. Likewise, the KCL equation for Node 1 is

$$i_a = i_b.$$

Combining these equations gives

$$i_a = i_c + i_d,$$

which is the KCL equation for the node at the bottom of the diagram. Now, with $i_b = 2$ A and $i_c = 3$ A, the preceding KCL equations indicate that $i_a = i_b = 2$ A and $i_d = i_b - i_c = 2$ A $- 3$ A $= -1$ A.

In Figure 1.7(c) we redraw Figure 1.6, showing the numerical values of v_a and i_d that we have deduced in earlier examples. From the figure, we see that

$$p_d = v_d i_d = 1 \text{ W}$$

and

$$p_a = -v_a i_a = -6 \text{ W}.$$

Therefore,

$$p_a + p_d = -5 \text{ W},$$

as claimed in the previous section.

1.3 Ideal Circuit Elements and Simple Circuit Analysis Examples

Circuit models of electrical systems and multiterminal electrical devices can be represented by a small number of idealized two-terminal circuit elements, which will be reviewed in this section. These elements have been defined to account for energy dissipation (resistor), injection (independent and dependent sources), and storage (capacitor and inductor) and are distinguished from one another by unique voltage–current, or v–i, relations.

Ideal resistor: An *ideal resistor* is a two-terminal circuit element having the v–i relation

$$v = Ri$$

known as *Ohm's law*. In this relation, $R \geq 0$ is a constant known as the *resistance*   **Ohm's law**
and v denotes the voltage drop in the direction of current i, as shown in Figure 1.8. Resistance R is measured in units of V/A = Ohms (Ω).

Resistors cannot inject energy into circuits, because Ohm's law $v = Ri$ and the absorbed power formula $p = vi$ imply that, for a resistor,

$$p = i^2 R = \frac{v^2}{R} \geq 0,$$

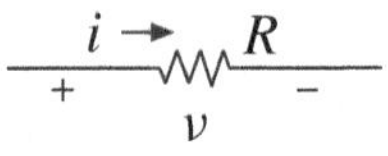

Figure 1.8   The circuit symbol for an ideal resistor.

since $R \geq 0$. Resistors primarily are used to represent energy sinks in circuit models of electrical systems.[5] The zigzag circuit symbol of the resistor is a reminder of the frictional energy loss of charge carriers moving through a resistor.

Example 1.15
Figure 1.9a shows a resistor carrying a 2 A current in the direction of a 4 V drop. Ohm's law $v = Ri$ indicates that the corresponding resistance value is

$$R = \frac{v}{i} = \frac{4\,\text{V}}{2\,\text{A}} = 2\,\Omega.$$

The resistor absorbs a power of $p = vi = 8\,\text{W}$ (or $i^2 R = 2^2 \times 2 = 8\,\text{W}$).

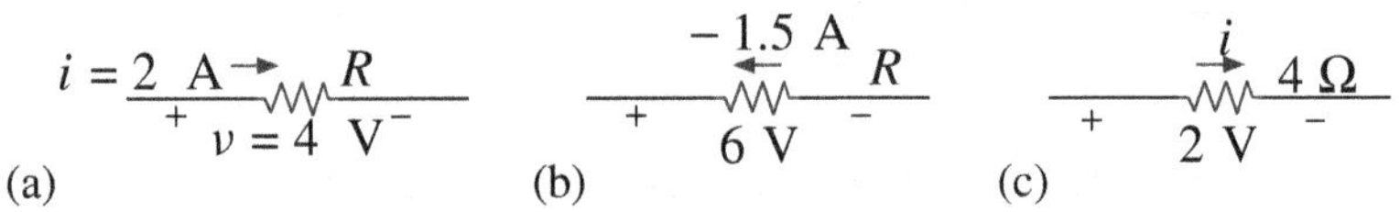

(a)   (b)   (c)

Figure 1.9   Ideal resistor examples.

Example 1.16
For the resistor shown in Figure 1.9b, and using Ohm's law,

$$R = \frac{v}{i} = \frac{6\,\text{V}}{1.5\,\text{A}} = 4\,\Omega.$$

Notice that to calculate R we used $i = 1.5$ A rather than -1.5 A, because the element current in the direction of the 6 V drop is 1.5 A–Ohm's law requires i to be the current in the direction of the voltage drop.

Example 1.17
For the resistor shown in Figure 1.9c, the element current is

$$i = \frac{v}{R} = \frac{2\,\text{V}}{4\,\Omega} = 0.5\,\text{A}.$$

---

[5]Negative resistance can be used to model certain energy sources. We will first encounter negative resistances at the end of Chapter 2 when we study the Thevenin and Norton equivalent circuits.

The special cases of an ideal resistor with $R = 0$ and $R = \infty$ are known as *short-circuit* and *open-circuit* elements, or *short* and *open*, respectively. Their special circuit symbols are shown in Figure 1.10.

$$\underline{\overline{\begin{matrix} short \\ v = 0 \end{matrix}}} \qquad \bullet\!\!—\!\!\begin{matrix} i = 0 \\ open \end{matrix}\!—\!\!\bullet$$

Figure 1.10   Circuit symbols for a "short" ($R = 0$) and an "open" ($R = \infty$).

For a *short*, $R = 0$, and therefore $v = 0$, independent of the current i; whereas, for an *open*, $R = \infty$, and therefore $i = 0$, independent of the voltage v. Consequently, the absorbed power vi is zero in both cases.[6] A short can conduct any externally driven current; likewise, an open can maintain any value of externally imposed voltage across its terminals.

Open and short

Independent voltage source: An *independent voltage source* is an element that maintains a specified potential difference v_s between its terminals, independent of the value of element current i. Its circuit symbol, shown in Figure 1.11, indicates the polarity[7] of voltage v_s. The value of the current i through the voltage source depends on the circuit containing the source and can be determined only by an analysis of the complete circuit.

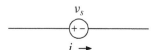

Figure 1.11   Circuit symbol for independent voltage source.

Example 1.18
For the circuit in Figure 1.12a with a 4 V voltage source, KVL and Ohm's law imply that

$$(8\,\Omega) \times i_R = 4\,\text{V};$$

so

$$i_R = \frac{4\,\text{V}}{8\,\Omega} = 0.5\,\text{A}.$$

---

[6]This is because in a short ($R = 0$) there is zero friction, while in an open ($R = \infty$) charge transport, and thus energy dissipation, cannot take place at all.

[7]More precisely, the plus and minus indicate that the potential difference between the plus and minus terminals is v_s. They do not indicate that the potential at the plus terminal is higher than the potential at the minus terminal. In other words, v_s may be positive or negative.

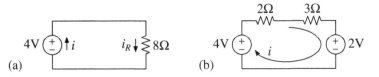

Figure 1.12   Circuit examples with independent voltage sources.

Since KCL requires that $i = i_R$ (for the given reference directions of i and i_R), the current of the 4 V source is $i = 0.5$ A.

Example 1.19

For the circuit shown in Figure 1.12b, a single current variable i is sufficient to represent all the element currents (indicated as a clockwise flow), since all four elements of the circuit are in series. In terms of i, KVL for the circuit is

$$4 = 2i + 3i + 2.$$

Therefore,

$$i = \frac{4 - 2}{2 + 3} = \frac{2}{5} = 0.4 \text{ A}.$$

Independent current source: An *independent current source* is an element that maintains a specified current flow i_s between its terminals, independent of the value of the element voltage v. Its circuit symbol, shown in Figure 1.13, indicates the reference direction[8] of the current i_s. The voltage v across the current source is affected by the circuit containing the source and can be determined only by analysis of the complete circuit.

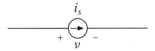

Figure 1.13   Circuit symbol for independent current source.

---

[8]The current may have a positive value, in which case positive charges flow in the direction of the arrow (or negative charges flow opposite the direction of the arrow), or it may have a negative value, in which case positive charges flow opposite the direction of the arrow.

Example 1.20

For the circuit in Figure 1.14a, with the 3 A current source, KCL and Ohm's law imply that

$$3\,A = \frac{v_R}{5\,\Omega},$$

since $\frac{v_R}{5\Omega}$ is the resistor current flowing from top to bottom. Hence,

$$v_R = 3 \times 5 = 15\,V.$$

Notice that KVL requires that $v = v_R = 15\,V.$

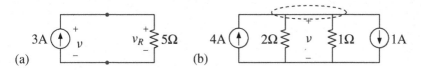

(a)   (b)

Figure 1.14   Circuits containing independent current sources.

Example 1.21

In Figure 1.14b the short at the top of the circuit, enclosed within the dashed oval, can be regarded as a single node, because the potential difference between any two points along a short is zero. Likewise, the bottom of the circuit can be regarded as a second node. The voltage drop in the circuit from top to bottom is denoted as v, which can be regarded as the element voltage of all four components of the circuit connected in parallel. Consequently, the $2\,\Omega$ and $1\,\Omega$ resistors conduct $\frac{v}{2}$ and $\frac{v}{1}$ ampere currents from top to bottom, respectively. Thus, the KCL equation for the top node can be written as

$$4 = \frac{v}{2} + \frac{v}{1} + 1,$$

from which we find

$$v = \frac{4-1}{\frac{1}{2}+1} = \frac{3}{1.5} = 2\,V.$$

Using $v = 2$ V, we now can calculate the resistor currents. For instance, a current of

$$\frac{2\,V}{2\,\Omega} = 1\,A$$

flows through the $2\,\Omega$ resistor from top to bottom.

Figure 1.15   Circuit symbols of dependent (or controlled) sources.

Dependent sources: Dependent voltage and current sources are represented by
the diamond-shaped symbols shown in Figure 1.15. These source elements have the
same characteristics as independent voltage and current sources, respectively, except
that v_s and i_s depend on a voltage or current v_x or i_y defined elsewhere in the circuit
containing the dependent source. The dependencies may take the forms

$$v_s = Av_x \text{ or } Bi_y$$
$$i_s = Cv_x \text{ or } Di_y,$$

where A and D are dimensionless constants and B and C are constants with units of
Ω and $\Omega^{-1} \equiv$ Siemens (S), respectively. A dependent voltage source specified as

$$v_s = Bi_y,$$

for instance, is a *current-controlled voltage source*.

Example 1.22
Figure 1.16 shows a circuit with a *voltage-controlled current source*

$$i_s = 2v_x,$$

where v_x denotes the voltage across a 2 Ω resistor. The KCL equation at
the top node is

$$i_1 + 2v_x = \frac{v_x}{2},$$

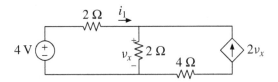

Figure 1.16   A circuit with a dependent current source. Voltage v_x regulates the
current supplied by the dependent current source into the circuit.

leading to the result

$$i_1 = -\frac{3}{2}v_x.$$

Using this result with

$$4 = 2i_1 + v_x$$

(which is the KVL equation of the loop on the left), we find the following solution:

$$v_x = -2\,\mathrm{V}$$

$$i_1 = 3\,\mathrm{A}.$$

Example 1.23

Figure 1.17 shows a circuit with a *current-controlled voltage source*

$$v_d = -\frac{1}{2}i_b,$$

where i_b is the current through a 2 Ω resistor. The KVL equation around the outer loop of the circuit is

$$3 = 2i_b + (-\frac{1}{2}i_b),$$

from which it follows that

$$i_b = 2\,\mathrm{A}.$$

Hence,

$$v_d = -\frac{1}{2}i_b = -1\,\mathrm{V}$$

and

$$v_b = 2i_b = 4\,\mathrm{V}.$$

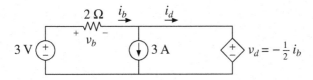

Figure 1.17   A circuit containing a current controlled voltage source v_d.

Also, the KCL equation at the top node requires that

$$i_b = 3 + i_d;$$

thus,

$$i_d = i_b - 3 = 2 - 3 = -1 \text{ A}.$$

Capacitor and inductor: Ideal *capacitors* and *inductors* are two-terminal circuit elements with v–i relations

$$i = C\frac{dv}{dt} \text{ (capacitor)}$$

$$v = L\frac{di}{dt} \text{ (inductor)},$$

where

$$C \geq 0 \text{ and } L \geq 0$$

are constants known as *capacitance* and *inductance*, respectively, and v denotes the voltage drop in the direction of current i, as shown in Figures 1.18a and 1.18b. The capacitance C is measured in units of A s/V = Farads (F) and the inductance L in units of V s/A = Henries (H).

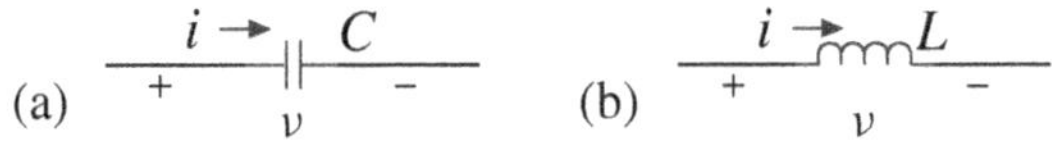

Figure 1.18   Circuit symbols for ideal capacitor (a) and inductor (b).

In the previous equations,

$$\frac{dv}{dt} \text{ and } \frac{di}{dt}$$

denote the time derivatives of v and i, respectively. In circuits with time-independent voltages and currents, the derivatives

$$\frac{dv}{dt} = 0 \text{ and } \frac{di}{dt} = 0,$$

and, consequently, capacitors and inductors behave as open (i.e., zero capacitor current) and short (i.e., zero inductor voltage) circuits, respectively. We refer to such circuits as DC circuits (DC for "direct current," meaning constant current) and use the term AC (AC for "alternating current") to describe circuits with time-varying voltage and currents. Capacitors and inductors play nontrivial roles only in AC circuits, where they respond differently than opens and shorts. **DC versus AC**

In AC circuits, capacitors and inductors represent energy storage elements, since they can re-inject the energy absorbed from charge carriers back to carrier motions. For instance, for a capacitor with capacitance C, the absorbed power is

$$p = vi = vC\frac{dv}{dt} = \frac{d}{dt}\left(\frac{1}{2}Cv^2\right),$$

and, therefore, the quantity

$$w \equiv \frac{1}{2}Cv^2$$

stands for the net absorbed energy of the capacitor, in units of J. When the absorbed power $p = vi = \frac{dw}{dt}$ is positive, w increases with time so that the capacitor is drawing energy from the circuit. Conversely, a negative $p = vi = \frac{dw}{dt}$ is associated with a decrease in w—that is, energy injection back into the circuit. This indicates that a capacitor does not dissipate its absorbed energy, but rather stores it in a form available for subsequent release when vi turns negative. We therefore will refer to $w = \frac{1}{2}Cv^2$ as the *stored energy* of the capacitor. It can likewise be argued that the stored energy of an inductor L carrying a current i is **Stored energy**

$$w = \frac{1}{2}Li^2.$$

Energy storage in a capacitor is associated with the separation of oppositely signed electrical charges between the two terminals of the element; the circuit symbol of the capacitor shows a pair of plates where these charge populations may be maintained separately in a physical capacitor.[9] Energy storage in an inductor is associated with the generation of magnetostatic fields in space due to the inductor current; the circuit symbol of the inductor is a representation of the helical configuration of physical inductors. The helical configuration is particularly efficient in generating intense magnetostatic fields and magnetic flux linkage.[10]

Analysis of AC circuits with capacitors and inductors will be delayed until Chapter 3. The physics of energy storage in capacitors and inductors typically is discussed in a first course in electromagnetics.

---

[9]The amount of charge stored on a capacitor plate is given as $q = Cv$, since the time-rate of change of q, namely, the derivative $\frac{dq}{dt} = C\frac{dv}{dt}$, matches the capacitor current $i = C\frac{dv}{dt}$.

[10]An inductor with current i generates a flux linkage of $\lambda = Li$, and the time-rate of change of λ, namely, $\frac{d\lambda}{dt} = L\frac{di}{dt}$, corresponds to the inductor voltage $v = L\frac{di}{dt}$. For a helical inductor coil, the flux linkage λ is the product of the magnetic flux ϕ generated by current i and the number of turns of the helix.

1.4 Complex Numbers

The analysis of resistive DC circuits—the main topic of Chapter 2—requires the use of only real numbers and real algebra. By contrast, calculating AC or time-varying response of circuits with capacitors and inductors may require solving differential equations, which often can be simplified with the use of complex numbers. We will first encounter complex numbers in Chapter 3, where we solve simple first-order differential equations describing RC and RL circuits with sinusoidal inputs. Complex numbers also will arise in the solution of second and higher order differential equations, irrespective of the input. Starting with Chapter 4, we will rely on complex arithmetic to provide an efficient solution to AC circuit problems where the signals are sinusoidal. More advanced circuit and system analysis methods (Fourier series, Fourier transform, and Laplace transform), discussed in succeeding chapters, will require a solid understanding of complex numbers, complex functions, and complex variables.

A review of complex numbers and arithmetic is provided in Appendix A at the end of this book. A short introduction to complex functions and complex variables is included as well. You should read Appendix A and then work the complex number exercises at the end of this chapter and Chapter 2 well before entering Chapters 3 and 4. It is important that you feel comfortable with complex addition, subtraction, multiplication, and division; understand the conversions between rectangular, polar, and exponential forms of complex numbers; and understand the relationship between trigonometric functions such as $\cos(\omega t)$ and complex exponentials $e^{j\omega t}$. In particular, *Euler's identity* and its implications will play a crucial role in Chapter 4 and the rest of this book.

EXERCISES

1.1   In the following circuit, determine R and v_s:

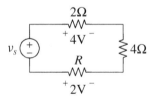

1.2   In the following circuit, determine i:

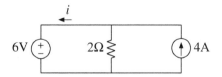

1.3 **(a)** In the following circuit, determine all of the unknown element and node voltages:

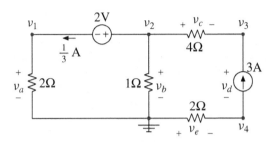

(b) What is the voltage drop in the above circuit from the reference to node 4?

(c) What is the voltage rise from node 2 to node 3?

(d) What is the voltage drop from node 1 to the reference?

1.4 **(a)** A volume of ionized gas filled with free electrons and protons can be modeled as a resistor. Consider such a resistor model supporting a 6 V potential difference between its terminals. We are told that in 1 s, 6.2422×10^{18} protons move through the resistor in the direction of the 6 V drop (say, from left to right) and 1.24844×10^{19} electrons move in the opposite direction. What is the net amount of electrical charge that transits the element in 1 s in the direction of the 6 V drop and what is the corresponding resistance R? Note that electrical charge q is 1.602×10^{-19} C for a proton and -1.602×10^{-19} C for an electron.

(b) Does a proton gain or lose energy as it transits the resistor? How many joules? Explain.

(c) Does an electron gain or lose energy as it transits the resistor? How many joules? Explain.

(d) A second resistor with 6 V potential difference conducts 1.87266×10^{19} electrons every second, but no proton is allowed to move through it. Compare the current, resistance, and absorbed power of the two resistors.

1.5   In the circuit pictured here, one of the independent voltage sources is injecting energy into the circuit, while the other one is absorbing energy. Identify the source that is injecting the energy absorbed in the circuit and confirm that the sum of all absorbed powers equals zero.

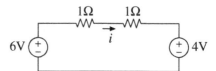

1.6   In the circuit pictured here, one of the independent current sources is injecting energy into the circuit, while the other one is absorbing energy. Identify the source that is injecting the energy absorbed in the circuit and confirm that the sum of all absorbed powers equals zero.

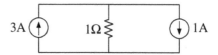

1.7   Calculate the absorbed power for each element in the following circuit and determine which elements inject energy into the circuit:

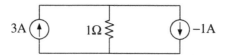

1.8   In the circuit given, determine i_x and calculate the absorbed power for each circuit element. Which element is injecting the energy absorbed in the circuit?

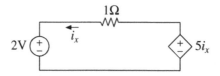

1.9   In the circuit given, determine v_x and calculate the absorbed power for each circuit element. Which element is injecting the energy absorbed in the circuit?

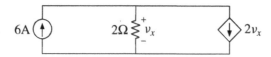

1.10   Some of the circuits shown next violate KVL or KCL and/or basic definitions
of two-terminal elements given in Section 1.3. Identify these ill-specified
circuits and explain the problem in each case.

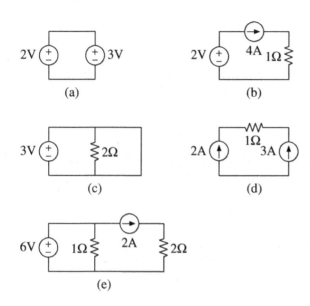

1.11   **(a)**   Let $A = 3 - j3$. Express A in exponential form.

(b)   Let $B = -1 - j1$. Express B in exponential form.

(c)   Determine the magnitudes of $A + B$ and $A - B$.

(d)   Express AB and A/B in rectangular form.

1.12   Do Exercise 1.11(a) through (d), but with $A = -3 - j3$ and $B = 1 + j2$.

1.13   **(a)**   Determine the rectangular forms of $e^{j0}, e^{j\frac{\pi}{2}}, e^{-j\frac{\pi}{2}}, e^{j\pi}, e^{-j\pi}$, and $e^{j2\pi}$.

(b)   Simplify $P = e^{j\pi} + e^{-j\pi}$, $Q = e^{j\frac{\pi}{2}} + e^{-j\frac{\pi}{2}}$, $R = 1 - e^{j\pi}$.

(c)   Show that $e^{j\frac{\pi}{2}m} = (-1)^{\frac{m}{2}}$.

1.14   **(a)**   Determine the rectangular forms of $7e^{j\frac{\pi}{4}}$, $7e^{-j\frac{\pi}{4}}$, $5e^{j\frac{3\pi}{4}}$, and $5e^{-j\frac{3\pi}{4}}$.

(b)   Simplify $P = 2e^{j\frac{5\pi}{4}} - 2e^{-j\frac{5\pi}{4}}$, $Q = 8e^{-j\frac{\pi}{4}} - 8e^{j\frac{\pi}{4}}$, and $R = \dfrac{e^{j\frac{3\pi}{4}}}{e^{-j\frac{\pi}{4}}}$.

1.15   **(a)**   Prove that $CC^* = |C|^2$.

(b)   Prove that $(C_1 C_2)^* = C_1^* C_2^*$.

1.16   (a)   Prove that $|C_1 C_2| = |C_1||C_2|$.

(b)   Prove that $|\frac{1}{C}| = \frac{1}{|C|}$.

(c)   Prove that $|\frac{C_1}{C_2}| = \frac{|C_1|}{|C_2|}$.

1.17   Show graphically on the complex plane that $|C_1 + C_2| \leq |C_1| + |C_2|$.

1.18   (a)   The function $f(t) = e^{j\frac{\pi}{4}t}$, for real-valued t, takes on complex values. Plot the values of $f(t)$ on the complex plane, for $t = 0, 1, 2, 3, 4, 5, 6$, and 7.

(b)   Repeat (a), but for the complex-valued function $g(t) = e^{(-\frac{1}{8} + j\frac{\pi}{4})t}$.

2

Analysis of Linear Resistive Circuits

Most analog signal processing systems are built with electrical circuits. Thus, the analysis and design of signal processing systems requires proficiency in *circuit analysis*, meaning the calculation of voltages and currents (or voltage and current waveforms) at various locations in a circuit. In this chapter we will describe a number of analysis techniques applicable to *linear resistive circuits* composed of resistors and some combination of independent and dependent sources. The techniques developed here will be applied in Chapter 3 to circuits containing operational amplifiers, capacitors, and inductors that can be used for signal processing purposes. Later, our basic techniques introduced here will be further developed in Chapter 4 for the analysis of linear circuits containing sinusoidal sources.

The topics to be covered in this chapter include resistor combinations and source transformations (analysis via circuit simplification), node-voltage and loop-current methods (systematic applications of KCL and KVL), and Thevenin and Norton equivalents of linear resistive networks and their interactions with external loads.

Strategies for circuit simplification; node-voltage and loop-current methods; linearity and superposition; coupling and available power of resistive networks

2.1   Resistor Combinations and Source Transformations

Analysis of resistive circuits frequently can be simplified via transformation to simpler equivalent circuits, using resistor combination and source transformation techniques. These approaches and some of their applications will be discussed in this section.

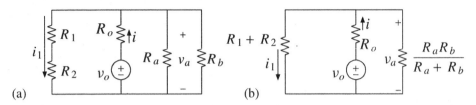

Figure 2.1 Resistors in series and parallel.

2.1.1 Resistor combinations and voltage and current division

Resistors R_1 and R_2 in Figure 2.1a carry the same current i_1 through a single branch of the circuit (leftmost branch) and therefore are said to be in *series*. Resistors R_a and R_b, on the other hand, support the same voltage v_a between the same pair of nodes and are said to be in *parallel*. We can analyze the circuit after simplifying it to the form shown in Figure 2.1b, using the series and parallel equivalents of these resistor pairs. We next will describe the simplification procedure and illustrate how to analyze the circuit.

In Figure 2.1a we recognize the total resistance of the left branch containing R_1 and R_2 as $R_1 + R_2$, since the voltage drop across the branch from top to bottom is

$$R_1 i_1 + R_2 i_1 = (R_1 + R_2) i_1 \equiv R_s i_1.$$

Thus, a single resistor

$$R_s = R_1 + R_2$$

Series equivalent

is the series equivalent of resistors R_1 and R_2, and replaces them in the simplified version of the circuit shown in Figure 2.1b. In general, the *series equivalent* of N resistors $R_1, R_2, \cdots, R_N$ (all carrying the same current on a single circuit branch) is

$$R_s = R_1 + R_2 + \cdots + R_N.$$

Likewise, we note the total resistance of the two parallel branches on the right of Figure 2.1a as $\frac{R_a R_b}{R_a + R_b}$, because the total current conducted from top to bottom through these branches is

$$\frac{v_a}{R_a} + \frac{v_a}{R_b} = v_a \left(\frac{1}{R_a} + \frac{1}{R_b} \right) \equiv \frac{v_a}{R_p}.$$

Thus, a single resistor

$$R_p = \left(\frac{1}{R_a} + \frac{1}{R_b} \right)^{-1} = \frac{R_a R_b}{R_a + R_b}$$

Parallel equivalent

is the parallel equivalent of resistors R_a and R_b, and replaces them in the simplified version of the circuit shown in Figure 2.1b. In general, the *parallel equivalent* of N

resistors $R_1, R_2, \cdots, R_N$ (all supporting the same voltage between the same pair of nodes) is

$$R_p = (\frac{1}{R_1} + \frac{1}{R_2} + \cdots + \frac{1}{R_N})^{-1}.$$

Example 2.1

Figure 2.2a replicates Figure 2.1a, but with numerical values for the resistors and the source. We next will solve for the unknown current i supplied by the source, using the technique of resistor combinations.

A simplified version of the circuit after series and parallel resistor combinations is shown in Figure 2.2b, where the series combination is

$$R_s = 1 + 5 = 6 \, \Omega$$

and the parallel combination is

$$R_p = \frac{3 \times 6}{3 + 6} = 2 \, \Omega.$$

Now notice that in Figure 2.2b the 6 Ω and 2 Ω resistors are in parallel, since they support the same voltage between the same pair of nodes. Thus, we replace them with their parallel equivalent

$$R_p = \frac{6 \times 2}{6 + 2} = \frac{12}{8} = 1.5 \, \Omega$$

and obtain the circuit shown in Figure 2.2c.

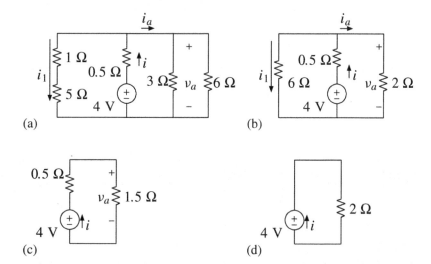

(a)   (b)   (c)   (d)

Figure 2.2 A resistive circuit (a) and its equivalents (b), (c), (d).

Finally, in Figure 2.2c we notice that the 0.5 Ω and 1.5 Ω resistors are in series and replace them by their series equivalent

$$R_s = 0.5 + 1.5 = 2\,\Omega$$

to obtain Figure 2.2d.

Clearly now, the source current in the circuit is

$$i = \frac{4\,\text{V}}{2\,\Omega} = 2\,\text{A}.$$

Suppose we had wanted to solve for v_a in Figure 2.2a. Our analysis then could have stopped with Figure 2.2c, which shows two resistors in series. This is a special case of the circuit shown in Figure 2.3a. Here we see that

$$i = \frac{v_s}{R_s} = \frac{v_s}{R_1 + R_2},$$

and, therefore,

$$v_1 = v_s \frac{R_1}{R_1 + R_2}$$

and

$$v_2 = v_s \frac{R_2}{R_1 + R_2}.$$

Voltage division

These *voltage division* equations tell us how the total voltage v_s across two resistors is divided between the elements.

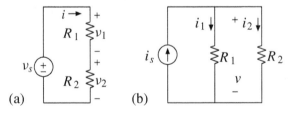

Figure 2.3   Voltage (a) and current (b) division.

Example 2.2

Apply voltage division to find v_a in Figure 2.2c.

Solution   The total voltage across the resistors in series is 4 V. Thus, applying voltage division, we find

$$v_a = 4\,\text{V}\,\frac{1.5}{0.5+1.5} = \frac{6}{2}\,\text{V} = 3\,\text{V}.$$

We can check this result by multiplying $i = 2\,\text{A}$ with $1.5\,\Omega$, which, of course, gives the same result.

Suppose we had wanted to find i_a in Figure 2.2a. The equivalent circuit in Figure 2.2b makes it clear that the current i splits into the two currents i_1 and i_a. The general situation is shown in Figure 2.3b. Here,

$$v = i_s R_p = i_s \frac{R_1 R_2}{R_1 + R_2},$$

and

$$i_1 = \frac{v}{R_1},\ i_2 = \frac{v}{R_2}.$$

Therefore,

$$i_1 = i_s \frac{R_2}{R_1 + R_2}$$

and

$$i_2 = i_s \frac{R_1}{R_1 + R_2}.$$

These *current-division* equations tell us how the total current i_s conducted by two resistors is split between them.

Current division

Example 2.3

Apply current division to find i_a in Figure 2.2b.

Solution   The parallel resistors carry a total current of $i = 2\,\text{A}$. (See Example 2.1.) Thus, using current division, we obtain

$$i_a = 2\,\text{A}\,\frac{6}{6+2} = \frac{12}{8}\,\text{A} = 1.5\,\text{A}.$$

We can check this result by dividing $v_a = 3\,\text{V}$ (see Example 2.2) with $2\,\Omega$, which gives the same result.

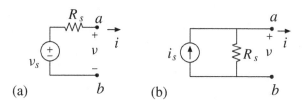

Figure 2.4 Two networks that are equivalent when $v_s = R_s i_s$. The term "network" refers to a circuit block with two or more external terminals.

2.1.2 Source transformations

In analyzing circuits, it sometimes is advantageous to replace a *voltage source in series with a resistor* by a *current source in parallel with a resistor* of the same value. Figure 2.4 shows these source–resistor combinations. It can be shown that, when terminals *a* and *b* are attached to another circuit, these source–resistor combinations will cause the same currents and voltages to occur in that circuit, so long as the source values are related by

$$v_s = R_s i_s \iff i_s = \frac{v_s}{R_s}.$$

In this case the source-resistor combinations are said to be *equivalent*, because they produce the same result when attached to a third circuit.

Proof: To prove that the source combinations in Figure 2.4 are equivalent for $v_s = R_s i_s$, we will find the expression for v in both circuits, in terms of the terminal current i (assumed to be carried by an external element or circuit connected between terminals *a* and *b*). Applying KVL in Figure 2.4a, we first find that $v_s = R_s i + v$, or

$$v = v_s - R_s i.$$

Applying KCL in Figure 2.4b, we next find that $i_s = \frac{v}{R_s} + i$, or

$$v = R_s i_s - R_s i.$$

Clearly, these expressions are identical for $v_s = R_s i_s$, in which case the two networks are equivalent because they apply the same voltage and inject the same current into an element or circuit attached to terminals *a* and *b*.

The next examples illustrate application of the *source transformation* method based on the equivalence of the networks shown in Figure 2.4 under the condition $v_s = R_s i_s$.

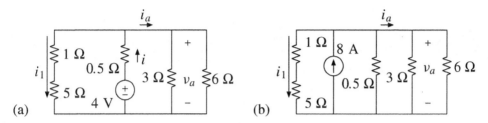

Figure 2.5 A source transformation example.

Example 2.4

Our goal is to find i_1 in Figure 2.5a.

We begin by exchanging the 4 V source in series with the 0.5 Ω resistor with a

$$\frac{4\,\text{V}}{0.5\,\Omega} = 8\,\text{A}$$

current source in parallel with a 0.5 Ω resistor. This exchange is permissible, as proven earlier, and is called a "source transformation." After the source transformation, the circuit in Figure 2.5a becomes the circuit shown in Figure 2.5b.

In this modified circuit, the current i_1 through the 6 Ω branch on the left can be easily calculated by current division. Since the parallel equivalent of the three resistors on the right is $\frac{1}{2.5}\,\Omega$, we find that

$$i_1 = 8\,\text{A}\frac{\frac{1}{2.5}}{\frac{1}{2.5}+6} = \frac{8}{1+15}\text{A} = 0.5\,\text{A}.$$

Example 2.5

In Figure 2.6 successive uses of source transformations and resistor combinations are illustrated to determine the unknown voltage v_{oc}. The simplified and equivalent versions of Figure 2.6a shown in Figures 2.6e and 2.6f are known as *Thevenin* and *Norton* equivalents (to be studied in Section 2.4). From either Figure 2.6e or 2.6f, we clearly see that

$$v_{oc} = 1\,\text{V}.$$

Resistor combinations, voltage and current division, and source transformations often are useful in the analysis of small circuits. But what if the circuit has thousands of loops and nodes? We need systematic ways of solving for voltages and currents in large circuits. In the next two sections we describe two procedures for systematic analysis of circuits: node analysis and loop analysis.

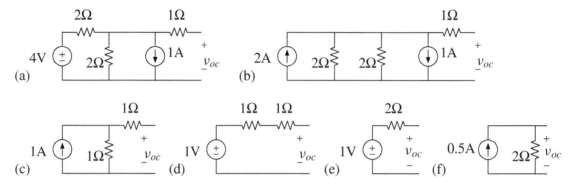

Figure 2.6 Simplification of circuit (a) by source transformations (a→b, c→d, and e→f) and resistor combinations (b→c and d→e) to its Thevenin (e) and Norton (f) equivalents. Note that in step b→c two current sources in parallel are combined into a single current source.

2.2 Node-Voltage Method

The node-voltage method is a systematic circuit analysis procedure that is based on the node-voltage concept (see Chapter 1 for a review) and application of KCL. The node-voltage method can be used for analyzing all possible resistive circuits, however complicated they may be. The popular and powerful circuit analysis software SPICE is based on the node-voltage method for that reason. The step-by-step procedure is as follows:

Node-voltage procedure

(1) In a circuit with $N + 1$ nodes, declare N node-voltage variables $v_1, v_2, \cdots, v_N$ with respect to a reference node (node 0), which is assigned a node voltage of zero.

(2) At each node with an unknown node voltage, write a KCL equation expressed in terms of node voltages.

(3) Solve the system of algebraic equations obtained in step 2 for the unknown node voltages.

Once all the node voltages are known, element voltages and currents easily can be calculated from the node voltages.

Example 2.6
Consider the circuit shown in Figure 2.7. The reference node is indicated by the ground symbol; v_1 and v_2 denote two unknown node voltages; and node voltage $v_3 = 3$ V has been directly identified, since there is an explicit 3 V rise from the reference to node 3 provided by an independent voltage source.

Following step 2 of the node-voltage method, we next construct KCL equations for nodes 1 and 2 where v_1 and v_2 were declared:

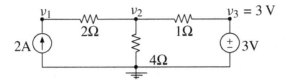

Figure 2.7 A node-voltage method example.

For node 1 we have

$$2 = \frac{v_1 - v_2}{2},$$

where $\frac{v_1-v_2}{2}$ denotes the current away from node 1 through the 2 Ω resistor.
Likewise, for node 2 we have

$$\frac{v_1 - v_2}{2} = \frac{v_2 - 0}{4} + \frac{v_2 - 3}{1},$$

or, equivalently,

$$0 = \frac{v_2 - v_1}{2} + \frac{v_2 - 0}{4} + \frac{v_2 - 3}{1},$$

where each term on the right corresponds to a current leaving node 2.
These KCL equations obtained from nodes 1 and 2 can be simplified to

$$v_1 - v_2 = 4$$
$$-2v_1 + 7v_2 = 12.$$

Their solution (step 3) is

$$v_1 = 8\,\text{V}$$
$$v_2 = 4\,\text{V}.$$

We can calculate the element voltages and currents in the circuit by using
these node voltages and $v_3 = 3$ V.

Example 2.7
The circuit shown in Figure 2.8 contains a voltage controlled current source
and three unknown node voltages marked as v_1, v_2, and v_3. We can apply
the node voltage method to the circuit by writing 3 KCL equations in which
each occurrence of voltage v_x controlling the dependent current source is
expressed in terms of node voltage v_3.

However, application of simple voltage division on the right side of the
circuit also shows that $v_3 = \frac{v_2}{2}$ as already marked on the circuit diagram.

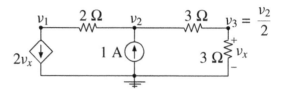

Figure 2.8 Another node-voltage example. Note that voltage v_x, which controls the dependent current source on the left, also can be written as node voltage v_3.

Therefore, we also should be able to solve this problem by writing down 2 KCL equations in terms of two unknowns v_1 and v_2. That approach is illustrated next:

We write the KCL equation for node 1 as

$$2\frac{v_2}{2} + \frac{v_1 - v_2}{2} = 0$$

where each term on the left side represents a current away from node 1, written in terms of the node-voltage unknowns. Notice how voltage-controlled current $2v_x$ has been represented as $2\frac{v_2}{2}$ in terms of node voltage $v_3 = \frac{v_2}{2}$.

The KCL equation for node 2 can be written as

$$\frac{v_1 - v_2}{2} + 1 = \frac{v_2 - 0}{6},$$

where the left side is the sum of currents into node 2 and the right side is a current away from node 2 through the 3 Ω resistor in series with a second 3 Ω resistor.

Simplifying these equations gives

$$v_1 + v_2 = 0$$
$$v_1 - 3v_2 = -6,$$

and solving them yields

$$v_1 = -3 \text{ V}$$
$$v_2 = 3 \text{ V}.$$

Also, $v_3 = \frac{v_2}{2} = 1.5$ V.

Example 2.7 illustrated how controlled sources are handled in the node-voltage method. Their contributions to the required KCL equations are entered after their values are expressed in terms of the node-voltage unknowns of the problem. The example also illustrated that if any of the unknown node voltages can be expressed readily in terms of other node voltages, the number of KCL equations to be formed can be reduced. The next example also illustrates such a reduction.

Example 2.8

Consider the circuit shown in Figure 2.9. In this circuit $v_1 = 2$ V has been directly identified, v_2 has been declared as an unknown node voltage, and it has been recognized at the outset that

$$v_3 = v_2 + 1$$

because of the 1 V rise from node 2 to node 3 provided by the 1 V independent voltage source.

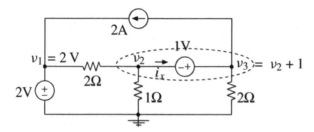

Figure 2.9 A node-voltage problem with a super-node.

To solve for the unknown node voltage v_2, we need to write the KCL equation for node 2. This equation can be expressed as

$$\frac{v_2 - 2}{2} + \frac{v_2 - 0}{1} + i_x = 0,$$

where we make use of a current variable i_x, which, according to the KCL equation written for node 3, is given by

$$i_x = 2 + \frac{(v_2 + 1) - 0}{2}.$$

Eliminating i_x between the two KCL equations, we obtain

$$\frac{v_2 - 2}{2} + \frac{v_2 - 0}{1} + 2 + \frac{(v_2 + 1) - 0}{2} = 0,$$

which is, in fact, a single KCL equation for a *super-node* in Figure 2.9, indicated by the dashed oval. (See the subsequent discussions of the super-node "trick.") Solving the equation, we obtain

$$v_2 = -0.75 \text{ V.}$$

Furthermore,

$$v_3 = v_2 + 1 = 0.25 \text{ V}.$$

The expression that we previously called the "super-node" KCL equation states that the sum of all currents out of the region in Figure 2.10, surrounded by the dashed oval, equals zero. Since there is no current flow into this region, the statement has the form of a KCL equation applied to the dashed oval. Thus, the dashed oval is an example of what is called a "super-node" in the node-voltage technique. Super-nodes can be defined around ordinary nodes connected by voltage sources, as in the previous example. When solving small circuits by hand, we find that clever application of the super-node concept sometimes can shorten the solution. We illustrate in the next example.

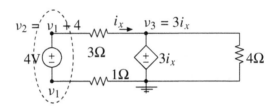

Figure 2.10   A node-voltage problem with a dependent source and a super-node.

Example 2.9
Figure 2.10 shows a circuit with a dependent source and a super-node. (See the dashed oval.) All node voltages in the circuit have been marked in terms of an unknown node voltage v_1 and current i_x that controls the dependent voltage source.

Using Ohm's law, we first note that current

$$i_x = \frac{v_2 - v_3}{3} = \frac{(v_1 + 4) - 3i_x}{3},$$

from which we obtain

$$i_x = \frac{v_1 + 4}{6}$$

in terms of unknown voltage v_1. We then can express the super-node KCL equation as

$$\frac{v_1}{1} + \frac{v_1 + 4}{6} = 0,$$

which can be solved for v_1 as

$$v_1 = -\frac{4}{7}\,\text{V}.$$

Thus,

$$v_2 = v_1 + 4 = \frac{24}{7}\,\text{V},$$

$$i_x = \frac{v_1 + 4}{6} = \frac{4}{7}\,\text{A},$$

and

$$v_3 = 3i_x = \frac{12}{7}\,\text{V}.$$

2.3 Loop-Current Method

The loop-current method is an alternative circuit analysis procedure based on system-atic application of KVL. When solving, by hand, small planar circuits[1] that contain primarily voltage sources, the loop-current method can be quicker than the node-voltage method. The next two examples will clarify what we mean by "loop current" and illustrate the main features of the method.

Example 2.10
In Figure 2.11a we see a single-loop circuit. All the element currents in the circuit can be represented in terms of a single *loop-current* variable i as shown in the diagram. A KVL equation for the loop is

$$3\,\text{V} = 2i + 4(2i) + 2\,\text{V};$$

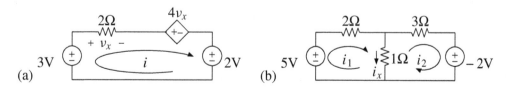

Figure 2.11 Loop-current method examples.

[1]Planar circuits are circuits that can be diagrammed on a plane with no element crossing another element.

therefore,

$$i = \frac{(3-2)\,\text{V}}{(2+8)\,\Omega} = 0.1\,\text{A}$$

and

$$v_x = 2i = 0.2\,\text{V}.$$

Example 2.11
Figure 2.11b shows a circuit with two-elementary loops (loops that contain no further loops within), which have been assigned distinct loop-current variables i_1 and i_2. In analogy with the previous problem, we consider i_1 to be the element current of the 5 V source and 2 Ω resistor on the left. Likewise, i_2 can be considered as the current of the 3 Ω resistor and -2 V source on the right. Furthermore, the current i_x through the 1 Ω resistor in the middle can be expressed in terms of loop currents i_1 and i_2; using KCL at the top node of the resistor, we see that

$$i_x = i_1 - i_2.$$

Clearly, all the element currents in the circuit are either directly known or can be calculated once the loop currents i_1 and i_2 are known. We will determine i_1 and i_2 by solving two KVL equations constructed around the two-elementary loops of the circuit.

The KVL equation for the loop on the left is

$$5 = 2i_1 + 1(i_1 - i_2),$$

where $1(i_1 - i_2) = 1i_x$ denotes the voltage drop across the 1 Ω resistor from top to bottom. Likewise, for the loop on the right we have

$$1(i_1 - i_2) = 3i_2 + (-2),$$

or, equivalently,

$$3i_2 + (-2) + 1(i_2 - i_1) = 0,$$

where $1(i_2 - i_1)$ denotes the voltage drop across the 1 Ω resistor from bottom to top. Rearranging the two equations, we obtain

$$3i_1 - i_2 = 5$$

and

$$-i_1 + 4i_2 = 2.$$

Solving these two equations in two unknowns yields

$$i_1 = 2\,\text{A}$$

and

$$i_2 = 1\,\text{A}.$$

Consequently,

$$i_x = i_1 - i_2 = 1\,\text{A},$$

and the voltage drop across the 1 Ω resistor from top to bottom is $1\,\Omega \times i_x = 1\,\text{V}$. The remaining element voltages also can be calculated in a similar way using the loop currents i_1 and i_2.

The preceding examples illustrate the main idea behind the loop-current method; namely, loop-current values are calculated for each elementary loop in the circuit. Then, if desired, element currents or voltages can be calculated from the loop currents. A step-by-step procedure for calculating the loop currents is as follows: **Loop-current procedure**

(1) In a planar circuit with N elementary loops, declare N loop-current variables $i_1, i_2, \cdots, i_N$,
(2) Around each elementary loop with an unknown loop current, write a KVL equation expressed in terms of loop currents.
(3) Solve the system of algebraic equations obtained in step 2 for the unknown loop currents.

Example 2.12
Consider the circuit shown in Figure 2.12. The circuit contains three elementary loops; therefore, there are three loop currents in the circuit. Two of these have been declared as unknowns i_1 and i_2, and the third one has been recognized as $i_3 = 2\,\text{A}$ to match the 2 A current source on the right. Since we have only two unknown loop currents i_1 and i_2, we need to write only two KVL equations. The KVL equation for loop 1 (where i_1 has been declared) is

$$14 = 2i_1 + 3i_x.$$

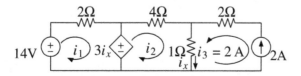

Figure 2.12 Loop-current example for a circuit containing a current source.

Likewise, the KVL equation for loop 2 (where i_2 has been declared) is

$$3i_x = 4i_2 + 1(i_2 + 2).$$

Note that i_x can be expressed as

$$i_x = i_2 + 2,$$

since the direction of loop currents i_2 and $i_3 = 2$ A coincide with the direction of i_x. Substituting for i_x in the two KVL equations and rearranging as

$$2i_1 + 3i_2 = 8$$

and

$$2i_2 = 4,$$

we are finished with the implementation of step 2. Clearly then, the solution (that is, step 3) is

$$i_2 = 2 \text{ A}$$

and

$$i_1 = 1 \text{ A}.$$

Example 2.13
Consider the circuit shown in Figure 2.13 with three loop currents declared as i_1, i_2, and i_3. Nominally, we need three KVL equations, one for each elementary loop. However, we note that loops 2 and 3 are separated by a 2 A current source and consequently,

$$i_3 = i_2 + 2,$$

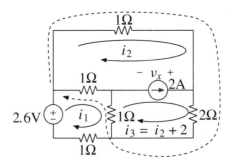

Figure 2.13 A loop-current example with a super-loop equation.

as already marked in the diagram. Therefore, in the KVL equations every occurrence of i_3 can be expressed as $i_2 + 2$ so that the equations contain only two unknowns, i_1 and i_2.

We start with loop 1 and obtain

$$1(i_1 - i_2) + 1(i_1 - (i_2 + 2)) + 1i_1 = 2.6,$$

expressing i_3 as $(i_2 + 2)$. The KVL equation for loop 2 is

$$1i_2 + v_x + 1(i_2 - i_1) = 0,$$

where

$$v_x = 2(i_2 + 2) + 1((i_2 + 2) - i_1)$$

is the voltage drop across the 2 A source as determined by writing the KVL equation around loop 3. Eliminating v_x between the two equations, we obtain

$$1i_2 + 2(i_2 + 2) + 1((i_2 + 2) - i_1) + 1(i_2 - i_1) = 0,$$

which is in fact the KVL equation around the dashed contour shown in Figure 2.13, which can be called a "super loop," in analogy with the super-node concept. The KVL equations for loop 1 and the dashed contour (super loop) simplify to

Super loop

$$3i_1 - 2i_2 = 4.6$$
$$-2i_1 + 5i_2 = -6.$$

Their solution is

$$i_1 = 1\,\text{A}$$
$$i_2 = -0.8\,\text{A},$$

and, consequently,

$$i_3 = i_2 + 2 = 1.2\,\text{A}.$$

Super loops such as the dashed contour in Figure 2.13 can be formed around elementary loops sharing a common boundary that includes a current source. In such cases, either of the elementary loop currents can be expressed in terms of the other (e.g., $i_3 = i_2 + 2$ in Example 2.13, earlier) and therefore, instead of writing two KVL equations around the elementary loops, it suffices to formulate a single KVL equation around the super loop.

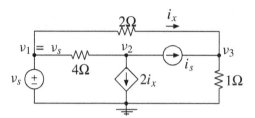

Figure 2.14 Circuit with arbitrary current and voltage sources i_s and v_s.

2.4 Linearity, Superposition, and Thevenin and Norton Equivalents

We might wonder how scaling the values of the independent sources in a resistive circuit will affect the value of some particular voltage or current. For example, will a doubling of the source values result in a doubling of the voltage across a specified resistor? More generally, we can ask how circuits *respond* to their source elements. By "circuit response," we refer to all voltages and currents excited in a circuit, and we consider whether the combined effect of all independent sources is the sum of the effects of the sources operating individually. We shall reach some general conclusions by examining these questions within the context of the node-voltage approach.

We first note that the application of KCL in node-voltage analysis always leads to a set of linear algebraic equations with node-voltage unknowns and *forcing terms* proportional to independent source strengths. For instance, for the circuit shown in Figure 2.14, the KCL equations to be solved are

$$v_1 = v_s$$

$$\frac{v_2 - v_1}{4} + 2\frac{v_1 - v_3}{2} + i_s = 0$$

$$\frac{v_3}{1} + \frac{v_3 - v_1}{2} = i_s,$$

in which the forcing terms v_s and i_s are due to the independent sources in the circuit. Dependent sources enter the KCL equations in terms of node voltages

$$2i_x = 2\frac{v_1 - v_3}{2},$$

(as for instance, in the second of the preceding equations), therefore, dependent sources do not contribute to the forcing terms. The consequence is that, node-voltage solutions are *linear combinations* of independent source strengths only, as in

$$v_1 = v_s$$

$$v_2 = -\frac{5}{3}v_s - \frac{4}{3}i_s$$

$$v_3 = \frac{1}{3}v_s + \frac{2}{3}i_s,$$

which is the solution to the KCL equations for the circuit of Figure 2.14.[2] Since we can calculate any voltage or current in a resistive circuit by using a linear combination of node voltages and/or independent source strengths (as we have seen in Section 2.2), the following general statement can be made:

Superposition principle: Let $f_1, f_2, \cdots, f_N$ be the complete set of independent source strengths (such as v_s and i_s in Figure 2.15a) in a resistive circuit. Then, any electrical response y in the circuit (such as voltage v in Figure 2.15a or some element current) can be expressed as

$$y = K_1 f_1 + K_2 f_2 + \cdots + K_N f_N,$$

where $K_1, K_2, \cdots, K_N$ are constant coefficients, unique for each response y.

Applying the superposition principle to the resistive circuit in Figure 2.15a, for instance, we can write

$$v = K_1 v_s + K_2 i_s,$$

where K_1 and K_2 are constant coefficients (to be determined in the next section). What this means is that voltage v is a *weighted linear superposition* of the independent sources v_s and i_s in the circuit. Likewise, any electrical response y in any resistive circuit is a weighted linear superposition of the independent sources $f_1, f_2, \cdots, f_N$.

Since any response in a resistive circuit can be expressed as a weighted linear superposition of independent sources, such circuits are said to be *linear*. We see from **Linearity** this property that a doubling of all independent sources will indeed double all circuit responses and that every circuit response is a sum of the responses due to the individual sources. We shall examine this latter point more closely in the next section.

The linearity property is not unique to resistive circuits; circuits containing resistors, capacitors, and inductors also can be linear, as we will see and understand in

---

[2]These results can be viewed as the solution of the matrix problem

$$\begin{bmatrix} 1 & 0 & 0 \\ \frac{3}{4} & \frac{1}{4} & -1 \\ -\frac{1}{2} & 0 & \frac{2}{3} \end{bmatrix} \begin{bmatrix} v_1 \\ v_2 \\ v_3 \end{bmatrix} = \begin{bmatrix} v_s \\ -i_s \\ i_s \end{bmatrix}$$

for the node-voltage vector on the left-hand side. Since the solution is the product of the source vector on the right with the inverse of the coefficient matrix on the left, the node-voltage values are linear combinations of the elements of the source vector and, hence, linear combinations of v_s and i_s.

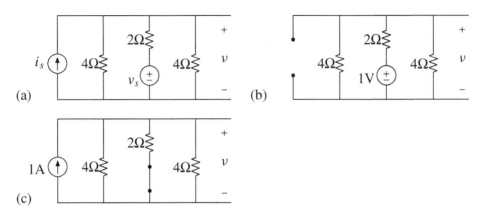

Figure 2.15 (a) A linear resistive circuit with two independent sources, (b) the same circuit with suppressed current source and $v_s = 1$ V, (c) the same circuit with suppressed voltage source and $i_s = 1$ A.

later chapters. On the other hand, circuits containing even a single nonlinear element, such as a diode, ordinarily will behave nonlinearly, meaning that circuit responses will not be weighted superpositions of independent sources in the circuit. A diode is nonlinear because its v–i relation cannot be plotted as a straight line. By contrast, resistors have straight-line v–i characteristics.

2.4.1 Superposition method

The superposition principle just introduced can be exploited in many ways in resistive circuit analysis.

For instance, as already discussed, voltage v in Figure 2.15a can be expressed as

$$v = K_1 v_s + K_2 i_s.$$

To determine K_1, we notice that, for $v_s = 1$ V and $i_s = 0$,

$$v = K_1 \times (1\text{ V}).$$

But with $v_s = 1$ V and $i_s = 0$, the circuit simplifies to the form shown in Figure 2.15b, since with $i_s = 0$ the current source becomes an effective open circuit. Analyzing the circuit of Figure 2.15b with the suppressed current source, we find that (using resistor combination and voltage division, for instance) $v = \frac{1}{2}$ V. Therefore, $v = K_1 \times (1\text{ V})$ implies that

$$K_1 = \frac{1}{2}.$$

Likewise, to determine K_2, we let $v_s = 0$ and $i_s = 1$ A in the circuit so that

$$v = K_2 \times (1\text{ A})$$

in the modified circuit with the suppressed voltage source, shown in Figure 2.15c. We find that in Figure 2.15c (using resistor combination and Ohm's law) $v = 1$ V. Therefore, $v = K_2 \times (1\,\text{A})$ implies that

$$K_2 = 1\frac{V}{A} = 1\,\Omega.$$

Hence, combining v_s and i_s with the weight factors K_1 and K_2, respectively, we find that in the circuit shown in Figure 2.15a,

$$v = \frac{1}{2}v_s + 1\,\Omega \times i_s.$$

Note that each term in this sum represents the contribution of an individual source to voltage v. Likewise, the superposition principle implies that any response in any resistive circuit is the sum of individual contributions of the independent sources acting alone in the circuit. We will use this notion in Example 2.14.

Example 2.14
In a linear resistive circuit with two independent current sources i_1 and i_2, a resistor voltage $v_x = 2$ V when

$$i_1 = 1\,\text{A} \quad \text{and} \quad i_2 = 0\,\text{A}.$$

But when

$$i_1 = -1\,\text{A} \quad \text{and} \quad i_2 = 3\,\text{A},$$

it is found that $v_x = 4$ V. Determine v_x if

$$i_1 = 0\,\text{A} \quad \text{and} \quad i_2 = 5\,\text{A}.$$

Solution   Because the circuit is linear, we can write

$$v_x = K_1 i_1 + K_2 i_2.$$

Using the given information, we then obtain

$$2 = K_1$$
$$4 = -K_1 + 3K_2.$$

Thus,

$$K_2 = \frac{4 + K_1}{3} = 2\,\Omega,$$

and we have

$$v_x = 5K_2 = 10\,\text{V}$$

for $i_1 = 0\,\text{A}$ and $i_2 = 5\,\text{A}$.

Example 2.15
Determine v_2 in Figure 2.7 from Section 2.2 using source suppression and superposition.

Solution   First, suppressing (setting to zero) the 2 A source in the circuit, we find that v_2 due to the 3 V source is

$$v_2 = 3\,\text{V}\frac{4\,\Omega}{1\,\Omega + 4\,\Omega} = \frac{12}{5}\,\text{V}.$$

To understand this expression, you must redraw Figure 2.7 with the 2 A source replaced by an open circuit and apply voltage division in the revised circuit. Second, suppressing the 3 V source, we calculate that v_2 due to the 2 A source is

$$v_2 = 2\,\text{A}\frac{4\,\Omega \times 1\,\Omega}{4\,\Omega + 1\,\Omega} = \frac{8}{5}\,\text{V}.$$

We obtain this expression by redrawing Figure 2.7 with the 3 V source set equal to zero, which implies that the voltage source is replaced by a short circuit. Finally, according to the superposition principle, the actual value of v_2 in the circuit is the sum of the two values just calculated, representing the individual contributions of each source. Hence,

$$v_2 = \frac{12}{5} + \frac{8}{5} = 4\,\text{V},$$

which agrees with the node-voltage analysis result obtained in Example 2.6 in Section 2.2.

2.4.2   Thevenin and Norton equivalents of resistive networks

Signal processing systems typically are composed of very large, interconnected circuits. For purposes of understanding how these circuits operate together, it often is advantageous to model one or more of these complex circuits by a much simpler circuit that is easier to understand and to work with mathematically. Here we are not suggesting a replacement of the physical circuit, but rather the introduction of a model whose mathematical behavior mimics that of the physical circuit.

Toward this end, consider the diagram shown in Figure 2.16a. The box on the left contains an arbitrary resistive network (e.g., the network shown in Figure 2.15a) including possibly a large number of independent and dependent sources. This is the network that we wish to model with a simpler circuit which will have equivalent

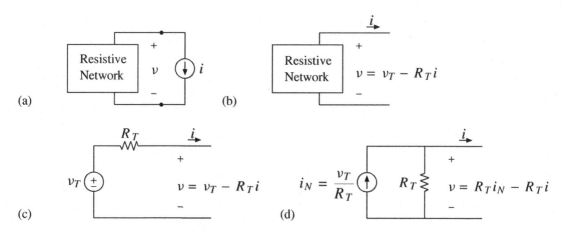

Figure 2.16   (a) A linear circuit containing an independent current source i and some linear resistive network, (b) the linear resistive network and its terminal v–i relation, (c) Thevenin equivalent of the network, and (d) Norton equivalent.

behavior as seen by the outside world. To incorporate the outside world, suppose that the network terminals are connected to a second circuit that draws current i and results in voltage v across the terminals. We have not shown the second circuit explicitly; instead, we have modeled the effect of the second circuit by the current source i.

Now, what might the simple replacement be for the resistive network? Because the network is resistive, its terminal voltage v will vary with current i in a linear form, say, as

$$v = v_T - R_T i,$$

where R_T and v_T are constants that depend on what is within the resistive network.[3] Both v_T and R_T may be either positive or negative.

Figure 2.16b emphasizes that the boxed resistive network is governed at its terminals by an equation having the preceding form. Any circuit having the same terminal v–i relation will be mathematically equivalent to the resistive network. The simplest such circuit is shown in Figure 2.16c and is known as the *Thevenin equivalent.*

Since we started by assuming that the box contains an arbitrary resistive network, this result means that *all* resistive networks can be represented in the simple form of Figure 2.16c! A second equivalent circuit is obtained as the source transform of the Thevenin equivalent (see Figure 2.16d) and is known as the *Norton equivalent.* Both

---

[3] A more rigorous argument for this relation is as follows: Since the circuit of Figure 2.16a is linear, the voltage v is a weighted linear superposition of all the independent sources within the box, as well as the independent current source i external to the box. Expressing the overall contributions of the sources within the box as v_T, and declaring a weighting coefficient $-R_T$ for the contribution of i, we obtain $v = v_T - R_T i$ with no loss of generality.

of these equivalent circuits[4] are useful for studying the coupling of resistive networks to external loads and other circuits.

We next will describe how to calculate the Thevenin voltage v_T, Norton current i_N, and Thevenin resistance R_T. We will then illustrate in the next section the use of the Thevenin equivalent in an important network coupling problem.

Thevenin voltage v_T: For $i = 0$, the terminal v–i relation of a resistive network, that is,

$$v = v_T - R_T i,$$

reduces to

$$v = v_T.$$

Thevenin voltage equals the open-circuit voltage

Thus, to find the Thevenin voltage v_T of a network, it is sufficient to calculate its output voltage v in the absence of an external load at the network terminals. The Thevenin voltage v_T, therefore, also is known as the "open-circuit voltage" of the network.

As an example, consider the network of Figure 2.15a, which is repeated in Figure 2.17a, with $v_s = 2$ V and $i_s = 1$ A. Using the result

$$v = \frac{1}{2}v_s + 1\,\Omega \times i_s$$

from the last section, with $v_s = 2$ V and $i_s = 1$ A, we calculate that $v = 2$ V is the open-circuit voltage of the network (i.e., the terminal voltage in the absence of an external load). Hence, Thevenin voltage of the network is

$$v_T = 2\,\text{V}.$$

Norton current i_N: We next will see that the Norton current of a linear network, defined as (see Figure 2.16d)

$$i_N \equiv \frac{v_T}{R_T},$$

also happens to be the current that flows through an external short placed between the network terminals (as in Figure 2.17b) in the direction of the voltage drop adopted for v and v_T. The termination shorts out the network output voltage v, and thus the relation $v = v_T - R_T i$ is reduced to

$$0 = v_T - R_T i,$$

implying that the short-circuit current of the network is

$$i = \frac{v_T}{R_T} = i_N.$$

---

[4]These two circuits are named after Leon C. Thevenin (1857-1926) of French Postes et Telegraphes and Edward L. Norton (1898-1983) of Bell Labs.

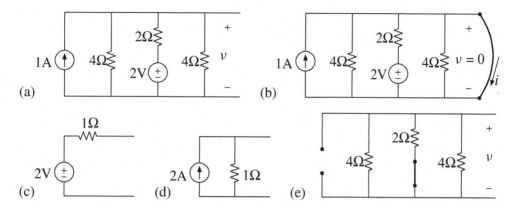

Figure 2.17 (a) A linear network with terminal voltage v, (b) the same network with an external short, (c) Thevenin equivalent, (d) Norton equivalent, and (e) the same network as (a) after source suppression.

Thus, to determine the Norton current of the network in Figure 2.17a, we place a short between the network terminals and calculate the short-circuit current i, as shown in Figure 2.17b. Applying KCL at the top node gives

Norton current equals the short-circuit current

$$1 = \frac{0 - 2}{2} + i,$$

because there is zero voltage across the two 4 Ω resistors. Therefore, the short-circuit current of the network is $i = 2$ A, and consequently,

$$i_N = 2\,\text{A}.$$

The Norton current of *any* linear network can be calculated as a short-circuit current.

Thevenin resistance R_T: One way to determine the Thevenin resistance R_T is to rewrite the formula for Norton current as

$$R_T = \frac{v_T}{i_N}$$

and use it with known values of v_T (open circuit voltage) and i_N (short circuit current). For the network shown in Figure 2.17a, for example,

$$R_T = \frac{v_T}{i_N} = \frac{2\,\text{V}}{2\,\text{A}} = 1\,\Omega;$$

the corresponding Thevenin and Norton equivalent circuits are as shown in Figures 2.17c and 2.17d, respectively.

The Thevenin resistance R_T also can be determined directly by a *source suppression method* without first finding the Thevenin voltage and Norton current. Before

Thevenin resistance

$R_T = \frac{v_T}{i_N}$

we describe the method, we observe that if all the independent source strengths in Figure 2.17a were halved, that is,

$$1 \, \text{A} \to 0.5 \, \text{A} \quad \text{and} \quad 2 \, \text{V} \to 1 \, \text{V},$$

the open-circuit voltage v_T and short-circuit current i_N of the network would also be halved. In other words,

$$v_T = 2 \, \text{V} \to 1 \, \text{V} \quad \text{and} \quad i_N = 2 \, \text{A} \to 1 \, \text{A}$$

because of the linearity of the network. However, the Thevenin resistance

$$R_T = \frac{v_T}{i_N} = \frac{2 \, \text{V}}{2 \, \text{A}} = 1 \, \Omega \to \frac{1 \, \text{V}}{1 \, \text{A}} = 1 \, \Omega$$

would remain unchanged. The Thevenin resistance would, in fact, remain unchanged even in the limiting case when all independent source strengths are suppressed to zero, as shown in Figure 2.17e. Indeed, the Thevenin resistance of the network in Figure 2.17e is the same as the Thevenin resistance of the original network shown in Figure 2.17a!

Source suppression method

This observation leads to the source suppression method for finding R_T:

(1) Replace all independent voltage sources in the network by short circuits and all independent current sources by open circuits.

(2) If the remaining network contains no dependent sources (as in Figure 2.17e), then R_T is the equivalent resistance, which we can determine usually by using series and/or parallel resistor combinations. (Note that in Figure 2.17e the parallel equivalent of the three resistors yields the correct Thevenin resistance $R_T = 1 \, \Omega$ obtained earlier.)

(3) If the remaining network contains dependent sources, or cannot be simplified by just series and parallel combinations, then R_T can be determined by the *test signal* method, illustrated in Example 2.17 to follow.

Example 2.16

Figure 2.18a shows a resistive network with a dependent source. Determine the Thevenin and Norton equivalents of the network.

Solution   To determine the Thevenin voltage v_T, we apply the node-voltage method to the circuit shown in Figure 2.18a. The unknown node voltages are v_T and $v_T - 2i_x$, where

$$i_x = \frac{v_T - 2i_x}{1}$$

is the current flowing down through the $1 \, \Omega$ resistor. We also have a super-node KCL equation

$$2 = i_x + \frac{v_T}{2}.$$

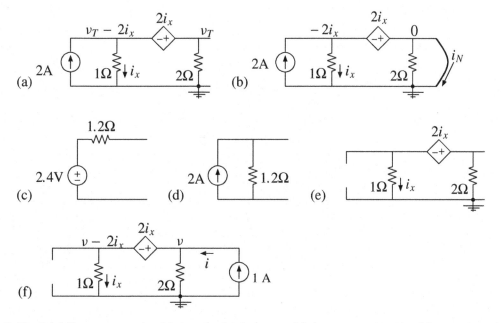

Figure 2.18   (a) A linear network with terminal voltage v_T, (b) the same network with an external short, (c) Thevenin equivalent, (d) Norton equivalent, (e) the same network with suppressed independent source, and (f) source suppressed network with an applied test source.

Solving these two equations in two unknowns, we find that

$$v_T = \frac{12}{5} = 2.4\,\text{V}.$$

We next proceed to find i_N by shorting the external terminals of the network as shown in Figure 2.18b. After placing the short at the network terminals, the voltage $v_T - 2i_x$ across the 1 Ω resistor has been reduced to $-2i_x$, and therefore the current through the resistor is

$$i_x = \frac{-2i_x}{1},$$

indicating that

$$i_x = 0.$$

Hence, neither resistor carries any current in Figure 2.18(b), and therefore,

$$i_N = 2\,\text{A}.$$

Finally,

$$R_T = \frac{v_T}{i_N} = \frac{12/5}{2} = \frac{6}{5} = 1.2 \, \Omega.$$

The Thevenin and Norton equivalents are shown in Figures 2.18c and 2.18d.

Test-signal method

The next example explains the test signal method for determining the Thevenin resistance in networks containing dependent sources.

Example 2.17

Figure 2.18e shows the network in Figure 2.18a with the independent source suppressed. The Thevenin resistance R_T of the original network is the equivalent resistance of the source-suppressed network, but the presence of a dependent source prevents us from determining R_T with series/parallel resistor combination methods. Instead, we can determine R_T by calculating the voltage response v of the source-suppressed network to a test current $i = 1$ A injected into the network, as shown in Figure 2.18f. R_T is then the equivalent resistance

$$R_T = \frac{v}{i} = \frac{v}{1 \, \text{A}}.$$

For a test current $i = 1$ A, we write KCL at the injection node to obtain

$$i = 1 \, \text{A} = \frac{v}{2} + i_x.$$

Also,

$$i_x = \frac{v - 2i_x}{1},$$

implying that

$$i_x = \frac{v}{3}.$$

Thus,

$$i = 1 \, \text{A} = \frac{v}{2} + \frac{v}{3} = \frac{5}{6}v,$$

from which we obtain

$$v = \frac{6}{5} = 1.2 \, \text{V}.$$

Hence,

$$R_T = \frac{v}{i} = \frac{1.2\,\text{V}}{1\,\text{A}} = 1.2\,\Omega,$$

as already determined in Example 2.16.

The Thevenin resistance in resistive networks with no dependent sources is always a positive quantity. However, in a network with dependent sources the Thevenin resistance can be zero or even negative (as happens in the next example). Remember that the Thevenin resistance is a quantity in a mathematical model—it is not a physical resistor.

Example 2.18
Determine the Thevenin equivalent of the network shown in Figure 2.19a.

Solution   Since the network contains no independent source, the open-circuit voltage $v_T = 0$ and the Thevenin equivalent is just the Thevenin resistance R_T.

To determine R_T, we will feed the network with an external 1 A current source, as shown in Figure 2.19b, and solve for the response voltage v at the network terminals. Then

$$R_T = \frac{v}{1\,\text{A}}.$$

The KCL equation at the top node is

$$\frac{v}{3} = v_x + 1,$$

and, via voltage division across the series resistors on the left, we also have

$$v_x = \frac{2}{2+1}v = \frac{2}{3}v.$$

Replacing v_x in the KCL equation with $\frac{2}{3}v$, we find that

$$v = -3\,\text{V}.$$

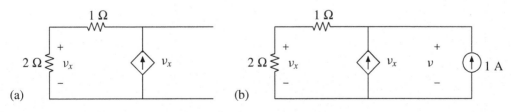

Figure 2.19   (a) A linear network with a dependent source, and (b) the same network excited by a 1 A test source to determine R_T.

Hence,

$$R_T = \frac{-3\,\text{V}}{1\,\text{A}} = -3\,\Omega.$$

2.5   Available Power and Maximum Power Transfer

Given a resistive network, it is important to ask how much power that network can transfer to an external load resistance. Ordinarily, we wish to size the load resistance so as to transfer the maximum power possible. For networks with positive Thevenin resistance R_T, there is an upper bound on the amount of power that can be delivered out of the network, a quantity known as the *available power* of the network.

To determine the general expression for the power available from a resistive network, we will examine the circuit shown in Figure 2.20. In this circuit an arbitrary resistive network is represented by its Thevenin equivalent on the left, and the resistor R_L on the right represents an arbitrary load.

Using voltage division, we find that the load voltage is

$$v_L = \frac{R_L}{R_T + R_L} v_T.$$

Hence, the absorbed load power

$$p_L = v_L i_L = \frac{v_L^2}{R_L}$$

is

$$p_L = \frac{R_L}{(R_T + R_L)^2} v_T^2.$$

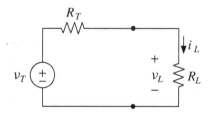

Figure 2.20   The model for the interaction of linear resistive networks with an external load R_L.

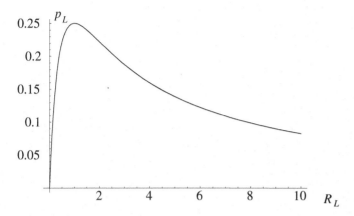

Figure 2.21   The plot of load power p_L in W versus load resistance R_L in Ω, for $V_T = 1$ V and $R_T = 1\,\Omega$.

For fixed values of v_T and $R_T > 0$ (i.e., for a specific network with specific values of v_T and R_T), this quantity will vary with load resistance R_L. From the expression it is easy to see that p_L vanishes for $R_L = 0$ and $R_L = \infty$, and is maximized for some positive value of R_L. (See Figure 2.21.) The maximum value of p_L is the available power p_a of the network.

To determine the value of R_L that maximizes p_L, we set the derivative of p_L with respect to R_L to zero, since the slope of the p_L curve vanishes at its maximum, as shown in Figure 2.21. The derivative is

$$\frac{dp_L}{dR_L} = \frac{(R_T + R_L)^2 - R_L 2(R_T + R_L)}{(R_T + R_L)^4} v_T^2.$$

Setting this to zero yields

$$\frac{R_T - R_L}{(R_T + R_L)^3} = 0$$

(for nonzero v_T). Thus, assuming that[5]

$$R_T > 0,$$

---

[5]With negative R_T (see Example 2.18 in the previous section), the transferred power p_L is maximized for

$$R_L = -R_T$$

at a value of ∞. In practice, such networks will violate the circuit model when $R_L = -R_T$ and will deliver only finite power to external loads.

the absorbed load power p_L will be maximum for

$$R_L = R_T.$$

Therefore, the available power p_a of a resistive network is the value of p_L for $R_L = R_T$. Evaluating p_L with $R_L = R_T$, we obtain

$$p_a = \frac{v_T^2}{4R_T}.$$

Available power

Also, since $\frac{v_T}{R_T} = i_N$,

$$p_a = \frac{i_N^2 R_T}{4}$$

is an alternative expression for the available power in terms of Norton current.

These expressions indicate that, to find the available power of any resistive network, all we need is its Thevenin or Norton equivalent.

Example 2.19
For the network of Figure 2.18a,

$$v_T = \frac{12}{5} \text{ V} \quad \text{and} \quad R_T = \frac{6}{5} \, \Omega.$$

Thus, for maximum power transfer, the load resistance should be selected as

$$R_L = R_T = \frac{6}{5} \, \Omega,$$

in which case the power transferred to the load will be

$$p_a = \frac{(12/5)^2}{4 \times 6/5} = \frac{12 \times 12}{24 \times 5} = 1.2 \, \text{W}.$$

Matched load

In summary, to achieve maximum power transfer from a resistive network, the external load resistance must be *matched* to the Thevenin resistance of the network. A matched load will absorb the available power of the network. For instance, the input resistance of an optimal speaker connected to an audio amplifier having an 8 Ω Thevenin resistance is 8 Ω. All other loads (speakers) not matching the Thevenin resistance of the amplifier will absorb less power than is available from the amplifier.

EXERCISES

2.1 In the following circuits, determine i_x:

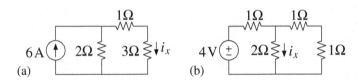

(a)　　(b)

2.2 In the following circuits, determine v_x and the absorbed power in the elements supporting voltage v_x:

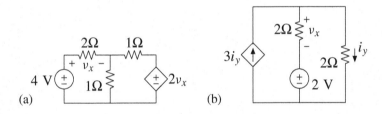

(a)　　(b)

2.3 In the circuit shown next, determine v_1, v_2, i_x, and i_y, using the node-voltage method. Notice that the reference node has not been marked in the circuit; therefore, you are free to select any one of the nodes in the circuit as the reference. The position of the reference (which should be shown in your solution) will influence the values obtained for v_1 and v_2, but not for i_x and i_y.

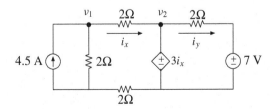

2.4   In the following circuit, determine the node voltages v_1, v_2, and v_3:

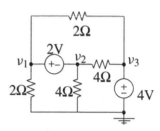

2.5   In the following circuit, determine node voltages v_1 and v_2:

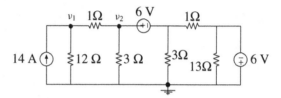

2.6   In the following circuits, determine loop currents i_1 and i_2:

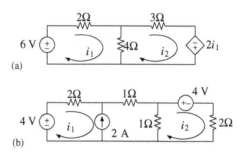

2.7   **(a)**   For the next circuit, obtain two independent equations in terms of loop-currents i_1 and i_2 and simplify them to the form

$$Ai_1 + Bi_2 = E$$
$$Ci_1 + Di_2 = F.$$

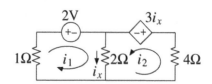

(b) Express the previous equations in the matrix form

$$\begin{pmatrix} A & B \\ C & D \end{pmatrix} \begin{pmatrix} i_1 \\ i_2 \end{pmatrix} = \begin{pmatrix} E \\ F \end{pmatrix}$$

and use matrix inversion or *Cramer's rule* to solve for i_1 and i_2.

2.8 By a sequence of resistor combinations and source transformations, the next circuit shown can be simplified to its Norton (bottom left) and Thevenin (bottom right) equivalents between nodes a and b. Show that

$$i_N = \frac{i}{2} + \frac{v}{R} \quad \text{and} \quad R_T = \frac{2}{3}R$$

and obtain the expression for v_T.

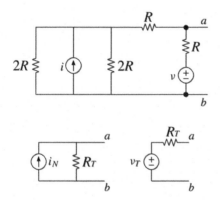

2.9 In the following circuit, it is observed that for $i = 0$ and $v = 1$ V, $i_L = \frac{1}{2}$ A, while for $i = 1$ A and $v = 0$, $i_L = \frac{1}{4}$ A.

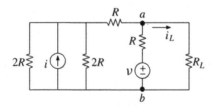

(a) Determine i_L when $i = 4$ A and $v = 2$ V. (You do not need the values of R and R_L to answer this part; just make use of the results of Problem 2.8.)

(b) Determine the values of resistances R and R_L.

(c) Is it possible to change the value of R_L in order to increase the power absorbed in R_L when $i = 4$ A and $v = 2$ V? Explain.

2.10   In the following circuit, find the open-circuit voltage and the short-circuit current between nodes a and b, and determine the Thevenin and Norton equivalents of the network between nodes a and b:

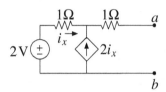

2.11   Determine the Thevenin equivalent of the following network between nodes a and b, and then determine the available power of the network:

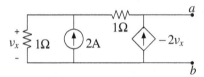

2.12   Determine i_x in Figure 2.11b, using source suppression followed by superposition.

2.13   In the next circuit, do the following:

(a) Determine v when $i_s = 0$.

(b) Determine v when $v_s = 0$.

(c) When $v_s = 4\,\text{V}$ and $i_s = 2\,\text{A}$, what is the value of v and what is the available power of the network? Hint: Make use of the results of parts (a) and (b) and the superposition method.

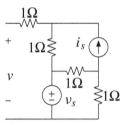

2.14 Consider the following circuit:

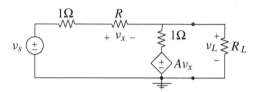

(a) Determine v_L, given that $v_s = 1$ V, $R = 1\,\mathrm{k\Omega}$, $R_L = 0.1\,\Omega$, and $A = 100$.

(b) Find an approximate expression for v_L that is valid when $R \gg 1\,\Omega$, $R_L \approx 1\,\Omega$, and $A \gg 1$.

2.15 Determine the Thevenin resistance R_T of the network to the left of R_L in the circuit shown in Problem 2.14. What is the approximate expression for R_T if $R \gg 1\,\Omega$ and $A \gg 1$?

2.16 For (a) through (e), assume that $A = 3 - j3$, $B = -1 - j1$, and $C = 5e^{-j\frac{\pi}{3}}$:

(a) Let $D = AB$. Express D in exponential form.

(b) Let $E = A/B$. Express E in rectangular form.

(c) Let $F = \frac{B}{C}$. Express F in exponential form.

(d) Let $G = (CD)^*$, where * denotes complex conjugation. Express G in rectangular and exponential forms.

(e) Let $H = (A + C)^*$. Determine $|H|$ and $\angle H$, the magnitude and angle of H.

3

Circuits for Signal Processing

---

*Op-amps,
capacitors,
inductors;
circuits and
systems
for signal
amplification,
addition,
subtraction,
integration;
LTI systems
and zero-input
and zero-state
response;
first-order
RC and RL
circuits and
ODE initial-value
problems;
steady-state
response;
nth-order
systems*

In this chapter, we begin taking the viewpoint that voltages and currents in electrical circuits may represent analog signals and that circuits can perform mathematical operations on these signals, such as addition, subtraction, scaling (amplification), differentiation and integration. Naturally, the type of math performed depends on the circuit and its components.

In Section 3.1 we describe a multiterminal circuit component called the operational amplifier and examine some of its simplest applications (amplification, addition, and differencing). We examine in Section 3.2 the use of capacitors and inductors (see Chapter 1 for their basic definitions) in circuits for signal differentiation and integration. In Section 3.3 we discuss system properties of differentiators and integrators and introduce the important concepts of system linearity and time-invariance. Analysis of linear and time-invariant (LTI) circuits with capacitors and inductors may require solving differential equations. In Section 3.4 we discuss solutions of first-order differential equations describing *RC* and *RL* circuits. Section 3.5 previews the analysis techniques for more complicated, higher-order LTI circuits and systems to be pursued in later chapters.

3.1   Operational Amplifiers and Signal Arithmetic

In Figure 3.1a the triangle symbol represents a multi-terminal electrical device known as an operational amplifier—the op-amp. As we shall see, op-amps are high-gain amplifiers that are useful components in circuits designed to process signals. The diagram in Figure 3.1a shows how an op-amp is "powered up," or *biased*, by raising

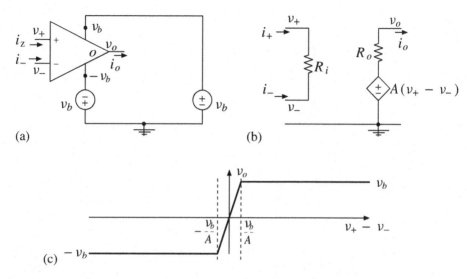

Figure 3.1 (a) Circuit symbol of the op-amp, showing five of its seven terminals and its biasing connections to a reference node, (b) equivalent linear circuit model of the op-amp showing only three terminals with node voltages v_+, v_-, and v_o, and (c) the dependence of v_o on differential input $v_+ - v_-$ and bias voltage v_b. The equivalent model shown in (b) is valid only for $v_+ - v_-$ between the dashed vertical lines.

the electrical potentials of two of its terminals[1] to DC voltages $\pm v_b$ with respect to some reference. Commercially available op-amps require v_b to be in the range of a few to about 15 V in order to function in the desired manner, as described here.

Glancing ahead for a moment, we see that Figure 3.2 shows two op-amp circuits, where (to keep the diagrams uncluttered) the biasing DC sources $\pm v_b$ are not shown. To understand what op-amps do in circuits, we find it necessary to know the relationship between the input voltages v_+, and v_-, and the output voltage v_o, defined at the three op-amp terminals ($+$, $-$, and o) shown in the diagrams. Because op-amps themselves are complicated circuits composed of many transistors, resistors, and capacitors, analysis of op-amp circuits would seem to be challenging. Fortunately, op-amps can be described by fairly simple equivalent circuit models involving the input and output voltages. Figure 3.1b shows the simplest op-amp model that is accurate enough for our purposes, which indicates that

$$v_o = A(v_+ - v_-) - R_o i_o.$$

It is because v_o depends on v_+ and v_- that the latter are considered as op-amp inputs and v_o is considered to be the op-amp output.

---

[1]Op-amps such as the Fairchild 741 have seven terminals. Two of these are for output offset adjustments, which will not be important to us. The discussion here will focus on the remaining five, shown in Figure 3.1a, two of which are used for powering up, or *biasing*, the device, as indicated in the figure.

**Input
and
output
terminals**

Terminal o is called the output terminal, and the terminals marked by $-$ and $+$ are referred to as *inverting* and *noninverting* input terminals. Resistors R_i and R_o in Figure 3.1b are called the input and output resistances of the op-amp, and A is a voltage gain factor (scales up the differential input $v_+ - v_-$) called the "open-loop gain." Typically, for an op-amp,

$$R_o \sim 1 - 10\,\Omega,$$

$$R_i \sim 10^6\,\Omega,$$

$$A \sim 10^6,$$

**Typical
op-amp
parameters**

where the symbol $\sim$ stands for "within an order of magnitude of." Very large values of A and R_i and a relatively small R_o are essential for op-amps to perform as intended in typical applications.

When the output terminal of an op-amp is left open—that is, when no connection is made to other components—as in Figure 3.1a, then $i_o = 0$ and, according to Figure 3.1b, v_o is just an amplified version of $v_+ - v_-$. However, an important detail that is not properly modeled by the simple op-amp equivalent circuit[2] is the saturation of v_o at $\pm v_b$. For the actual physical op-amp, the output v_o cannot exceed v_b in magnitude. Assuming small i_o, Figure 3.1c shows the variation of the output voltage v_o as a function of the differential input voltage $v_+ - v_-$. Only for

$$|v_+ - v_-| < \frac{v_b}{A}$$

**Linear
vs
saturated**

is the variation of v_o with $v_+ - v_-$ linear, as described by the model in Figure 3.1b. Otherwise, the op-amp is said to be *saturated* or behaving nonlinearly.

Clearly, then, to maintain an op-amp in the desired linear regime—that is, to ensure that $|v_+ - v_-| < \frac{v_b}{A}$—it is necessary to keep the *differential input* $v_+ - v_-$ extremely small. Typically, with $A \sim 10^6$ and $v_b \approx 10$ V, we must have $|v_+ - v_-| < 10\,\mu$V. When this condition is achieved, the op-amp input currents i_+ and i_- through the input resistance $R_i \sim 10^6\,\Omega$ will have magnitudes less than $\frac{10\,\mu\text{V}}{10^6\,\Omega} = 10\,$pA, which is an exceedingly small (in fact, negligible) current. Hence, for all practical purposes, for an op-amp operating in the linear regime (i.e., between the dashed vertical lines in Figure 3.1c), we have

$$v_+ \approx v_-,$$

$$i_+ \approx 0,$$

$$i_- \approx 0,$$

to within microvolts and picoamps.

---

[2]The equivalent circuit model of an op-amp should not be confused with the actual structure of the interior of the op-amp. The Fairchild 741 op-amp circuit consists of 20 transistors, 11 resistors, and one capacitor. The equivalent circuit in Figure 3.1b is a *model* that merely describes the relationship of the signals at the op-amp terminals under normal (linear) operating conditions.

These conditions are fundamental approximations that tremendously simplify the analysis of op-amp circuits *known* to be in linear operation. As you will discover later, if you are having difficulty in analyzing an op-amp circuit, it is probably because you have forgotten to apply one of the previous conditions—so memorize them! These equations are known as *ideal op-amp approximations*. The reason for this terminology is, $v_+ = v_-$ and $i_+ = i_- = 0$ would be exactly true for an ideal op-amp with $R_i \to \infty$ and $A \to \infty$. Notice that there is no ideal op-amp approximation that constrains the output voltage v_o. Instead, as we shall see, the value of v_o depends upon the circuit in which the op-amp is embedded.

Ideal op-amp approximations

Applying the ideal op-amp approximations to the analysis of circuits containing nonideal (i.e., real, "off-the-shelf") op-amps amounts to ignoring voltage and current terms in our KVL and KCL equations having magnitudes less than $\sim 10\,\mu V$ and $10\,pA$. This will result in negligible errors in the calculation of the larger voltages and currents in a circuit that typically will interest us.

Alternatively, we can analyze linear op-amp circuits by substituting for each op-amp the equivalent circuit model shown in Figure 3.1b and writing down the exact KVL and KCL equations in the resulting circuit diagrams. In such calculations, we use $R_o \sim 1\,\Omega - 10\,\Omega$, $R_i \sim 10^6\,\Omega$, and $A \sim 10^6$ instead of the op-amp approximations introduced earlier. This approach can be slightly more accurate (if exact values of R_o, R_i, and A are known), but requires far more effort than using the simple ideal op-amp relations.

In the following discussion of well-known linear op-amp circuits, we will illustrate both of these approaches.

3.1.1 Voltage follower and noninverting amplifier

Figures 3.2a and 3.2b show two important linear op-amp circuits known as the *voltage follower* and the *noninverting amplifier*. Clearly, the circuits are related, because the voltage follower of Figure 3.2a is a special case of the amplifier of Figure 3.2b, with $R_1 = 0$ and $R_2 = \infty$. We next will analyze the circuits by using the ideal op-amp approximations and later justify the analysis by showing that for small inputs the circuits operate in the linear regime.

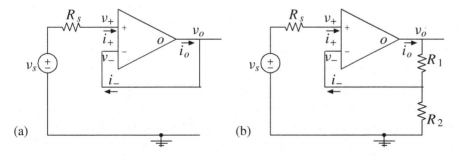

Figure 3.2   (a) A voltage-follower (or buffer), and (b) a noninverting amplifier.

Starting with the voltage-follower circuit and using the ideal op-amp approximations, we can argue as follows: Since $i_+ \approx 0$, we can ignore the voltage drop across resistor R_s and thus claim that

$$v_+ \approx v_s.$$

Also, since $v_- \approx v_+$, it follows that

$$v_- \approx v_s.$$

Finally, in the voltage follower circuit

$$v_o = v_-,$$

and, as a consequence,

$$v_o \approx v_s.$$

Now, for the noninverting amplifier of Figure 3.2b, $v_o \neq v_-$ because of the voltage drop across R_1. Instead, we can write v_- in terms of v_o as

$$v_- \approx v_o \frac{R_2}{R_1 + R_2},$$

because $i_- \approx 0$ and, as a consequence, v_o is distributed across R_1 and R_2 in accordance with voltage division. Given that $v_- \approx v_s$, this leads to

$$v_o \approx (1 + \frac{R_1}{R_2})v_s.$$

Clearly, then, the circuit in Figure 3.2b is a voltage amplifier with a *voltage gain*

$$G \equiv \frac{v_o}{v_s} \approx (1 + \frac{R_1}{R_2}).$$

The amplifier is called noninverting because the gain is positive, and this preserves the algebraic sign of the amplified voltage.

According to these results, a noninverting amplifier multiplies the input voltage v_s by a gain G that is independent of the source resistance R_s (so long as it is not infinite), whereas the voltage follower is simply a noninverting amplifier with unity gain ($G = 1$). What makes both of these circuits very useful is the fact that their gain G

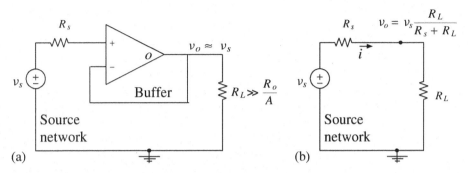

Figure 3.3 (a) Feeding a load through a buffer (see Exercise Problem 3.1 for a detailed discussion of this circuit), versus (b) a direct connection between the source network and the load.

remains essentially unchanged when the circuits are terminated by an external load—for example, by some finite resistance R_L—as shown in Figure 3.3a, so long as the load resistance exceeds a value on the order of $\frac{R_o}{A}$. (See the discussion that follows.) The voltage follower in particular can be used as a *unity gain* buffer between a source circuit and a load, as shown in Figure 3.3a, to prevent "loading down" the source voltage. That is, the entire source voltage v_s appears across the load resistance R_L. **Buffering**

Ordinarily, if two circuits are to be connected to one another, then making a direct connection changes the behavior of the two circuits. This is illustrated in Figure 3.3b, where connecting R_L to the terminals of the source network reduces the voltage at those terminals. By connecting the load resistance to the source network through a voltage follower, as in Figure 3.3a, R_L does not draw current from the source network (instead, the current is supplied by the op-amp) and the full source voltage appears across the load resistance. More generally, the connection of one circuit to another through a voltage follower allows both circuits to continue to operate as designed.

The preceding ideal op-amp analysis did not provide us with detailed information such as the range of values of R_s for which the gain G will be independent of R_s. To obtain such information, we would need to insert the op-amp equivalent model of Figure 3.1b into the circuit (to replace the op-amp symbol) and then reanalyze the circuit without making any assumptions about $v_\pm$ and $i_\pm$. We will lead you through such a calculation in Exercise Problem 3.2 to confirm that the noninverting amplifier gain G is well approximated by $1 + \frac{R_1}{R_2}$, so long as $R_s \ll R_i$ and $\frac{R_o}{A} \ll R_L$. Such calculations also are demonstrated in the next section in our discussion of the inverting amplifier.

We will close this section with a discussion of *negative feedback*, the magic behind how a voltage follower circuit keeps itself within the linear operating regime. Notice how the output terminal of the op-amp in the voltage follower is *fed back* to its own inverting input. That connection configures the circuit with a negative feedback loop, ensuring that $v_+ - v_-$ remains between the dashed lines in Figure 3.1c, so long as $|v_s| < v_b$. Let us see how. **Negative feedback**

Assume for a moment that $v_o = v_b$, which requires having $v_+ - v_- > 0$ in the model curve in Figure 3.1c. But in the circuit in Figure 3.2a, $v_o = v_b$ implies that $v_+ - v_- = v_s - v_b < 0$ if $|v_s| < v_b$. Since $v_+ - v_- > 0$ and $v_+ - v_- < 0$ are contradictory, $v_o = v_b$ is clearly not possible when $|v_s| < v_b$. Neither is $v_o = -v_b$ for similar reasons. In that case $v_+ - v_- < 0$, according to Figure 3.1c, whereas, according to the circuit, $v_+ - v_- = v_s + v_b > 0$ if $|v_s| < v_b$. The upshot is that if $|v_s| < v_b$, then the only noncontradictory situation in a voltage follower is when $|v_o| < v_b$—that is, being in a linear operation.

To see that negative feedback is *essential* to the linearity of the voltage follower, consider what happens when we connect the op-amp output to the *noninverting* input, as in Figure 3.4—the circuit in Figure 3.4 appears identical to a voltage follower, except for the use of *positive* feedback! Then a variety of outcomes for v_o that deviate from the voltage follower behavior are possible. For instance, a saturated solution of $v_o = v_b$ is possible with $v_s = 0$, since, with $v_o = v_b$, Figure 3.1c implies $v_+ - v_- > 0$, which is consistent with $v_+ - v_- = v_b - 0 = v_b$ obtained from Figure 3.4.

All op-amp circuits discussed in this chapter and elsewhere in this textbook use negative feedback—some fraction of v_o is always *added* to v_- by an appropriate connection between the output and inverting input—and thus the described circuits operate linearly when excited with sufficiently small inputs. For instance, the noninverting amplifier adds a fraction $\frac{R_2}{R_1+R_2}$ of v_o to v_- and, therefore, operates linearly for $|v_s| < \frac{v_b}{G}$.

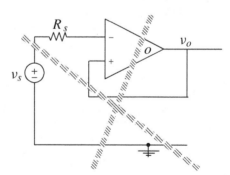

Figure 3.4   Not a voltage follower, because of positive feedback.

Example 3.1

In the linear op-amp circuit shown in Figure 3.5, determine the node voltages v_o, v_1, and v_2 assuming that $V_{s1} = 3\,\text{V}$, $V_{s2} = 4\,\text{V}$, and $R_{s1} = R_{s2} = 100\,\Omega$.

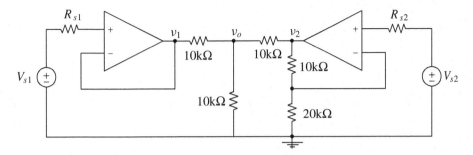

Figure 3.5 A circuit with two op-amps.

Solution   The left-end of the circuit is a voltage follower; therefore,

$$v_1 \approx V_{s1} = 3\,\text{V}.$$

At the right-end we notice a noninverting amplifier with a gain of

$$G = 1 + \frac{10\,\text{k}\Omega}{20\,\text{k}\Omega} = 1.5;$$

therefore,

$$v_2 \approx GV_{s2} = (1.5)(4\,\text{V}) = 6\,\text{V}.$$

To obtain v_o, we write the KCL equation at the middle node as

$$\frac{v_o - 3}{10\,\text{k}} + \frac{v_o}{10\,\text{k}} + \frac{v_o - 6}{10\,\text{k}} = 0,$$

giving

$$v_o = 3\,\text{V}.$$

Example 3.2
Both sources of Example 3.1 are doubled so that now $V_{s1} = 6\,\text{V}$, and $V_{s2} = 8\,\text{V}$. What is the new value of v_o? Assume that the biasing voltage is $v_b = 15$ V for both op-amps.

Solution   Since the circuit in Figure 3.5 is linear, a doubling of both inputs causes a doubling of the response v_o. Hence, the new value of v_o is 6 V. This is the correct result, because the new values of v_1 and v_2, namely, 6 V and 12 V, are both below the saturation level of 15 V, and therefore, the circuit remains in linear operation.

Example 3.3

Would the circuit in Figure 3.5 remain in linear operation if the source voltage values of Example 3.1 were tripled? Once again, assume that $v_b = 15$ V.

Solution   No, the circuit would enter into a nonlinear saturated mode, because with a tripled value for V_{s2} the response v_2 could not triple to 18 V (since v_2 cannot exceed the biasing voltage of 15 V).

3.1.2   Inverting amplifier

The op-amp circuit shown in Figure 3.6a employs negative feedback and is known as an *inverting amplifier*. Ideal op-amp analysis of the circuit—justified because of negative feedback—proceeds as follows.

Because the noninverting terminal in Figure 3.6a is in contact with the reference, the corresponding node voltage is $v_+ = 0$. Therefore, the ideal op-amp approximation $v_- \approx v_+$ implies that $v_- \approx 0$. Thus, the current through resistor R_s toward the inverting terminal can be calculated as

$$\frac{v_s - v_-}{R_s} \approx \frac{v_s - 0}{R_s} = \frac{v_s}{R_s}.$$

The current through resistor R_f away from the inverting terminal is, likewise,

$$\frac{v_- - v_o}{R_f} \approx \frac{0 - v_o}{R_f} = -\frac{v_o}{R_f}.$$

Since $i_- \approx 0$, KCL applied at the inverting terminal gives

$$\frac{v_s}{R_s} \approx -\frac{v_o}{R_f} + 0.$$

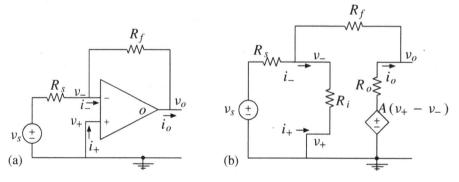

Figure 3.6   (a) An op-amp circuit (inverting amplifier), and (b) the same circuit where the op-amp is shown in terms of its equivalent circuit model.

Hence,

$$v_o \approx -\frac{R_f}{R_s} v_s,$$

which shows that the circuit is a voltage amplifier with a voltage gain of

$$G = \frac{v_o}{v_s} \approx -\frac{R_f}{R_s}.$$

Because the gain is negative, the amplifier is said to be *inverting*.

Let us next verify the preceding result by using the more accurate equivalent-circuit approach, where we replace the op-amp by its equivalent linear model, introduced earlier in Figure 3.1b. Making this substitution, we find that the noninverting amplifier circuit takes the form shown in Figure 3.6b. Applying KCL at the inverting terminal (where v_- is defined) gives

$$\frac{v_- - v_s}{R_s} + \frac{v_-}{R_i} + \frac{v_- - v_o}{R_f} = 0.$$

Likewise, the KCL equation for the output terminal is

$$\frac{v_o - v_-}{R_f} + \frac{v_o - A(0 - v_-)}{R_o} = 0.$$

From the second equation, we obtain

$$v_- = v_o \frac{\frac{1}{R_f} + \frac{1}{R_o}}{\frac{1}{R_f} - \frac{A}{R_o}} \approx -v_o \frac{\frac{1}{R_o}}{\frac{A}{R_o}} = -\frac{v_o}{A},$$

assuming that $R_o \ll R_f$ and $A \gg 1$. Substituting $v_- \approx -\frac{v_o}{A}$ into the first KCL equation gives

$$\frac{-\frac{v_o}{A} - v_s}{R_s} + \frac{-\frac{v_o}{A}}{R_i} + \frac{-\frac{v_o}{A} - v_o}{R_f} \approx 0,$$

which implies that

$$v_o\left(\frac{1}{AR_s} + \frac{1}{AR_i} + \frac{1}{AR_f} + \frac{1}{R_f}\right) \approx -\frac{v_s}{R_s}.$$

Clearly, with $AR_s \gg R_f$, $AR_i \gg R_f$, and $A \gg 1$, the first three terms within the parentheses on the left can be neglected to yield

$$v_o \approx -\frac{R_f}{R_s} v_s.$$

This is the result that we obtained earlier by using the ideal op-amp approximation.

The advantages of the (more laborious) equivalent-circuit approach just illustrated is that it provides us with the validity conditions for the result (which are not obtained with the ideal approximation method). The previous exercise showed that the validity conditions are

$$A \gg 1$$

and

$$R_o \ll R_f \ll A R_s, \; A R_i.$$

Since for typical op-amps $A \sim 10^6$, $R_o \sim 10 \, \Omega$, and $R_i \sim 10^6 \, \Omega$, these conditions are readily satisfied if R_s and R_f are chosen in the kΩ range. For instance, with $R_f = 20 \, \text{k}\Omega$ and $R_s = 5 \, \text{k}\Omega$,

$$v_o \approx -\frac{R_f}{R_s} v_s = -4 v_s$$

would be a valid result, whereas our simple gain formula would not be accurate for $R_f = 20 \, \Omega$ and $R_s = 5 \, \Omega$. Importantly, our detailed analysis has shown that the inverting amplifier gain $G \approx -\frac{R_f}{R_s}$ is not sensitive to the exact values of A, R_i, or R_o; it is sufficient that A and R_i be very large and R_o quite small. Op-amps are intentionally designed to satisfy these conditions.

To examine how the inverting amplifier gain may depend on a possible load R_L connected to the output terminal, we find it sufficient to calculate the Thevenin resistance R_T of the equivalent circuit shown in Figure 3.6b. The calculation can be carried out by use of the circuit shown in Figure 3.7, where the source v_s of Figure 3.6b has been suppressed and a 1 A current has been injected into the output terminal to implement the test current method discussed in Chapter 2. Using straightforward steps

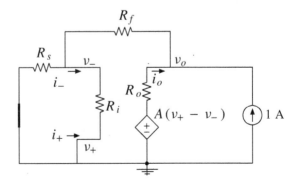

Figure 3.7 Test current method is applied to determine the Thevenin resistance of the equivalent circuit of an inverting amplifier.

and the assumptions given, we find that

$$v_o \approx \frac{R_o}{1 + A\frac{R_s}{R_s+R_f}} \sim \frac{R_o}{A}$$

for $R_s \sim R_f$. Thus, the Thevenin resistance of the equivalent circuit is $R_T \sim \frac{R_o}{A}$ in typical usage, just as for a voltage follower. (See Exercise Problem 3.2.) The amplifier gain will not be sensitive to R_L so long as $R_L \gg \frac{R_o}{A}$, and, in addition, the total current conducted by R_L remains within specified limits (typically, less than a few tens of mA, depending on the specific op-amp being used).

3.1.3   Sums and differences

A variation of the op-amp inverting amplifier is shown in Figure 3.8a. This circuit works as an adder or a summer. Since $v_- \approx v_+ = 0$, the total current from the left toward the minus terminal of the op-amp is approximately

$$\frac{v_1 - 0}{R_1} + \frac{v_2 - 0}{R_2}.$$

Equating this current to the current $\frac{0-v_o}{R_f}$ through R_f toward the output node, we obtain

$$v_o \approx -\left(\frac{R_f}{R_1}v_1 + \frac{R_f}{R_2}v_2\right).$$

Clearly, the circuit sums the input voltages v_1 and v_2, with respective weighting coefficients $-\frac{R_f}{R_1}$ and $-\frac{R_f}{R_2}$. The circuit can be modified in a straightforward way to combine three or more inputs in a similar manner. Furthermore, because of the low Thevenin resistance of the inverting amplifier, the weighted sum will appear in full strength across any load that is reasonably large.

The circuit shown in Figure 3.8b forms the difference, $v_1 - v_2$, between voltages v_1 and v_2. This becomes apparent when we note that $v_- \approx v_+ \approx \frac{v_1}{2}$ (obtained by

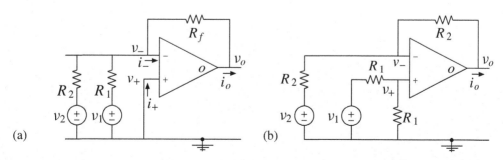

(a)   (b)

Figure 3.8   (a) An adder circuit, and (b) differencing circuit.

applying voltage division to v_1, since $i_+ \approx 0$), so that the KCL equation at the inverting input node is

$$\frac{v_2 - \frac{1}{2}v_1}{R_2} \approx \frac{\frac{1}{2}v_1 - v_o}{R_2}.$$

Hence,

$$v_o \approx v_1 - v_2.$$

3.2   Differentiators and Integrators

In the previous section we discussed resistive op-amp circuits for signal amplification, summation, and differencing. In this section we shall see that, by including capacitors or inductors, we can build op-amp circuits for signal differentiation and integration. These operations are, of course, relevant for processing time-varying signals (or give rise to them) and therefore, this will be our first exposure to so-called AC circuits.

3.2.1   Differentiator circuits

Capacitors and inductors as differentiators: Since the v–i relations for capacitors and inductors (see Chapter 1) are

$$i = C\frac{dv}{dt} \quad \text{(Capacitor)}$$
$$v = L\frac{di}{dt} \quad \text{(Inductor)},$$

capacitors are natural differentiators of their voltage inputs and inductors differentiate their current inputs.

Example 3.4

The capacitor voltage in the circuit shown in Figure 3.9 is

$$v(t) = 5\cos(100t) \text{ V}.$$

Determine the capacitor's current response.

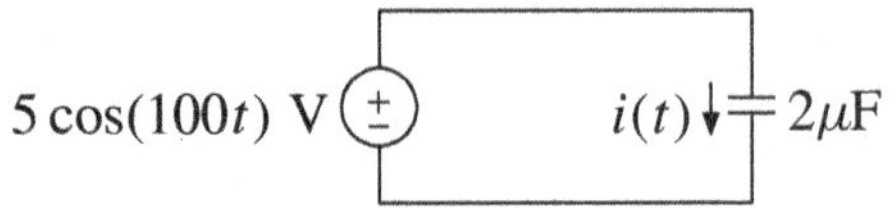

Figure 3.9   A capacitor circuit with an imposed capacitor voltage signal and a current response $i(t)$ proportional to the time derivative of the imposed voltage.

Solution

$$i(t) = C\frac{dv}{dt} = 2\,\mu\text{F} \times \frac{d}{dt}[5\cos(100t)\,\text{V}]$$

$$= 2 \times 5 \times (-100)\sin(100t)\,\mu\text{A} = -\sin(100t)\,\text{mA}.$$

Example 3.5

The 1 H inductor shown in Figure 3.10a responds to the ramp current input shown in Figure 3.10b, with a *unit-step* voltage $v(t) = u(t)$ shown in Figure 3.10c. The unit-step voltage output is 0 for $t < 0$, when the input $i(t)$ is constant (with zero value), and 1 for $t > 0$, when

$$i(t) = t\,\text{A},$$

and, thus,

$$L\frac{di}{dt} = 1\,\text{V}.$$

Op-amp differentiators: The op-amp circuit shown in Figure 3.11a converts the current-response of the capacitor (which is proportional to the derivative of its voltage) into an output voltage and therefore functions as a voltage differentiator. To understand this behavior, note that $v_- \approx 0$ because of the ideal op-amp approximation $v_- \approx v_+ = 0$. Therefore, the capacitor current, left-to-right, is approximately

$$C\frac{d}{dt}(v_s(t) - 0) = C\frac{dv_s}{dt}.$$

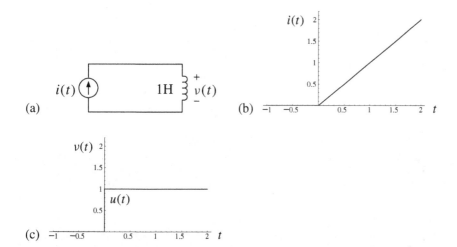

Figure 3.10 (a) An inductor circuit, (b) current input to the inductor, and (c) voltage response of the inductor described by a unit-step function, $u(t)$.

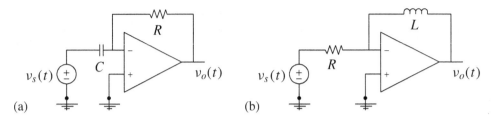

Figure 3.11 (a) A differentiator, and (b) a differentiator with an inductor.

Because $i_- \approx 0$, this capacitor current is very nearly matched to the resistor current

$$\frac{0 - v_o(t)}{R} = -\frac{v_o(t)}{R}$$

directed toward the output terminal of the op-amp. Therefore,

$$v_o(t) = -RC\frac{dv_s}{dt}.$$

We leave it as an exercise for you to derive the output formula

$$v_o(t) = -\frac{L}{R}\frac{dv_s}{dt}$$

for the differentiator circuit shown in Figure 3.11b.

3.2.2 Integrator circuits

Capacitors and inductors as integrators: Consider the circuit shown in Figure 3.12a, where a capacitor responds to an applied current $i(t)$ with the voltage $v(t)$. Integrating the capacitor current

$$i(t) = C\frac{dv}{dt}$$

between two points in time, say, from a to b, we get

$$\int_a^b i(t)dt = \int_a^b C\frac{dv}{dt}dt = C\int_a^b dv = C[v(b) - v(a)],$$

which indicates that

$$v(b) = v(a) + \frac{1}{C}\int_a^b i(t)dt.$$

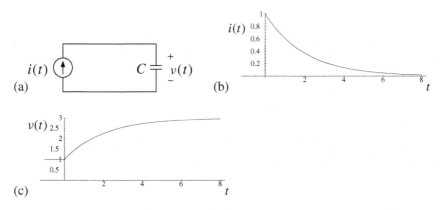

(a)

(b)

(c)

Figure 3.12   (a) Capacitor as an integrator, and (b) input and output signals of the integrator circuit discussed in Example 3.6.

Replacing t with τ under the integral sign, and then replacing a and b with t_o and t, respectively, we obtain

$$v(t) = v(t_o) + \frac{1}{C} \int_{t_o}^{t} i(\tau)d\tau,$$

an explicit formula for the capacitor voltage in terms of the input current $i(t)$.

Clearly, the result implies that if we know the capacitor voltage at some initial instant t_o, then we can determine any subsequent value of the capacitor voltage in terms of a *definite integral* of the input current from time t_o to the time t of interest. We will use the term *initial value* or, occasionally, *initial state*, to refer to $v(t_o)$.   **Initial state**

Example 3.6
In the circuit shown in Figure 3.12a, suppose that $C = 1$ F and $v(0) = 1$ V. Let

$$i(t) = e^{-\frac{1}{2}t} \text{ A}$$

for $t > 0$ as shown in Figure 3.12b. Calculate the voltage response $v(t)$ for $t > 0$.

Solution   Using the general expression for $v(t)$ obtained previously, with $t_o = 0$, $v(0) = 1$ V, $i(\tau) = e^{-\frac{1}{2}\tau}$ A, and $C = 1$ F, we find that

$$v(t) = 1 + \frac{1}{1} \int_{0}^{t} (e^{-\frac{1}{2}\tau} \text{ A})d\tau = (1 + \frac{e^{-\frac{1}{2}t} - 1}{-\frac{1}{2}}) \text{ V} = 3 - 2e^{-\frac{1}{2}t} \text{ V}.$$

The plot of the response $v(t)$ is shown in Figure 3.12c.

Notice in Figures 3.12b and 3.12c that $i(t) = 0$ A and $v(t) = 1$ V for $t < 0$. These values of the capacitor current and voltage for $t < 0$ are consistent with one another, since a constant $v(t)$ implies zero $i(t)$, according to

$$i(t) = C\frac{dv}{dt}.$$

For $t > 0$ the curves in the figure also satisfy this same relation, because the voltage curve was computed in Example 3.6 to satisfy this very same constraint.

A comparison of the $i(t)$ and $v(t)$ curves in Figures 3.12b and 3.12c leads to the following important observation: Even though the input curve $i(t)$ is discontinuous at $t = 0$, the output curve $v(t)$ does not display a discontinuity. For reasons to be explained next, the following general statement can be made about a capacitor voltage:

Capacitor voltage can't "jump"

The voltage response of a capacitor to a *practical* input current must be continuous.

Explanation: Recall from Chapter 1 that the capacitor is an energy storage device, with the stored energy varying with capacitor voltage as

$$w(t) = \frac{1}{2}Cv^2(t).$$

Furthermore, a capacitor stores net electrical charge on its plates, an amount

$$q(t) = Cv(t)$$

on one plate assigned positive polarity and $-q(t)$ on the other. Therefore, any time discontinuity in capacitor voltage would lead to accompanying discontinuities in stored charge and energy, which would require infinite current in order to add a finite amount of charge and energy in zero time. Such discontinuous changes are naturally prohibited in practical circuits where current amplitudes remain bounded, consistent with the fact that

$$|i(t)| = |C\frac{dv}{dt}| = C|\frac{dv}{dt}| < \infty$$

when $v(t)$ is a continuous function.

In the following example, we will make use of the *continuity* of capacitor voltage to calculate the response of a capacitor to a *piecewise continuous* input current:

Example 3.7
Calculate the voltage response of a 2 F capacitor to the discontinuous current input shown in Figure 3.13a, for $t > 0$. Assume that $v(0) = 0$ V.

Solution For the period $0 < t < 1$, where t is in units of seconds, the current input is $i(t) = 1$ A. Therefore, for that period,

$$v(t) = v(0) + \frac{1}{2}\int_0^t 1\,d\tau = 0 + \frac{t-0}{2} = \frac{t}{2}\,\text{V}.$$

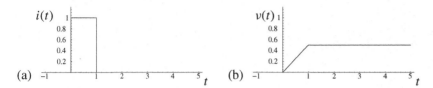

Figure 3.13   (a) A piecewise continuous input current $i(t)$, and (b) response $v(t)$ of a 2 F capacitor discussed in Example 3.6.

Now, even though this result is only for the interval $0 < t < 1$, we still can use it to calculate

$$v(1) = \frac{1}{2}\,\text{V}$$

because of the continuity of the capacitor voltage $v(t)$ at $t = 1$. Thus, for $t > 1$, where the current $i(t) = 0$, we find that

$$v(t) = v(1) + \frac{1}{2}\int_1^t 0\,d\tau = v(1) = \frac{1}{2}\,\text{V}.$$

Notice that $v(t)$ grows from zero to $\frac{1}{2}$ V (as shown in Figure 3.13b) during the time interval $0 < t < 1$ because of a steady current flow that charges up the capacitor within that time interval. After $t = 1$, the current stops so that no more charging or discharging takes place and the capacitor voltage remains fixed at the level of $v(1) = \frac{1}{2}$ V forever.

While a capacitor is an integrator of its input current, an inductor, shown in Figure 3.14, produces a current response that is related to the integral of the voltage applied across its terminals. This can be seen by integrating

$$v(t) = L\frac{di}{dt}$$

to yield

$$i(t) = i(t_o) + \frac{1}{L}\int_{t_o}^t v(\tau)d\tau,$$

Figure 3.14   An inductor acts as an integrator of its voltage input $v(t)$.

in analogy with the voltage response of a current-driven capacitor. Also, by analogy, it should be clear that the current-response of an inductor to a *practical* voltage input must be continuous.

Example 3.8
A voltage input

$$v(t) = u(t) \, \text{V}$$

is applied to a 3 H inductor, where $u(t)$ represents the unit step function introduced in Figure 3.10c. Find the current response of the inductor for $t > 0$ if $i(t) = 0$ for $t < 0$.

Solution   The applied input $v(t) = u(t)$ V is zero for $t < 0$, which is consistent with $i(t)$ being zero in the same time period.

Now, for $t > 0$ the input is $v(t) = u(t) = 1$V, and therefore, the current response is

$$i(t) = i(0) + \frac{1}{3} \int_0^t 1 \, d\tau = i(0) + \frac{t}{3} = \frac{t}{3} \text{A}.$$

where, by the continuity of the inductor current, $i(0) = 0$. We can express the overall response curve for all t as

$$i(t) = \frac{t}{3} u(t) \, \text{A},$$

making use of the unit step function notation.

Op-amp integrators: If we were to swap the positions of the capacitor and resistor in the differentiator circuit of Figure 3.11a, we would obtain the integrator circuit shown in Figure 3.15a. To see that Figure 3.15a is an integrator, let us analyze the circuit, using the ideal op-amp approximations.

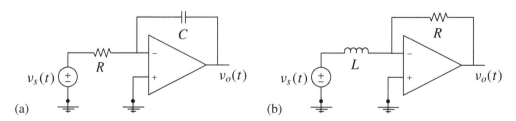

Figure 3.15   An op-amp integrator with (a) a capacitor and (b) an inductor.

The resistor current is approximately

$$\frac{v_s(t)}{R},$$

flowing from left to the right, and the capacitor current is

$$C\frac{d}{dt}(0 - v_o(t)) = -C\frac{dv_o}{dt},$$

in the same direction. Matching these currents, we find that

$$v_s(t) \approx -RC\frac{dv_o}{dt}.$$

The implication is (in analogy with capacitor and inductor integrators) that

$$v_o(t) \approx v_o(t_o) - \frac{1}{RC}\int_{t_o}^{t} v_s(\tau)d\tau.$$

We leave it as an exercise for you to verify that for the integrator circuit shown in Figure 3.15b,

$$v_s(t) \approx -\frac{L}{R}\frac{dv_o}{dt},$$

and therefore,

$$v_o(t) \approx v_o(t_o) - \frac{R}{L}\int_{t_o}^{t} v_s(\tau)d\tau.$$

3.3   Linearity, Time Invariance, and LTI Systems

The differentiator and integrator circuits of Section 3.2 can be viewed abstractly as *analog systems*. Such systems convert their time-continuous input signals $f(t)$ to **Analog** output signals $y(t)$ according to some rule that defines the system. For instance, for **system** the integrator system shown in Figure 3.16, the input–output rule has the form

$$y(t) = y(t_o) + \int_{t_o}^{t} f(\tau)d\tau.$$

For another analog system, the rule may be specified in terms of a differential equation for the output $y(t)$ that depends on input $f(t)$.

This book is concerned mainly with signal processing systems that can be implemented with lumped-element electrical circuits, and, in particular, with *linear* and *time-invariant* (LTI) systems such as differentiators, integrators, and *RC* and *RL* circuits to be examined in the next section. The purpose of this section is to describe **LTI** what we mean by an LTI system and to introduce some key terminology to be used **systems** throughout the rest of the book.

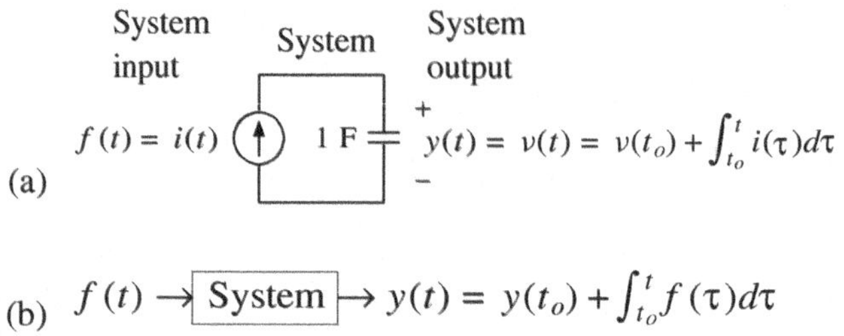

System   System
input   output
System

$$f(t) = i(t) \quad \text{1 F} \quad y(t) = v(t) = v(t_o) + \int_{t_o}^t i(\tau)d\tau$$

(a)

(b) $f(t) \rightarrow \boxed{\text{System}} \rightarrow y(t) = y(t_o) + \int_{t_o}^t f(\tau)d\tau$

Figure 3.16 (a) An integrator system with an input $i(t)$ and output $v(t)$, and (b) a symbolic representation of the same system in terms of the input and output signals designated as $f(t)$ and $y(t)$, respectively.

We already have seen that an integrator output having the form

$$y(t) = y(t_o) + \int_{t_o}^t f(\tau)d\tau,$$

Input and initial state

for $t > t_o$, depends not only on the input $f(t)$ from time t_o to t, but also on $y(t_o)$, the *initial state* of the integrator. In Figure 3.16, $y(t_o)$ is the initial voltage across the capacitor, which is proportional to the initial charge on the capacitor. So we think of this as the initial state. Notice that the contributions of $y(t_o)$ and $\{f(\tau), t_o < \tau < t\}$ to $y(t)$ are *additive*, and each contribution, namely, $y(t_o)$ and $\int_{t_o}^t f(\tau)d\tau$, vanish with vanishing $y(t_o)$ and $f(t)$, respectively.

Zero-input response is independent of the input and zero-state response is independent of the initial state

Thus, for $f(t) = 0$—that is, with *zero input*—the integrator output is just

$$y(t_o),$$

while for $y(t_o) = 0$—that is, under the *zero state* condition—the output is only

$$\int_{t_o}^t f(\tau)d\tau.$$

Therefore, it is convenient to think of $y(t_o)$ and $\int_{t_o}^t f(\tau)d\tau$ as *zero-input* and *zero-state components* of the output, respectively, and, conversely, to think of the output as a sum of zero-input and zero-state response components:

$$y(t) = \underbrace{y(t_o)}_{\textit{Zero-input response}} + \underbrace{\int_{t_o}^t f(\tau)d\tau}_{\textit{Zero-state response}}$$

In a very general sense, and using the same terminology, we can state that:

Linearity

A system is said to be *linear* if its output is a sum of distinct zero-input and zero-state responses that vary linearly with the initial state of the system and linearly with the system input, respectively.

The zero-state response is said to vary linearly with the system input—known as *zero-state linearity*—if (using the symbolic notation of Figure 3.16b), under zero-state conditions,

$$f_1(t) \longrightarrow \boxed{\text{Linear}} \longrightarrow y_1(t)$$

and

$$f_2(t) \longrightarrow \boxed{\text{Linear}} \longrightarrow y_2(t),$$

imply that

$$K_1 f_1(t) + K_2 f_2(t) \longrightarrow \boxed{\text{Linear}} \longrightarrow K_1 y_1(t) + K_2 y_2(t),$$

for any arbitrary constants K_1 and K_2. In other words, with zero-state linear systems, a **Zero-state** *weighted sum of inputs produces a similarly weighted sum of corresponding zero-state* **linearity** *outputs*, consistent with the *superposition principle*.

Example 3.9
Verify that an integrator system

$$i(t) \longrightarrow \boxed{\text{Integ}} \longrightarrow v(t_o) + \int_{t_o}^{t} i(\tau)d\tau$$

is zero-state linear.

Solution   Assuming a zero initial condition, $v(t_o) = 0$, we can express the integrator outputs caused by inputs $i_1(t)$ and $i_2(t)$ as

$$v_1(t) = \int_{t_o}^{t} i_1(\tau)d\tau$$

and

$$v_2(t) = \int_{t_o}^{t} i_2(\tau)d\tau,$$

respectively. With a new input

$$i_3(t) \equiv K_1 i_1(t) + K_2 i_2(t),$$

which is a linear combination of the original inputs, the zero-state output is

$$v_3(t) = \int_{t_o}^{t} (K_1 i_1(\tau) + K_2 i_2(\tau))d\tau.$$

But we can rewrite this as

$$v_3(t) = K_1 \int_{t_o}^{t} i_1(\tau)d\tau + K_2 \int_{t_o}^{t} i_2(\tau)d\tau = K_1 v_1(t) + K_2 v_2(t),$$

which is a linear combination of the original outputs (with the same coefficients in the linear combination), implying that the system is zero-state linear.

Example 3.10

Identify the zero-input and zero-state response components of the output of the op-amp differentiator

$$v_s(t) \longrightarrow \boxed{\text{Diff}} \longrightarrow -RC\frac{dv_s}{dt}$$

and show that the system is zero-state linear.

Solution For zero-input, $v_s(t) = 0$ and the system output is also zero. So, the system has no dependence on initial conditions and the zero-input response is identically zero. Thus, the system output consists entirely of the zero-state component. The system outputs caused by inputs $v_{s1}(t)$ and $v_{s2}(t)$ can be expressed as

$$v_1(t) = -RC\frac{dv_{s1}}{dt}$$

and

$$v_2(t) = -RC\frac{dv_{s2}}{dt},$$

respectively. With the input

$$v_{s3}(t) \equiv K_1 v_{s1}(t) + K_2 v_{s2}(t),$$

the zero-state output is

$$v_3(t) = -RC\frac{d(K_1 v_{s1}(t) + K_2 v_{s2}(t))}{dt}$$

$$= -K_1 RC\frac{dv_{s1}}{dt} - K_2 RC\frac{dv_{s2}}{dt}$$

$$= K_1 v_1(t) + K_2 v_2(t),$$

which is a linear combination of the original outputs. Hence, the system is zero-state linear.

The foregoing examples focused on the concept of zero-state linearity. *Zero-input linearity* is a similar concept: A system is zero-input linear if the aforementioned type of superposition principle applies to the calculation of the zero-input response. That is, for systems that are zero-input linear, a linear combination of initial states (initial conditions) produces an output that is the linear combination of the outputs caused by the two sets of initial conditions operating individually. For example, hypothetical systems with zero-input responses

Zero-input linearity

$$y(t) = 3y(t_o)^2$$

and

$$y(t) = 1 + y(t_o)$$

would not be zero-input linear. This is easily seen because doubling the initial condition does not double the response due to the initial condition, which is a necessary component of linearity. (Choose $K_1 = 2$ and $K_2 = 0$ in the linear combination of initial states.) A third system with zero-input response

$$y(t) = y(t_o)e^{-(t-t_o)}$$

would be zero-input linear. The integrator and differentiator examined in Examples 3.9 and 3.10 also are zero-input linear.

Example 3.11

Is the system

$$f(t) \longrightarrow \boxed{\text{System}} \longrightarrow y(t_o) + f^2(t)$$

defined for $t > t_o$ linear?

Solution   Clearly, the system zero-input response $y(t_o)$ satisfies zero-input linearity. However, the zero-state response $f^2(t)$ is nonlinear, and therefore the system is not linear. To observe violation of zero-state linearity, simply double the input from $f(t)$ to $2f(t)$, and notice that the zero-state response is *not doubled*. Rather, it is *quadrupled* from $f^2(t)$ to $4f^2(t)$! That can't happen in a linear system. In a linear system a doubled input always must produce a doubled output; it can't be anything else.

Integrators and differentiators, as well as the RC and RL circuits examined in the next section, satisfy an additional system property known as time-invariance. Such systems that are both linear and *time-invariant* are referred to as LTI systems, for short. In time-invariant systems, *delayed inputs cause equally delayed outputs*, in the sense that if

Time-invariance and LTI systems

$$f(t) \longrightarrow \boxed{\text{Time-inv.}} \longrightarrow y(t),$$

then

$$f(t - t_o) \longrightarrow \boxed{\text{Time-inv.}} \longrightarrow y(t - t_o)$$

for zero initial states and arbitrary t_o.

All electrical circuits constructed with constant valued components (with the exception of source elements) are necessarily time-invariant, since in the description of such circuits the choice of time origin $t = 0$ is totally arbitrary. By contrast, a circuit containing a time-dependent resistor, for example, would not be time-invariant.

Example 3.12

Determine the zero-state response of an integrator system

$$i(t) \longrightarrow \boxed{\text{Integ}} \longrightarrow v(t_o) + 2 \int_{t_o}^{t} i(\tau)d\tau$$

with inputs

$$i_1(t) = at, \, t > 0$$

and

$$i_2(t) = a(t - 3), \, t > 3 \, \text{s},$$

and show that responses to $i_1(t)$ and $i_2(t)$ are consistent with time-invariance. The constant a is arbitrary.

Solution   The zero-state response to input $i_1(t)$ can be expressed as

$$v_1(t) = 2 \int_0^t i_1(\tau)d\tau = 2 \int_0^t a\tau d\tau = a \, \tau^2\big|_{\tau=0}^{t} = at^2, \, t > 0.$$

Likewise, the zero-state response to input $i_2(t)$ is

$$v_2(t) = 2 \int_3^t i_2(\tau)d\tau = 2 \int_3^t a(\tau - 3)d\tau$$

$$= a \, (\tau^2 - 6\tau)\big|_{\tau=3}^{t} = a(t - 3)^2, \, t > 3 \, \text{s}.$$

Since

$$v_2(t) = a(t - 3)^2, \, t > 3 \, \text{s}$$

is the same as

$$v_1(t - 3), \, t - 3 > 0,$$

the results are consistent with time-invariance of the system: A delayed input causes an equally delayed zero-state output.

As we study the properties of first-order RC and RL circuits in the next section, we will make use of the systems terminology introduced in this section. The terminology and associated concepts will, in fact, play an essential role throughout the rest of this book.

3.4   First-Order *RC* and *RL* Circuits

Figure 3.17 shows a source network with an open circuit voltage $v_s(t)$ and a Thevenin resistance R that are connected at time t_o across a capacitor C. For $t > t_o$, the loop current $i(t)$ and resistor voltage $Ri(t)$ easily can be deduced from the capacitor voltage $v(t)$. Therefore, we will focus our efforts on how to find $v(t)$ for $t > t_o$. The problem of determining $v(t)$ for $t > t_o$ can be viewed as an LTI system problem with a system input

$$f(t) = v_s(t),$$

output

$$y(t) = v(t),$$

and an input–output relation corresponding to the solution of a first-order *ordinary differential equation* (ODE) derived and examined next.

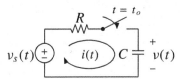

Figure 3.17   A first-order *RC* circuit that constitutes an LTI system.

We can identify the ODE that governs the RC circuit shown in Figure 3.17 for $t > t_o$ by writing the KVL equation around the loop. Since the loop current can be expressed as

$$i(t) = C\frac{dv}{dt}$$

in terms of the capacitor voltage $v(t)$, the KVL equation is

$$v_s(t) = RC\frac{dv}{dt} + v(t).$$

This equation can be rearranged as

$$\frac{dv}{dt} + \frac{1}{RC}v(t) = \frac{1}{RC}v_s(t),$$

which is a first-order linear ODE with constant coefficients that describes the circuit for $t > t_o$.

This differential equation is called "first-order" and "ordinary" because it contains only the first ordinary derivative of its *unknown* $v(t)$, instead of higher-order derivatives or partial derivatives. It is said to have *constant coefficients* because the coefficient $\frac{1}{RC}$ in front of both $v(t)$ and the forcing function $v_s(t)$ do not vary with time. This is true because the circuit components R and C are assumed to have constant values. The ODE also is said to be *linear*, because a linear combination of inputs applied to the ODE produces a solution that is the linear combination of the individual outputs.[3] Furthermore, the ODE also satisfies the zero-input linearity condition, as we will see in the next section.

The linearity and the constant coefficients of the preceding ODE guarantee that the RC circuit of Figure 3.17 constitutes an LTI system for $t > t_o$. The same system properties also apply to all resistive circuits containing a single capacitor, because we can represent all such circuits as in Figure 3.17 by using Thevenin equivalents.

3.4.1 *RC*-circuit response to constant sources

Figure 3.18 is a special case of Figure 3.17, with $t_o = 0$ and

$$v_s(t) = V_s,$$

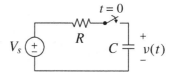

Figure 3.18   An *RC* circuit with a switch that closes at $t = 0$ and a DC voltage source V_s.

---

[3]**Verification of linearity:** Assume that

$$\frac{dv}{dt} + \frac{1}{RC}v(t) = \frac{1}{RC}f(t)$$

and

$$\frac{dw}{dt} + \frac{1}{RC}w(t) = \frac{1}{RC}g(t)$$

are true and therefore that $v(t)$ and $w(t)$ are different solutions of the same ODE with different inputs $f(t)$ and $g(t)$. A weighted sum of the equations, with coefficients K_1 and K_2, can be expressed as

$$\frac{d(K_1v + K_2w)}{dt} + \frac{1}{RC}(K_1v(t) + K_2w(t)) = \frac{1}{RC}(K_1f(t) + K_2g(t)),$$

which implies that the solution of the ODE, with an input $K_1f(t) + K_2g(t)$, is the linear superposition $K_1v(t) + K_2w(t)$ of solutions $v(t)$ and $w(t)$, obtained with inputs $f(t)$ and $g(t)$, respectively. Hence, superposition works, and the ODE is said to be linear.

a DC input. Clearly, the capacitor response $v(t)$ for $t > 0$ is a solution of the linear ODE

$$\frac{dv}{dt} + \frac{1}{RC}v(t) = \frac{1}{RC}V_s,$$

which satisfies an initial condition

$$v(0) = v(0^-),$$

where $v(0^-)$ stands for the capacitor voltage just before the switch is closed. The continuity of capacitor voltage discussed in Section 3.2.2 allows us to evaluate $v(t)$ at $t = 0$ and requires that we match $v(0)$ to $v(0^-)$, a voltage value established by the past charging/discharging activity of the capacitor.

To solve the ODE initial value problem just outlined, we first note that

$$v(t) = V_s$$

is a *particular solution* of the ODE, meaning that it fits into the differential equation. However, we must find a solution that also matches the initial value at $t = 0^-$. So, unless $V_s = v(0^-)$, *this* particular solution is not viable.

Next, we note that, with no loss generality, an entire family of solutions of the ODE can be written as

$$v(t) = v_h(t) + V_s,$$

where it is required that

$$\frac{d}{dt}(v_h(t) + V_s) + \frac{1}{RC}(v_h(t) + V_s) = \frac{1}{RC}V_s,$$

or, equivalently,

$$\frac{dv_h}{dt} + \frac{1}{RC}v_h(t) = 0.$$

But the last ODE—known as the *homogeneous* form of the original ODE—can be integrated directly to obtain a *homogeneous* (or complementary) solution

$$v_h = Ae^{-\frac{t}{RC}}$$

Margin notes:

ODE initial-value problem

Particular, homogeneous, and general solutions

where A is an arbitrary constant.[4] Thus,

$$v(t) = Ae^{-\frac{t}{RC}} + V_s$$

is the family of *general solutions* of the ODE, where A can be any constant.

Imposing the initial condition

$$v(0) = v(0^-)$$

on the general solution, we find that

$$v(0) = A + V_s = v(0^-),$$

so that

$$A = v(0^-) - V_s.$$

Therefore, the solution to the initial-value problem posed earlier, which is the capacitor voltage in the RC-circuit for $t > 0$, is

$$v(t) = [v(0^-) - V_s]e^{-\frac{t}{RC}} + V_s.$$

This solution simultaneously satisfies the ODE and also matches the initial condition at $t = 0^-$.

This result has an interesting structure that makes it easy to remember. The first term decays to zero as $t \to \infty$, leaving the second term, which is a DC solution (i.e., the solution after the voltages and currents are no longer changing). But, under DC conditions, the capacitor in Figure 3.18 becomes an open circuit, taking the full value of the DC source voltage V_s. Thus, the component V_s in the solution for $v(t)$ should be viewed as the *final* value of $v(t)$, as opposed to its *initial* value $v(0^-)$. The transition from initial to final DC state for $v(t)$ occurs as an exponential decay of the difference $v(0^-) - V_s$ between the two states, with the rate of decay controlled by a *time-constant RC*.

RC time-constant

---

[4]**Verification:** The homogeneous ODE implies that

$$\frac{dv_h}{v_h} = -\frac{1}{RC}dt,$$

which in turn integrates into

$$\ln v_h = -\frac{1}{RC}t + K,$$

where K is an arbitrary integration constant. Hence,

$$e^{\ln v_h} = v_h = e^{(-\frac{1}{RC}t + K)} = Ae^{-\frac{1}{RC}t},$$

as claimed, where $A \equiv e^K$ is another arbitrary constant.

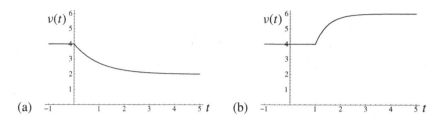

Figure 3.19   (a) Response function for $v(0^-) = 4\,\text{V}$, $V_s = 2\,\text{V}$, $RC = 1\,\text{s}$, and (b) for $v(1^-) = 4\,\text{V}$, $V_s = 6\,\text{V}$, $RC = \frac{1}{2}\,\text{s}$, and 1 s time delay in closing the switch.

Figure 3.19a shows $v(t)$ for the case $v(0^-) = 4\,\text{V}$, $V_s = 2\,\text{V}$, $R = 2\,\Omega$, and $C = \frac{1}{2}\,\text{F}$. The time constant of the decay is

$$RC = (2\,\Omega) \times (\frac{1}{2}F) = 1\,\text{s},$$

which is the amount of time it takes for

$$v(0^-) - V_s$$

to drop to

$$(v(0^-) - V_s)e^{-1} \approx 0.37(v(0^-) - V_s).$$

We derived this result for $v(t)$, assuming a switch closing time of $t_o = 0$. For an arbitrary switching time t_o, the original result can be shifted in time to obtain

$$v(t) = [v(t_o^-) - V_s]e^{-\frac{t-t_o}{RC}} + V_s$$

for $t > t_o$. Here $v(t_o^-)$ denotes the initial state of the capacitor voltage just before the switch is closed at $t = t_o$. Figure 3.19b depicts $v(t)$ for the case with $t_o = 1\text{s}$, $v(1^-) = 4\,\text{V}$, $V_s = 6\,\text{V}$, $R = 1\,\Omega$, and $C = \frac{1}{2}\,\text{F}$. Notice that now the RC time constant is 0.5 s, which is one-half the value assumed in Figure 3.19a.

In the next set of examples we will make use of the general results obtained above.

Example 3.13

Consider the circuit shown in Figure 3.20a. The switch in the circuit is moved at $t = 0$ from position a to position b, bringing the capacitor into the left side of the network. Assuming that the capacitor is in the DC state when the switch is moved, calculate the capacitor voltage $v(t)$ for $t > 0$.

Solution   For $t < 0$, the capacitor is in the DC state and behaves like an open circuit. Therefore, the 2 A source current in the circuit flows through the 1 Ω resistor on the right, generating a 2 V drop from top to bottom.

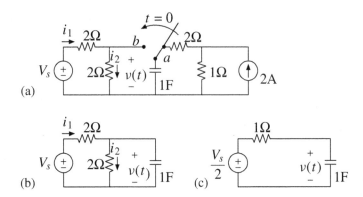

Figure 3.20 (a) A switched RC circuit, (b) capacitor circuit for $t > 0$, and (c) equivalent circuit.

Because the $2\,\Omega$ resistor in series with the capacitor conducts no current, there is no voltage drop across the $2\,\Omega$ resistor. Thus, the capacitor voltage $v(t)$ matches the 2 V drop across the $1\,\Omega$ resistor, giving

$$v(0^-) = 2\,\text{V}.$$

Figure 3.20b shows the capacitor circuit for $t > 0$. The resistive network across the capacitor can be replaced by its Thevenin equivalent, yielding the equivalent circuit in Figure 3.20c. Using the equivalent circuit, we see that as $t \to \infty$, $v(t) \to \frac{V_s}{2}$, which is the final state of the capacitor when it becomes an open circuit. We also note that the RC time constant is $1\,\Omega \times 1\,\text{F} = 1\,\text{s}$. Hence, for $t > 0$,

$$v(t) = [v(0^-) - \frac{V_s}{2}]e^{-t} + \frac{V_s}{2}.$$

Notice that the expression for $v(t)$ in Example 3.13 also can be written as

$$v(t) = \underbrace{v(0^-)e^{-t}}_{\text{Zero-input}} + \underbrace{\frac{V_s}{2}(1 - e^{-t})}_{\text{Zero-state}}.$$

Zero-input and zero-state components

Clearly, the foregoing zero-input and zero-state components vary linearly with the initial state $v(0^-)$ and input V_s, respectively. Therefore, the zero-input and zero-state linearity conditions are satisfied and the circuit constitutes a linear system (as claimed, but not shown, earlier on).

Example 3.14
Calculate the currents $i_1(t)$ and $i_2(t)$ in the circuit shown in Figure 3.20a.

Solution   From the figure, we see that for $t < 0$

$$i_1(t) = i_2(t) = \frac{V_s}{4\,\Omega}.$$

For $t > 0$,

$$i_1(t) = \frac{V_s - v(t)}{2}$$

and

$$i_2(t) = \frac{v(t)}{2}.$$

Substituting the expression for $v(t)$ from the previous equation, we obtain

$$i_1(t) = -\frac{v(0^-)}{2}e^{-t} + \frac{V_s}{4}(1 + e^{-t})$$

and

$$i_2(t) = \frac{v(0^-)}{2}e^{-t} + \frac{V_s}{4}(1 - e^{-t})$$

for $t > 0$. The voltage waveform $v(t)$ and the current waveforms $i_1(t)$ and $i_2(t)$ are plotted in Figure 3.21, for the case $v(0^-) = 2\,\text{V}$ (as in Example 3.13) and $V_s = 8\,\text{V}$.

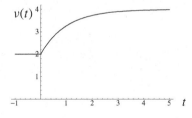

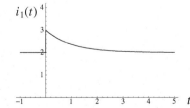
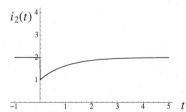

Figure 3.21   Capacitor voltage and resistor current responses in the circuit shown in Figure 3.20.

Notice that the capacitor voltage curve in Figure 3.21 is continuous across $t = 0$ (the switching instant in Examples 3.13 and 3.14), but the curves for $i_1(t)$ and $i_2(t)$ are discontinuous. Clearly, it is impossible to assign unique values to $i_1(0)$ and $i_2(0)$. Instead, we note that $i_1(0^-) = 2\,\text{A}$, $i_1(0^+) = 3\,\text{A}$ and $i_2(0^-) = 2\,\text{A}$, $i_2(0^+) = 1\,\text{A}$, where $i_1(0^-)$ and $i_1(0^+)$, for instance, refer to the limiting values of $i_1(t)$ as $t = 0$ is approached from the left and right, respectively. All solutions in the circuit for $t > 0$ can be found using the initial-state $v(0^-)$ of the capacitor voltage, without knowledge of $i_1(0^-)$, $i_2(0^-)$, etc., as we found out explicitly in Example 3.14. In this sense the initial capacitor voltage $v(0^-)$ fully describes the initial state of the entire RC circuit for $t > 0$.

3.4.2 *RL*-circuit response to constant sources

Figure 3.22 shows an RL circuit with a DC current source I_s. Since the inductor voltage drop is $L\frac{di}{dt}$ in the direction of $i(t)$, the KCL equation at the top node of the circuit can be written as

$$I_s = \frac{L\frac{di}{dt}}{R} + i(t).$$

So, the ODE that describes the inductor current in the circuit is

$$\frac{di}{dt} + \frac{R}{L}i(t) = \frac{R}{L}I_s.$$

The solution to this ODE for $t > 0$ can be expressed as

$$i(t) = [i(0^-) - I_s]e^{-\frac{t}{L/R}} + I_s,$$

by analogy with the RC-circuit solution developed in the previous section.

$\frac{L}{R}$ **time-constant**

Clearly, the inductor current $i(t)$ in the RL circuit evolves from an initial value of $i(0^-)$ to a final DC value of I_s as $t \to \infty$ and the inductor turns into an effective short circuit. The time-constant of exponential variation is $\frac{L}{R}$. If the input I_s is applied, beginning at some delayed time $t = t_o$, then a delayed version of the solution,

$$i(t) = [i(t_o^-) - I_s]e^{-\frac{t-t_o}{L/R}} + I_s,$$

is pertinent for $t > t_o$.

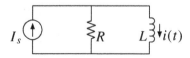

Figure 3.22   An *RL* circuit with DC input I_s.

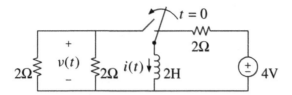

Figure 3.23 An *RL* circuit with a make-before-break type switch.

Example 3.15

Consider the circuit shown in Figure 3.23 where the switch moves from right to left and the inductor is connected into both sides of the circuit at the single instant $t = 0$. Determine the inductor current $i(t)$ and voltage $v(t)$ for $t > 0$. Assume that $\frac{di}{dt} = 0$ for $t < 0$.

Solution   From the figure, we see that

$$i(0^-) = 2\,\text{A},$$

because the inductor is effectively a short circuit for $t < 0$ (since $\frac{di}{dt} = 0$). For $t > 0$, the inductor finds itself in the source-free segment on the left. Hence,

$$I_s \equiv \lim_{t \to \infty} i(t) = 0.$$

Also, the equivalent resistance in parallel with the inductor is $1\,\Omega$, and therefore the exponential decay time-constant is

$$\frac{L}{R} = \frac{2\,\text{H}}{1\,\Omega} = 2\,\text{s}.$$

Thus, for $t > 0$, the inductor current is

$$i(t) = i(0^-)e^{-\frac{t}{L/R}} = 2e^{-0.5t}\,\text{A}.$$

Next, we notice that half of this current flows upward through each resistor on the left, and therefore the voltage $v(t)$ is $-(2\,\Omega) \times \frac{i(t)}{2}$, or

$$v(t) = -2e^{-0.5t}\,\text{V}.$$

The resistors will dissipate the initial energy

$$w = \frac{1}{2}2\,\text{H} \times (2\,\text{A})^2 = 4\,\text{J}$$

stored in the inductor, and all signals in the circuit vanish as $t \to \infty$.

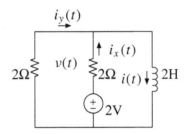

Figure 3.24   An *RL* circuit with a 2 V DC input.

Example 3.16

Consider the circuit shown in Figure 3.24. We will assume that

$$i(0^-) = 0$$

and calculate $i(t)$ for $t > 0$. First, we notice that the inductor becomes an effective short circuit as $t \to \infty$, so the currents

$$i_x(t) \to 1\,\text{A} \quad \text{and} \quad i_y(t) \to 0.$$

Therefore, the final inductor current is 1 A. The Thevenin resistance of the network seen by the inductor is 1 Ω (just replace the voltage source with a short circuit and combine the two parallel resistors); therefore,

$$\frac{L}{R} = 2\,\text{s}.$$

Using these values, as well as $i(0^-) = 0$, we write

$$i(t) = [0 - 1]e^{-0.5t} + 1\,\text{A} = 1 - e^{-0.5t}\,\text{A}.$$

This is a zero-state response of the circuit, because we started with a zero initial-state, $i(0^-) = 0$.

What if the initial state of the circuit described were $i(0^-) = 2\,\text{A}$? In that case the response $i(t)$ would be the superposition of the previous expression and the zero-input solution of the circuit for $i(0^-) = 2\,\text{A}$. But, we already have found the zero-input solution. Under the zero-input condition, the voltage source in Figure 3.24 is replaced with a short circuit and the circuit reduces to the circuit analyzed in Example 3.15 for the same initial current $i(0^-) = 2\,\text{A}$. Therefore, superposing the answers in Examples 3.15 and 3.16, we get

$$i(t) = 2e^{-0.5t}\,\text{A} + (1 - e^{-0.5t})\,\text{A} = 1 + e^{-0.5t}\,\text{A}.$$

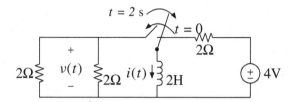

Figure 3.25 In this circuit the switch is moved twice, at $t = 0$ and $t = 2$ s.

Now, what if $i(0^-)$ were 4 A? No problem! In this case we can double the zero-input response just calculated, since the response is linear in $i(0^-)$, and add it to the zero-state response. The answer is

$$i(t) = 4e^{-0.5t}\,\text{A} + (1 - e^{-0.5t})\,\text{A} = 1 + 3e^{-0.5t}\,\text{A}.$$

Example 3.17
As a final example, we will calculate the inductor current $i(t)$ in the circuit shown in Figure 3.25. This is the same circuit as in Example 3.15, but, at $t = 2$ s, the switch is returned back to its original position. Therefore, the inductor current $i(t)$ evolves until $t = 2$ s, exactly as determined in Example 3.15, namely,

$$i(t) = 2e^{-0.5t}\,\text{A}.$$

So, the inductor current just before the switch is moved again is

$$i(2^-) = 2e^{-1}\,\text{A}.$$

As $t \to \infty$, the inductor current will build up from this value to a final current value of 2 A, with a time constant of

$$\frac{2\,\text{H}}{2\,\Omega} = 1\,\text{s}.$$

Notice that the time constant is different than before, because the inductor sees a different Thevenin resistance after $t = 2$ s. Therefore, for $t > 2$ s, the current variation is

$$i(t) = (2e^{-1} - 2)e^{-(t-2)} + 2\,\text{A}.$$

The complete current waveform is shown in Figure 3.26.

3.4.3   *RC*- and *RL*-circuit response to time-varying sources

As we have found, the capacitor voltage in *RC* circuits and inductor current in *RL* circuits are described by linear first-order ODEs of the form

$$\frac{dy}{dt} + ay(t) = bf(t),$$

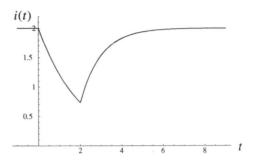

Figure 3.26   Inductor current waveform for the circuit shown in Figure 3.25.

where $y(t)$ denotes a capacitor voltage (in RC circuits) or inductor current (in RL circuits), a and b are circuit-dependent constants, and $f(t)$ is an input source or signal. For instance, in the RC circuit shown in Figure 3.27,

$$a = b = \frac{1}{RC}$$

and

$$f(t) = v_s(t)$$

for $t > 0$.

In RC and RL circuits with time varying inputs (as in Figure 3.27), the general solution of

$$\frac{dy}{dt} + ay(t) = bf(t)$$

for $t > 0$ can be written as

$$y(t) = Ae^{-at} + y_p(t),$$

where Ae^{-at} is a solution of the homogeneous ODE

$$\frac{dy_h}{dt} + ay_h = 0$$

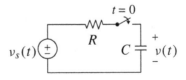

Figure 3.27   A first-order circuit with an arbitrary time-varying input $f(t) = v_s(t)$ applied at $t = 0$.

	Source function $f(t)$	Particular solution of $\frac{dy}{dt} + ay(t) = bf(t)$
1	constant D	constant K
2	Dt	$Kt + L$ for some K and L
3	De^{pt}	Ke^{pt} if $p \neq -a$ Kte^{pt} if $p = -a$
4	$\cos(\omega t)$ or $\sin(\omega t)$	$H\cos(\omega t + \theta)$, where H and θ depend on ω, a, and b

Table 3.1 Suggestions for particular solutions of $\frac{dy}{dt} + ay(t) = bf(t)$ with various source functions $f(t)$.

and $y_p(t)$ is a particular solution of the original ODE (like those suggested in Table 3.1 for inputs $f(t)$ having various forms). Since

$$y(0) = A + y_p(0),$$

it follows that

$$y(t) = [y(0) - y_p(0)]e^{-at} + y_p(t)$$

$$= \underbrace{y(0)e^{-at}}_{\text{Zero-input}} + \underbrace{y_p(t) - y_p(0)e^{-at}}_{\text{Zero-state}},$$

where $y(0) = y(0^-)$, an initial capacitor voltage or an inductor current. In the second line of the preceding equation, we have grouped the solution into its zero-input response (due to initial condition) and zero-state response (due to input).

Our solutions in the next set of examples will be specific applications of this general result for first-order *RC* and *RL* circuits.

For reference, the result can be generalized further to

$$y(t) = [y(t_o) - y_p(t_o)]e^{-a(t-t_o)} + y_p(t)$$

for $t > t_o$, where $y(t_o)$ is an initial-state (capacitor voltage or inductor current) defined at $t = t_o$. In this case the zero-state response for $t > t_o$ is

$$y_p(t) - y_p(t_o)e^{-a(t-t_o)},$$

while the zero-input response is

$$y(t_o)e^{-a(t-t_o)}.$$

Example 3.18

Find the capacitor voltage $y(t)$ in an RC circuit described by

$$\frac{dy}{dt} + y(t) = e^{-2t}$$

for $t > 0$. Assume a zero initial state—that is, $y(0^-) = 0$.

Solution   Since

$$f(t) = e^{-2t},$$

we can try, according to Table 3.1, a particular solution

$$y_p(t) = Ke^{-2t}$$

where K is a constant to be determined. Replacing $y(t)$ in the ODE by Ke^{-2t}, we find that

$$-2Ke^{-2t} + Ke^{-2t} = e^{-2t},$$

which implies that

$$K = -1.$$

Therefore, a particular solution of the ODE is

$$y_p(t) = -e^{-2t}.$$

The particular solution does not match the initial condition (otherwise we would be finished), and so we proceed to the general solution (sum of the homogeneous and particular solutions):

$$y(t) = Ae^{-t} - e^{-2t}.$$

Applying the initial condition yields

$$y(0) = A - 1,$$

implying that

$$A = y(0) + 1.$$

But since $y(0) = y(0^-) = 0$, it follows that

$$A = 1.$$

Hence, the zero-state solution for $t > 0$ is

$$y(t) = e^{-t} - e^{-2t}.$$

Notice that the zero-state response determined in Example 3.18 consists of two transient functions, e^{-t} and e^{-2t}. The term *transient* describes functions that vanish as $t \to \infty$.

Transient function

Example 3.19
What is the zero-state solution of

$$\frac{dy}{dt} + y(t) = e^{-2t}$$

for $t > -1$ s, assuming that $y(-1^-) = 0$?

Solution   This problem is similar to Example 3.18, except that a solution is sought for $t > -1$ s, with $y(-1^-) = 0$. Evaluating the general solution

$$y(t) = Ae^{-t} - e^{-2t}$$

(from Example 3.18) at $t = -1$ s, we find

$$y(-1) = Ae - e^2 = e(A - e).$$

Since we require

$$y(-1) = 0,$$

we conclude that

$$A = e.$$

Hence, the zero-state solution for the period $t > -1$ is

$$y(t) = e^{-(t-1)} - e^{-2t}.$$

Example 3.20
Let $R = 1 \, \Omega$, $C = 1 \text{F}$, and

$$v_s(t) = \cos(t)$$

in the *RC* circuit shown earlier in Figure 3.27. Then, for $t > 0$, the capacitor voltage $v(t)$ will be the solution to the ODE

$$\frac{dv}{dt} + v(t) = \cos(t).$$

Determine $v(t)$ for $t > 0$, assuming zero initial state—that is, $v(0^-) = 0$.

Solution   Since the ODE input is $\cos(t)$, a particular solution according to Table 3.1, should have the form

$$v_p(t) = H \cos(t + \theta) = A \cos(t) - B \sin(t),$$

where[5]

$$A = H \cos \theta \ \text{ and } \ B = H \sin \theta.$$

Therefore,

$$\frac{dv_p}{dt} = -A \sin(t) - B \cos(t).$$

Substituting $v_p(t)$ and $\frac{dv_p}{dt}$ into the ODE, we obtain

$$(-A \sin(t) - B \cos(t)) + (A \cos(t) - B \sin(t)) = \cos(t),$$

or, equivalently,

$$(A - B) \cos(t) - (A + B) \sin(t) = \cos(t).$$

This identity imposes the two constraints

$$A - B = 1 \ \text{ and } \ A + B = 0,$$

yielding

$$A = \frac{1}{2} \ \text{ and } \ B = -\frac{1}{2}.$$

Now, substituting the values for A and B into $A = H \cos \theta$ and $B = H \sin \theta$ gives

$$\frac{1}{2} = H \cos \theta \ \text{ and } \ -\frac{1}{2} = H \sin \theta.$$

It follows that

$$\frac{\sin \theta}{\cos \theta} = \tan \theta = -1,$$

and so

$$\theta = -45°$$

---

[5] Here, we are making use of the trig identity $\cos(a + b) = \cos a \cos b - \sin a \sin b$.

and

$$H = \frac{1/2}{\cos(-45°)} = \frac{1}{\sqrt{2}},$$

or equivalently, $\theta = 135°$ and $H = -\frac{1}{\sqrt{2}}$. Therefore, a particular solution of the ODE is

$$v_p(t) = H \cos(t + \theta) = \frac{1}{\sqrt{2}} \cos(t - 45°).$$

Consequently, the general solution can be written as

$$v(t) = Ae^{-t} + \frac{1}{\sqrt{2}} \cos(t - 45°),$$

where the first term is the homogeneous solution with arbitrary constant A. Employing the initial condition, $v(0) = 0$, gives

$$v(0) = A + \frac{1}{\sqrt{2}} \cos(-45°) = A + \frac{1}{2},$$

which implies that

$$A = -\frac{1}{2}.$$

Thus, the zero-state solution for $t > 0$ is

$$v(t) = -\frac{1}{2}e^{-t} + \frac{1}{\sqrt{2}} \cos(t - 45°).$$

Clearly, the first term $-\frac{1}{2}e^{-t}$ is transient and the second term $\frac{1}{\sqrt{2}} \cos(t - 45°)$ is non-transient.

Example 3.21
Given that

$$y(0^-) = 1,$$

what is the zero-input response of

$$\frac{dy}{dt} + ay(t) = f(t)?$$

Is the solution transient?

Solution   To find the zero-input response we set $f(t) = 0$ in the ODE to obtain

$$\frac{dy}{dt} + ay(t) = 0.$$

The general solution of this homogeneous ODE is

$$y(t) = Ae^{-at}.$$

Since

$$y(0) = A = y(0^-) = 1,$$

we conclude that

$$y(t) = e^{-at}$$

for $t > 0$. This solution is transient only if $a > 0$. A first-order RC circuit with negative R will display a non-transient zero-state response. Note: A negative Thevenin resistance is possible for a network with dependent sources.

In Examples 3.18 through 3.21, we saw that both the zero-input and zero-state responses of 1st-order ODEs can include non-transient as well as transient functions. The part of a system response remaining after the transient components have vanished is called the system *steady-state* response. Of course, if nothing remains after the transients vanish, then the steady-state response is, trivially, zero. Such was the case in Example 3.18, where we found that the system response was composed entirely of transient functions. In Example 3.20, on the other hand, the steady-state response was $\frac{1}{\sqrt{2}} \cos(t - 45°)$.

Steady-state response

Example 3.22
Suppose that

$$\frac{dv}{dt} + v(t) = \cos(t)$$

is valid for *all* t. Then, what is the steady-state component of the response $v(t)$?

Solution   From Example 3.20, we know that a particular solution to the ODE is the co-sinusoid

$$v_p(t) = \frac{1}{\sqrt{2}} \cos(t - 45°),$$

which is non-transient. The specification of an initial condition is irrelevant, because no matter when (for what time) an initial condition is specified, the homogeneous solution has the form Ae^{-t}, which is transient and vanishes with increasing t. Therefore, the steady-state component of $v(t)$ is the co-sinusoidal solution $v_p(t)$ given above.

3.5 *n*th-Order LTI Systems

As we saw in Section 3.4, RC and RL circuits containing a single energy storage element (a capacitor C or inductor L) are described by first-order linear ODEs. Similarly, linear circuits with n distinct energy storage elements are described by nth-order ODEs. For instance, the parallel RLC circuit shown in Figure 3.28 is a second-order LTI system described by second-order ODEs determined in Example 3.23.

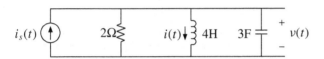

Figure 3.28 A parallel RLC circuit with a current source input $i_s(t)$.

Example 3.23
Determine the ODEs for the inductor current $i(t)$ and capacitor voltage $v(t)$ in the parallel RLC circuit shown in Figure 3.28.

Solution   The KCL equation for the top node is

$$i_s(t) = \frac{v(t)}{2} + i(t) + 3\frac{dv}{dt},$$

where we have

$$v(t) = 4\frac{di}{dt},$$

using the $v - i$ relation for the 4 H inductor. Eliminating $v(t)$ in the KCL equation, we get

$$i_s(t) = 2\frac{di}{dt} + i(t) + 12\frac{d^2i}{dt^2},$$

or

$$\frac{d^2i}{dt^2} + \frac{1}{6}\frac{di}{dt} + \frac{1}{12}i(t) = \frac{1}{12}i_s(t),$$

which is the ODE for the inductor current $i(t)$. Taking the derivative of both sides of the ODE and making the substitution

$$\frac{di}{dt} = \frac{v(t)}{4},$$

we next obtain

$$\frac{d^2}{dt^2}(\frac{v(t)}{4}) + \frac{1}{6}\frac{d}{dt}(\frac{v(t)}{4}) + \frac{1}{12}\frac{v(t)}{4} = \frac{1}{12}\frac{di_s}{dt},$$

which implies

$$\frac{d^2v}{dt^2} + \frac{1}{6}\frac{dv}{dt} + \frac{1}{12}v(t) = \frac{1}{3}\frac{di_s}{dt}.$$

This is the ODE describing the capacitor voltage. Notice that the ODEs for $i(t)$ and $v(t)$ are identical except for their right-hand sides. Thus, the forms of the homogeneous solutions of the ODEs are identical.

**nth-order
linear ODE
with
constant
coefficients**

As the order n of an LTI circuit or system increases, obtaining the governing ODEs of the form

$$\frac{d^n y}{dt^n} + a_1\frac{d^{n-1}y}{dt^{n-1}} + \cdots + a_n y(t) = b_o\frac{d^m f}{dt^m} + b_1\frac{d^{m-1}f}{dt^{m-1}} + \cdots + b_m f(t)$$

and solving them becomes increasingly more involved and difficult. Fortunately, there are efficient, alternative ways of analyzing LTI circuits and systems that do not depend on the formulation and solution of differential equations. The details of such methods, which are particularly useful when n is large, depend on whether or not the system zero-input response is transient.

Here is the central idea: In an LTI system with a *transient* zero-input response, the steady-state response to a co-sinusoidal input applied at $t = -\infty$ will itself be a co-sinusoid and will not depend on an initial state. (See Example 3.22.) Therefore, in such systems—known as *dissipative* LTI systems—the zero-state response to a super-position of co-sinusoidal inputs can be written as a superposition (because of linearity) of co-sinusoidal responses. This superposition method for zero-state response calcula-tions in dissipative LTI systems is known as the *Fourier method* and will be described in detail beginning in Chapter 5. An extension of the method, known as the *Laplace method*, is available for *non-dissipative* LTI systems where the zero-input response is not transient.

**Dissipative
LTI systems
and
Fourier
method**

Our plan for learning how to handle nth-order circuit and system problems is as follows. In Chapters 5 through 7 we will study the Fourier method for zero-state response calculations in dissipative LTI systems. This is a very powerful method because, as we will discover in Chapter 7, *any* practical signal that can be generated in the lab can be expressed as a weighted superposition of $\sin(\omega t)$ and $\cos(\omega t)$ signals with different ω's. Since the Fourier method requires that we know the system steady-state response to co-sinusoidal inputs, we need an efficient method for calculating such responses in circuits and systems of arbitrary complexity (arbitrary order n). We will develop such a method in Chapter 4. The discussion of the Laplace method for handling non-dissipative system problems will be delayed until Chapter 11.

We close this chapter with two simple examples on *zero-input response* in *n*th-order systems. A more complete coverage of the same topic will be provided in Chapter 11 after we learn the Laplace method.

Example 3.24

Determine the zero-input solution of the second-order ODE

$$\frac{d^2 y}{dt^2} + 3\frac{dy}{dt} + 2y(t) = f(t), \quad t > 0,$$

and discuss whether or not the system is dissipative.

Solution   To find the zero-input solution, we set

$$f(t) = 0,$$

and solve the homogeneous ODE

$$\frac{d^2 y}{dt^2} + 3\frac{dy}{dt} + 2y(t) = 0.$$

This equation can be satisfied by

$$y(t) = e^{st}$$

with certain allowed values for s. To find the permissible s we insert e^{st} for $y(t)$ in the ODE and obtain

$$s^2 e^{st} + 3s e^{st} + 2e^{st} = 0,$$

which implies

$$(s^2 + 3s + 2)e^{st} = 0,$$

and thus,

$$s^2 + 3s + 2 = (s + 1)(s + 2) = 0.$$

Clearly, the permissible values for s are

$$s = -1 \text{ and } s = -2.$$

The zero-input solution of the ODE generally is a weighted superposition of e^{-t} and e^{-2t}. In other words,

$$y(t) = Ae^{-t} + Be^{-2t},$$

where A and B are constants chosen so that $y(t)$ satisfies prescribed initial conditions. For example, the initial state may be specified as the values of $y(t)$ and its first derivative at $t = 0$, so that A and B can be found by solving

$$y(0) = A + B$$

and

$$\left. \frac{dy}{dt} \right|_{t=0} = -A - 2B.$$

The system is dissipative, because the zero-input solution is transient.

In general, we can construct solutions of any nth-order homogeneous ODE

$$\frac{d^n y}{dt^n} + a_1 \frac{d^{n-1} y}{dt^{n-1}} + \cdots + a_n y(t) = 0$$

by superposing functions proportional to $e^{s_i t}$, where the constants s_i correspond to the roots of the polynomial

$$s^n + a_1 s^{n-1} + \cdots + a_n,$$

Characteristic polynomial

known as the *characteristic polynomial* of the ODE. For instance, the characteristic polynomial of

$$\frac{d^2 y}{dt^2} + 3 \frac{dy}{dt} + 2y(t) = 0,$$

used in Example 3.24, is

$$s^2 + 3s + 2 = (s + 1)(s + 2).$$

Example 3.25
Repeat Example 3.24 with the ODE

$$\frac{d^2 y}{dt^2} + \frac{dy}{dt} - 2y(t) = f(t).$$

Solution   The characteristic polynomial is

$$s^2 + s - 2 = (s - 1)(s + 2).$$

Hence, permissible values for s are 1 and -2, and the zero-input solution is of the form

$$y(t) = Ae^t + Be^{-2t}.$$

Because the first continually increases as $t \to \infty$, the zero-input response is non-transient and the system must be non-dissipative.

From the previous examples, it should be clear that whether or not an LTI circuit is dissipative depends on the algebraic sign of the roots of its characteristic polynomial.[6] However, even before examining the characteristic polynomial, we can recognize a circuit as dissipative if it contains at least one current-carrying resistor and contains no dependent sources. That is true, because such a circuit would have no new source of energy under zero-input conditions (i.e., with the independent sources suppressed) and would dissipate, as heat, whatever energy it may have stored in its capacitors and inductors, via current flowing through the resistor.

Resistive linear circuits with no controlled sources are guaranteed to be dissipative

EXERCISES

3.1   **(a)**   In Figure 3.3a, given that $i_+ \approx 0$, what happens to the current $\frac{v_o}{R_L}$ in the circuit? Hint: the answer is related to the answer of part (b).

(b)   For $v_s = 1$ V, $R_s = 50\ \Omega$, and $R_L = 1$ kΩ, what is the power absorbed by resistor R_L in the circuit shown in Figure 3.3a and where does that power come from?

3.2   **(a)**   Confirm that substitution of the linear op-amp model of Figure 3.1b into the noninverting amplifier circuit of Figure 3.2b leads to the following circuit diagram:

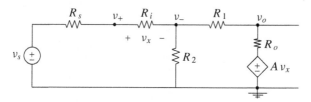

(b)   Assuming that $A \gg 1$, $R_i \gg R_s$, and $R_i \gg R_o$, show that $v_o \approx (1 + R_1/R_2)v_s$ in the equivalent circuit model shown in part (a).

(c)   Determine the short circuit current i_x in the following circuit:

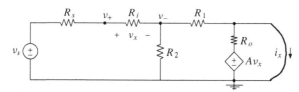

(d)   What is the Thevenin resistance $R_T = v_o/i_x$ of the equivalent circuit model above? Use the results from parts (b) and (c).

---

[6]In later chapters, we commonly will encounter situations where the roots of the characteristic polynomial are complex, as in Example 3.23, where the polynomial is $s^2 + \frac{1}{6}s + \frac{1}{12}$. In such cases, we will learn that the system is dissipative if the *real parts* of the roots are negative.

3.3   In the next circuit shown, determine the node voltage $v_o(t)$. You may assume that the circuit behaves linearly and make use of the ideal op-amp approximations ($v_+ \approx v_-$ and $i_+ \approx i_- \approx 0$).

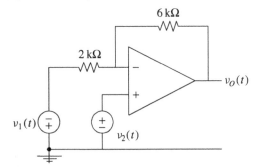

3.4   In the following circuit, determine the node voltage v_o, using the ideal op-amp approximations and assuming that $R_a = R_b = 1\,k\Omega$:

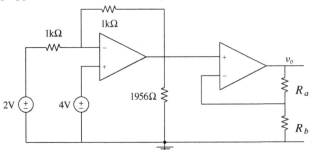

3.5   Repeat Problem 3.4 for $R_a = 0$ and $R_b = \infty$.

3.6   In the circuit shown next, determine the voltage v_x, assuming linear operation.

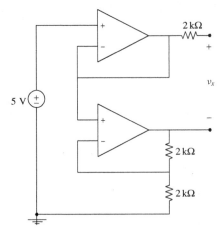

3.7   **(a)** In the following circuit, determine the capacitor current $i(t)$:

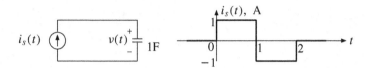

(b) In the next circuit, determine and plot the capacitor voltage $v(t)$. Assume that $v(t) = 0$ for $t < 0$.

3.8   In the following circuit, determine the output $v_o(t)$, using the ideal op-amp approximations:

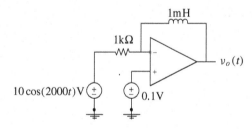

3.9   In the following circuit, determine $v_o(t)$:

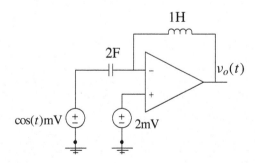

3.10   Using KCL and the v–i relations for resistors and capacitors, show that the voltage $v(t)$ in the following circuit satisfies the ODE

$$3\frac{dv}{dt} + \frac{1}{2}v(t) = i_s(t).$$

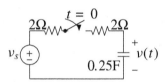

3.11   In the next circuit, $v(t) = 2$ V for $t < 0$. Determine $v(t)$ for $t > 0$ after the switch is closed, and identify the zero-state and zero-input components of $v(t)$. In the circuit, v_s denotes a DC voltage source (time-independent).

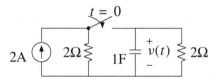

3.12   In the next circuit, $v(t) = 0$ for $t < 0$. Determine $v(t)$ for $t > 0$ after the switch is closed.

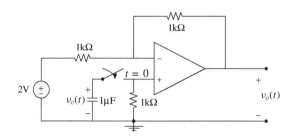

3.13   Assuming linear operation and $v_c(0) = 1$ V, determine $v_o(t)$ at $t = 1$ ms in the following circuit:

3.14 Determine the ODE that describes the inductor current $i(t)$ in the next circuit. Hint: Apply KVL using a loop current $i(t)$ such that $v(t) = 2\frac{di}{dt}$.

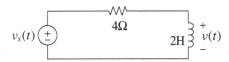

3.15 In the circuit that follows, find $i(t)$ for $t > 0$ after the switch is closed. Assume that $i(t) = 0$ for $t < 0$.

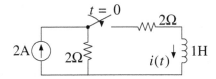

3.16 The circuit shown next is in DC steady state before the switch flips at $t = 0$. Find $v_L(0^-)$ and $i_L(0^-)$, as well as $i_L(t)$ and $v_L(t)$, for $t > 0$.

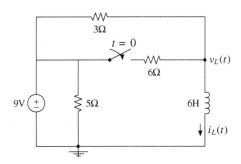

3.17 Obtain the second-order ODE describing the capacitor voltage $v(t)$ in the series RLC circuit shown next. Hint: Proceed as in Problem 3.14 and use $i(t) = 2\frac{dv}{dt}$ for the loop current.

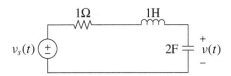

3.18   A second-order linear system is described by

$$\frac{d^2v}{dt^2} + 3\frac{dv}{dt} + 2v(t) = \cos(2t).$$

Confirm that the transient function

$$v_h(t) = Ae^{-t} + Be^{-2t}$$

is the homogeneous solution of the ODE and that its particular solution can be expressed as

$$v_p(t) = H\cos(2t + \theta).$$

Determine the values of H and θ. Hint: See Example 3.20 in Section 3.4.3.

3.19   **(a)** Show that $\frac{e^{j2t} + e^{-j2t}}{2} = \cos(2t)$.

        (b) Express $\frac{e^{j4t} - e^{-j4t}}{2j}$ in terms of a sine function.

        (c) Given that $-4(e^{j3t} + e^{-j3t}) = A\cos(3t + \theta)$, determine $A > 0$ and θ.

        (d) Express $P = \text{Re}\{2e^{j\frac{\pi}{3}}e^{j5t}\}$ in terms of a cosine function.

3.20   Let $f(x) = x$, where x is a complex variable.

        (a) Sketch the surface $|f(x)|$ over the 2-D complex plane. Describe in words what the surface looks like.

        (b) Describe in words the appearance of the surface $\angle f(x)$.

3.21   Let $f(x) = x - (2 + j3)$. Sketch the surface $|f(x)|$ over the 2-D complex plane and describe in words what the surface looks like.

4

Phasors and Sinusoidal Steady State

Suppose that we wish to calculate the voltage $v(t)$ in Figure 4.1a, where the input source is co-sinusoidal. In Chapter 3 we learned how to do this by writing and then solving a differential equation. The solution involved doing a large amount of algebra, using trigonometric identities. It turns out that there is a far simpler method for calculating just the steady-state response $v(t)$ in Figure 4.1a. The trick is to calculate V in Figure 4.1b by treating it as a DC voltage in a resistive circuit having a 1 V source and resistors with values $-j\,\Omega$ and $1\,\Omega$. We then can use voltage division to obtain

Steady-state AC-circuit calculations by complex arithmetic; phasors and impedances; average power and impedance matching; resonance

$$V = (1\,\text{V})\frac{1}{-j+1} = \frac{1}{1-j}\,\text{V}.$$

Now, in this calculation j represents the *imaginary unit* and, thus, V is complex. If you were to convert V to polar form by using your engineering calculator, you would see an output like

$$(0.707107, 45°),$$

indicating a magnitude 0.707107 and an angle of 45° for the complex number V. Surprisingly, this magnitude and angle are the correct magnitude and angle of $v(t)$ in Figure 4.1a so that $v(t)$ is simply

$$v(t) = 0.707107\cos(t + 45°)\,\text{V}.$$

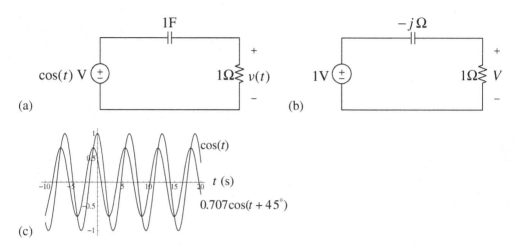

Figure 4.1 (a) An *RC* circuit with a cosine input, (b) the equivalent phasor circuit, and (c) input and output signals of the circuit.

Figure 4.1c compares the input signal, cos(t), with the output signal, 0.707107 cos(t + 45°). Notice that the output signal has the same frequency and shape as the input, but it is attenuated (reduced in amplitude) and shifted in phase.

   In this chapter we will learn why this trick with complex numbers works (even though DC circuits do not have imaginary resistances, and a capacitor is not a resistor). The trick is known as the *phasor method*, where V stands for the phasor of $v(t)$ and 1 V is the phasor of cos(t) V. We will see that the phasor method is a simple technique for determining the steady-state response of LTI circuits to sine and cosine inputs. For brevity, we will refer to such responses as "sinusoidal steady-state." We will explain the phasor method in Section 4.1: We will see why the capacitor of Figure 4.1a was converted to an imaginary resistance known as *impedance* and learn how to use impedances to calculate sinusoidal steady-state responses in LTI circuits. Section 4.2 is mainly a sequence of examples demonstrating how to apply the phasor method using the analysis techniques of Chapter 2, including source transformations, loop and node analyses, and Thevenin and Norton equivalents. Sections 4.3 and 4.4 describe power calculations in sinusoidal steady-state circuits and also discuss a phenomenon known as *resonance*.

4.1   Phasors, Co-Sinusoids, and Impedance

The phasor method requires a working knowledge of addition, subtraction, multiplication, and division of complex numbers and an understanding that, given any complex number C,

$$\text{Re}\{C\} = \frac{C + C^*}{2},$$

$$\text{Im}\{C\} = \frac{C - C^*}{j2}.$$

Furthermore, given any real ϕ, you should be able to recite Euler's formula in your sleep:

$$e^{\pm j\phi} = \cos\phi \pm j\sin\phi,$$

as well as its corollaries

$$\text{Re}\{e^{j\phi}\} = \cos\phi = \frac{e^{j\phi} + e^{-j\phi}}{2}$$

and

$$\text{Im}\{e^{j\phi}\} = \sin\phi = \frac{e^{j\phi} - e^{-j\phi}}{j2}.$$

Finally, you should be adept at converting between Cartesian and polar-form representation of complex numbers. After completing your review of complex numbers in Appendix A, read the rest of this section to learn about phasors and the phasor method.

4.1.1 Phasors and co-sinusoids

Consider the signal $f(t)$, defined as

$$f(t) \equiv \text{Re}\{Fe^{j\omega t}\},$$

where ω is a real constant and F is a complex constant that can be written in exponential form as

$$F = |F|e^{j\theta}.$$

Then what kind of a signal is $f(t)$ and how can we plot it?

Let's find out. Substituting the polar form of F into the formula for $f(t)$, and using the identity $\text{Re}\{e^{j\phi}\} = \cos\phi$, we get

$$f(t) = \text{Re}\{|F|e^{j\theta}e^{j\omega t}\} = |F|\text{Re}\{e^{j(\omega t + \theta)}\} = |F|\cos(\omega t + \theta).$$

So,

$$f(t) = \text{Re}\{Fe^{j\omega t}\} = |F|\cos(\omega t + \theta)$$

Cosine signal

is a *cosine signal* with

Amplitude,
phase,
phase shift,
radian frequency,
and phasor

- *amplitude* $|F|$,
- *phase* $\omega t + \theta$,
- *phase shift* $\theta = \angle F$, and
- *radian frequency* ω.

The complex constant $F = |F|e^{j\theta}$, which has a magnitude that is the amplitude of $f(t)$ and a phase that is the phase shift of $f(t)$, will be called the *phasor* of $f(t)$.

Example 4.1

The cosine signal

$$f(t) = \operatorname{Re}\{6e^{j\frac{\pi}{3}}e^{j2t}\} = 6\cos(2t + \frac{\pi}{3})$$

is plotted in Figure 4.2a. Its phasor is

$$F = 6e^{j\frac{\pi}{3}}.$$

The location of the phasor F in the complex plane is marked in Figure 4.2b. The magnitude, $|F| = 6$, is the peak value of the signal $f(t) = 6\cos(2t + \frac{\pi}{3})$ in Figure 4.2a. The angle of phasor F in Figure 4.2b is the phase shift, $\frac{\pi}{3} = 60°$, of the signal $f(t)$.

Example 4.2

The signal

$$g(t) = 3\cos(2t - \frac{\pi}{4}),$$

with frequency $\omega = 2\,\frac{\text{rad}}{\text{s}}$, has phasor

$$G = 3e^{-j\frac{\pi}{4}} = 3\angle - \frac{\pi}{4} = 3\angle - 45°.$$

See the phasor G and the signal $g(t)$ in Figure 4.2.

Period

The *period* of signals $f(t)$ and $g(t)$, with frequency $\omega = 2\,\frac{\text{rad}}{\text{s}}$, is $T = \frac{2\pi}{\omega} = \pi$ s, which is the time interval between the successive peaks (crests) of the $f(t)$ and $g(t)$ curves shown in Figure 4.2a.

Example 4.3

The phasor of

$$v(t) = 2\cos(6t - \frac{\pi}{2})$$

is

$$V = 2e^{-j\frac{\pi}{2}} = -j2 = 2\angle - 90°.$$

Signal $v(t)$ has frequency $\omega = 6\,\frac{\text{rad}}{\text{s}}$, and its period is $\frac{2\pi}{6} = \frac{\pi}{3}$ s.

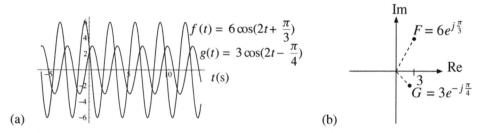

(a)   (b)

Figure 4.2 (a) Plots of cosine signals $f(t) = 6\cos(2t + \frac{\pi}{3})$ and $g(t) = 3\cos(2t - \frac{\pi}{4})$ versus t, and (b) the locations of the corresponding phasors $F = 6e^{j\frac{\pi}{3}}$ and $G = 3e^{-j\frac{\pi}{4}}$ in the complex plane. Note that the signal $f(t)$ "leads" (peaks earlier than) $g(t)$ by $\frac{\pi}{3} + \frac{\pi}{4}$ rad $= 105°$, because the angle of F is 105° greater than the angle of G in the complex plane. Equivalently, $g(t)$ "lags" (peaks later than) $f(t)$ by 105°. Also, the amplitude of $g(t)$ is half as large as the amplitude of $f(t)$, because the magnitude of phasor G is half the magnitude of phasor F.

Since

$$\sin \phi = \cos(\phi - \frac{\pi}{2})$$

for all real ϕ (a trig identity that you should be able to visualize), the foregoing $v(t)$ also can be expressed as $2\sin(6t)$. Therefore,

$$V = 2e^{-j\frac{\pi}{2}} = -j2$$

is also the phasor of signal $2\sin(6t)$. In fact, using Euler's identity, we obtain

$$v(t) = \text{Re}\{Ve^{j6t}\} = \text{Re}\{-j2e^{j6t}\}$$
$$= \text{Re}\{-j2(\cos(6t) + j\sin(6t))\} = 2\sin(6t),$$

as well as

$$v(t) = \text{Re}\{Ve^{j6t}\} = \text{Re}\{2e^{-j\frac{\pi}{2}}e^{j6t}\}$$
$$= 2\text{Re}\{e^{j(6t - \frac{\pi}{2})}\} = 2\cos(6t - \frac{\pi}{2}).$$

In general, the phasor of a *sine signal*   **Sine signal**

$$|F| \sin(\omega t + \phi) = |F| \cos(\omega t + \phi - \frac{\pi}{2})$$

is

$$|F|e^{j(\phi - \frac{\pi}{2})} = -j|F|e^{j\phi}.$$

For instance, the phasor of

$$i(t) = 6\sin(20t - \frac{\pi}{4})$$

is

$$I = -j6e^{-j\frac{\pi}{4}},$$

which is also the same as

$$6e^{-j\frac{\pi}{2}}e^{-j\frac{\pi}{4}} = 6e^{-j\frac{3\pi}{4}} = 6\angle -135°.$$

Example 4.4

$w(t) = 5\sin(5t + \frac{\pi}{3})$. What is the phasor W of $w(t)$?

Solution   The phasor of

$$5\cos(5t + \frac{\pi}{3})$$

is

$$5\angle 60°.$$

Therefore, the phasor of

$$5\sin(5t + \frac{\pi}{3})$$

is

$$-j(5\angle 60°) = (1\angle -90°)(5\angle 60°) = 5\angle -30°.$$

Hence, $W = 5\angle -30°$.

Example 4.5

A cosine signal $p(t)$, with frequency $3\,\frac{rad}{s}$, has phasor $P = j7$. What is $p(t)$?

Solution

$$p(t) = \text{Re}\{j7e^{j3t}\} = \text{Re}\{7e^{j(3t+\frac{\pi}{2})}\} = 7\cos(3t + \frac{\pi}{2}).$$

Alternatively,

$$P = j7 = 7\angle 90° \Rightarrow p(t) = 7\cos(3t + 90°),$$

since $\omega = 3\frac{\text{rad}}{\text{s}}$. As another alternative,

$$p(t) = \text{Re}\{j7e^{j3t}\} = 7\text{Re}\{j(\cos(3t) + j\sin(3t))\}$$
$$= 7\text{Re}\{j\cos(3t) - \sin(3t)\} = -7\sin(3t).$$

All versions of $p(t)$ just given are equivalent.

We can simplify certain mathematical operations on *co-sinusoidal* signals (meaning signals that are expressed either as a cosine or a sine) by working with the phasor   **Co-sinusoids** representation

$$\text{Re}\{Fe^{j\omega t}\}.$$

In particular, linear combinations of sinusoids and their derivatives, all having the same frequency ω, easily can be calculated by the use of phasors.

4.1.2 Superposition and derivatives of co-sinusoids

In this section, we will state and prove two principles concerning co-sinusoids and their phasors and also demonstrate their applications in circuit analysis.

Superposition principle: The weighted superposition

$$k_1 f_1(t) + k_2 f_2(t)$$

of co-sinusoids

$$f_1(t) = \text{Re}\{F_1 e^{j\omega t}\}$$

and

$$f_2(t) = \text{Re}\{F_2 e^{j\omega t}\}$$

with phasors F_1 and F_2 is also a co-sinusoid with the phasor

$$k_1 F_1 + k_2 F_2.$$

Proof   The superposition principle claims that

$$k_1 f_1(t) + k_2 f_2(t) = \text{Re}\{(k_1 F_1 + k_2 F_2)e^{j\omega t}\}.$$

To prove this claim, we expand its right-hand side as

$$\text{Re}\{(k_1 F_1 + k_2 F_2)e^{j\omega t}\} = \text{Re}\{k_1 F_1 e^{j\omega t}\} + \text{Re}\{k_2 F_2 e^{j\omega t}\}$$
$$= k_1 \text{Re}\{F_1 e^{j\omega t}\} + k_2 \text{Re}\{F_2 e^{j\omega t}\} = k_1 f_1(t) + k_2 f_2(t).$$

Therefore, the claim is correct.[1] QED

Example 4.6
Suppose that as shown in Figure 4.3

$$i_1(t) = 3\cos(3t) \text{ A}$$

and

$$i_2(t) = -4\sin(3t) \text{ A}$$

denote currents flowing into a circuit node. Determine the amplitude and phase shift of the current $i_3(t) = i_1(t) + i_2(t)$.

Solution   The phasor of $i_1(t)$ is $I_1 = 3$ A, while the phasor of $i_2(t)$ is $I_2 = j4$ A. Since

$$i_3(t) = i_1(t) + i_2(t),$$

the superposition principle implies that

$$I_3 = I_1 + I_2.$$

Hence,

$$I_3 = 3 + j4 = 5e^{j\tan^{-1}(\frac{4}{3})} \text{ A},$$

and therefore,

$$i_3(t) = 5\cos(3t + \tan^{-1}(\tfrac{4}{3}))\text{A}.$$

Clearly, the amplitude of $i_3(t)$ is 5 A and its phase shift is $\tan^{-1}(\frac{4}{3}) \approx 53.13°$. Signals $i_1(t), i_2(t)$, and $i_3(t) = i_1(t) + i_2(t)$ are shown in Figure 4.4.

Example 4.7
Let

$$v_1(t) = 3\sin(5t) \text{ V}$$

---

[1]Notice that the proof does not hold for $f_1(t)$ and $f_2(t)$ with different frequencies ω_1 and ω_2. Therefore, the superposition principle is valid only for co-sinusoids having the same frequency ω.

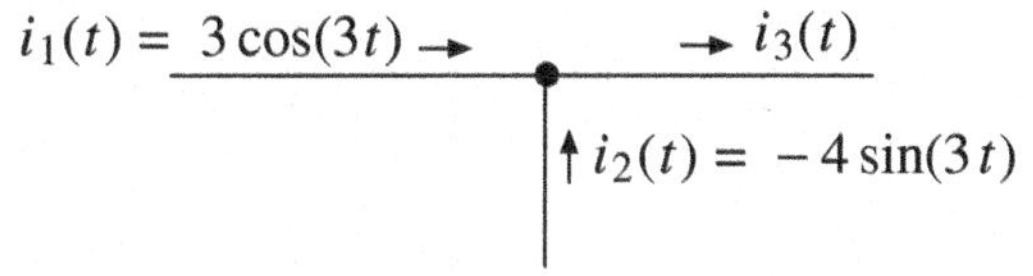

$$i_1(t) = \ 3\cos(3t) \rightarrow \qquad \rightarrow i_3(t)$$

$$\uparrow i_2(t) = -4\sin(3t)$$

Figure 4.3 A circuit node where three branches with currents $i_1(t)$, $i_2(t)$, and $i_3(t)$ meet.

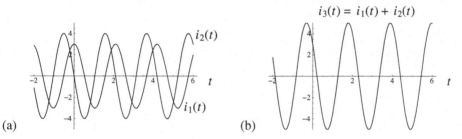

(a)   (b)

Figure 4.4 Signal waveforms of Example 4.6: (a) $i_1(t)$ and $i_2(t)$, and (b) $i_3(t) = i_1(t) + i_2(t)$.

and

$$v_2(t) = 2\cos(5t - \frac{\pi}{4}) \text{ V}$$

denote the two element voltages in the circuit shown in Figure 4.5. Calculate the amplitude of the third element voltage $v_3(t)$.

Solution   Since the KVL equation for the loop indicates that

$$v_3(t) = v_1(t) + v_2(t),$$

we apply the superposition principle, giving

$$V_3 = V_1 + V_2 = -j3 + 2e^{-j\frac{\pi}{4}} \text{ V}.$$

Using a calculator we easily find that $|V_3| \approx 4.635$ V, which is the amplitude of voltage $v_3(t)$.

As the previous examples illustrate, we can write KVL and KCL circuit equations for co-sinusoidal signals in terms of the signal phasors. Specifically,   **Phasor KVL and KCL**

$$\left\{ \sum V_{\text{drop}} = \sum V_{\text{rise}} \right\}_{\text{loop}}$$

and

$$\left\{ \sum I_{\text{in}} = \sum I_{\text{out}} \right\}_{\text{node}},$$

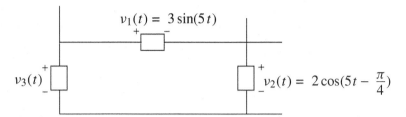

Figure 4.5 An elementary loop with three circuit elements. All elements carry co-sinusoidal voltages.

where $V_{\text{drop (rise)}}$ denotes a phasor voltage drop (rise) and $I_{\text{in (out)}}$ denotes a phasor current flowing into (out of) a node.

Derivative principle

(i) The derivative $\frac{df}{dt}$ of co-sinusoid

$$f(t) = \text{Re}\{Fe^{j\omega t}\}$$

is a co-sinusoid with phasor $j\omega F$.

(ii) The nth derivative $\frac{d^n f}{dt^n}$ is also a co-sinusoid with phasor $(j\omega)^n F$.

Proof   The claim of statement (i) is

$$\frac{df}{dt} = \text{Re}\{j\omega Fe^{j\omega t}\}.$$

To prove this claim, we calculate[2] the derivative of co-sinusoid $f(t)$ as

$$\frac{df}{dt} = \frac{d}{dt}\text{Re}\{Fe^{j\omega t}\} = \text{Re}\{F\frac{d}{dt}e^{j\omega t}\} = \text{Re}\{j\omega Fe^{j\omega t}\},$$

which indeed has the phasor $j\omega F$. Statement (ii) follows by induction from statement (i), because the nth derivative is the first derivative of the $(n-1)$st derivative.

Example 4.8
Apply the superposition and derivative rules to find a particular solution of the ODE

$$\frac{df}{dt} + 4f(t) = 2\cos(4t).$$

---

[2]Note: $\frac{d}{dt}\text{Re}\{Fe^{j\omega t}\} = \text{Re}\{F\frac{d}{dt}e^{j\omega t}\}$ is justified because $\text{Re}\{Fe^{j\omega t}\} = \frac{Fe^{j\omega t}+F^*e^{-j\omega t}}{2}$ and $\text{Re}\{F\frac{d}{dt}e^{j\omega t}\} = \frac{F\frac{d}{dt}e^{j\omega t}+F^*\frac{d}{dt}e^{-j\omega t}}{2}$.

Solution   Since the right-hand side of the equation is a co-sinusoid, the preceding phasor rules imply that the left-hand side can be a simple sum of co-sinusoids $\frac{df}{dt}$ and $4f(t)$, with phasors $j\omega F$ and $4F$, respectively. Here, F is the phasor of a co-sinusoid $f(t)$ with frequency $\omega = 4\,\frac{\text{rad}}{\text{s}}$. Thus, equating the phasors of the left- and right-hand sides of the equation, we get

$$j4F + 4F = 2,$$

from which it follows that

$$F = \frac{2}{4 + j4} = \frac{2}{4\sqrt{2}e^{j\frac{\pi}{4}}} = \frac{1}{2\sqrt{2}}e^{-j\frac{\pi}{4}}.$$

The corresponding co-sinusoid

$$f(t) = \frac{1}{2\sqrt{2}}\cos(4t - \frac{\pi}{4})$$

is a particular solution of the ODE. (You easily can confirm this by substituting the result into the ODE.) This solution is also the steady-state component of the zero-state solution of the ODE. Notice that, by using phasors, we simplified the solution tremendously, requiring no trigonometric identities.

Example 4.9
Determine the particular solution of

$$\frac{d^2y}{dt^2} + 3\frac{dy}{dt} + 2y = 5\sin(6t).$$

Solution   Equating the phasors of both sides (using $\omega = 6\,\frac{\text{rad}}{\text{s}}$), we have

$$(j6)^2Y + 3(j6)Y + 2Y = -j5.$$

So,

$$Y = \frac{-j5}{-34 + j18} \approx 0.130\angle 117.9°,$$

giving the particular solution

$$y(t) = 0.130\cos(6t + 117.9°).$$

These examples illustrate how phasors can be used to find steady-state solutions of linear constant-coefficient ODEs with co-sinusoidal inputs. But, phasors also can be used to completely avoid having to deal with differential equations in the analysis of dissipative LTI circuits with co-sinusoid inputs. The key is to establish phasor $V\text{–}I$ relations for inductors and capacitors in such circuits, as illustrated in the next example.

Example 4.10

Consider an inductor with some co-sinusoidal current

$$i(t) = \text{Re}\{Ie^{j\omega t}\}$$

and a voltage drop

$$v(t) = \text{Re}\{Ve^{j\omega t}\}$$

in the direction of the current. (See Figure 4.6a.) Express the voltage phasor V in terms of the current phasor I.

Solution   We can do this by replacing each co-sinusoid in the inductor v–i relation

$$v(t) = L\frac{di}{dt}$$

by its phasor. Since the phasor of $\frac{di}{dt}$ is $j\omega I$, the result is

$$V = j\omega L\, I.$$

Similar V–I relations also can be established for capacitors and resistors embedded in dissipative LTI circuits. For a capacitor (see Figure 4.6b), the v–i relation

$$i(t) = C\frac{dv}{dt}$$

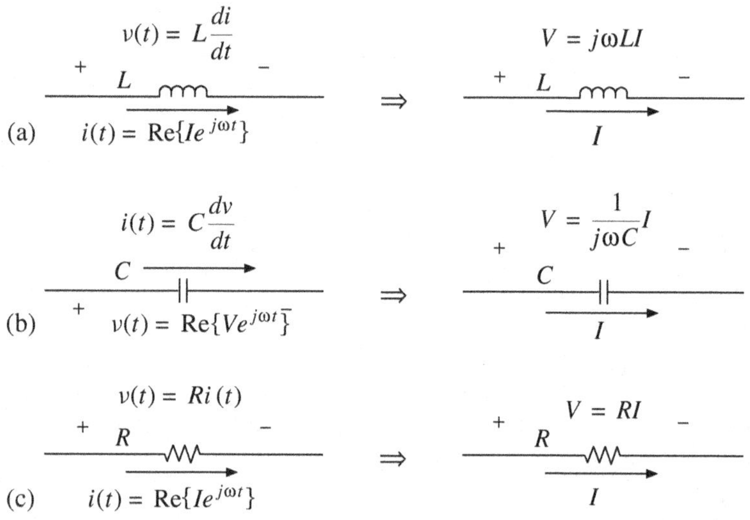

Figure 4.6   Elements with co-sinusoidal signals and their phasor $V - I$ relations: (a) an inductor, (b) a capacitor, and (c) a resistor.

implies that

$$I = j\omega C \, V \;\; \text{or} \;\; V = \frac{1}{j\omega C} \, I.$$

For a resistor (see Figure 4.6c), Ohm's law,

$$v(t) = Ri(t),$$

implies that

$$V = RI.$$

In the next section, we will discuss the implications of these phasor V–I relations for the analysis of LTI circuits with co-sinusoid inputs.

Phasor V–I relations

4.1.3 Impedance and the phasor method

In the previous section, we learned that phasor V–I relations for inductors L, capacitors C, and resistors R carrying co-sinusoidal signals have the form

$$V = ZI, \qquad \cdot$$

with

$$Z \equiv \begin{cases} j\omega L & \text{for inductors} \\ \dfrac{1}{j\omega C} & \text{for capacitors ,} \\ R & \text{for resistors} \end{cases}$$

where ω is the signal frequency. The parameter Z is known as *impedance* and is measured in units of ohms, since $Z = \frac{V}{I}$ is a voltage-to-current ratio, just like ordinary resistance $R = \frac{v(t)}{i(t)}$. Unlike resistance, however, impedance is, in general a complex quantity. Its imaginary part is known as *reactance*. Inductors and capacitors have reactances ωL and $-\frac{1}{\omega C}$, respectively, but the reactance of a resistor is zero. The real part of the impedance is known as *resistance*. Inductors and capacitors have zero resistance—they are purely *reactive*.[3]

Impedance

---

[3] An alternative form of the phasor V–I relation is

$$I = YV,$$

where

$$Y \equiv \frac{1}{Z}$$

is known as *admittance* and is measured in Siemens ($S = \Omega^{-1}$). The real and imaginary parts of admittance are known as *conductance* and *susceptance*, respectively.

Using the phasor $V-I$ relation

$$V = ZI,$$

and the phasor KVL and KCL equations

$$\left\{ \sum V_{\text{drop}} = \sum V_{\text{rise}} \right\}_{\text{loop}}$$

and

$$\left\{ \sum I_{\text{in}} = \sum I_{\text{out}} \right\}_{\text{node}}$$

discussed in the previous section, it is possible to express all of the mathematical constraints (KVL, KCL, and voltage-current relations for the elements) for LTI circuits (with co-sinusoidal inputs) in an algebraic form. The next example demonstrates this possibility.

Example 4.11

In the circuit shown in Figure 4.7a, we can write

$$2\cos(2t) = i_1(t) + i_2(t),$$

which is the KCL equation for the top node,

$$2i_2(t) + \frac{1}{2}\frac{di_2}{dt} = v_c(t),$$

the KVL equation for the right loop, and

$$i_1(t) = \frac{1}{2}\frac{dv_c}{dt},$$

giving us three coupled differential equations in three unknowns. Since the source term in this set of equations is a co-sinusoid, $2\cos(2t)$, the corresponding phasor equations—which we obtain using $\omega = 2$ rad/s (the source frequency) and the superposition and derivative principles—are

$$2 = I_1 + I_2,$$

$$V_c = 2I_2 + jI_2,$$

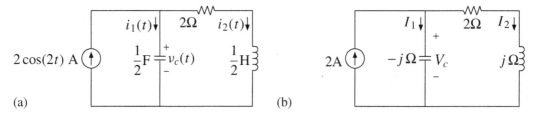

(a)   (b)

Figure 4.7 (a) An *RLC* circuit with a cosine input, and (b) the equivalent phasor circuit.

and

$$I_1 = jV_c,$$

respectively.

Notice that we could have obtained the phasor equations above directly from the phasor equivalent circuit shown in Figure 4.7b, without ever writing the differential equations, by applying phasor KCL and KVL and the V–I relations pertinent to inductors, capacitors, and resistors. In Figure 4.7b each element of Figure 4.7a has been replaced by the corresponding imped-ance calculated at the source frequency $\omega = 2$ rad/s. For example, $\frac{1}{2}$ H is replaced by $j(2\,\frac{\text{rad}}{\text{s}})(\frac{1}{2}\,\text{H}) = j\,\Omega$, and each co-sinusoid by the corre-sponding phasor. The solution of the phasor equations (obtained either way) gives

$$I_1 = 2 + j1\,\text{A} = \sqrt{5}e^{j0.464}\,\text{A},$$
$$I_2 = -j\,\text{A} = 1e^{-j\frac{\pi}{2}}\,\text{A},$$

and

$$V_c = -jI_1 = \sqrt{5}e^{j(-1.107)}\,\text{V},$$

so that

$$i_1(t) = \sqrt{5}\cos(2t + 0.464)\,\text{A},$$
$$i_2(t) = 1\cos(2t - \frac{\pi}{2})\,\text{A},$$

and

$$v_c(t) = \sqrt{5}\cos(2t - 1.107)\,\text{V}$$

are the steady-state co-sinusoidal currents and capacitor voltage in the orig-inal circuit. This can be confirmed by substituting these current expressions into the preceding set of differential equations.

In Example 4.11 we demonstrated the basic procedure for calculating the steady-state response of dissipative LTI circuits to co-sinusoidal inputs. However, we no longer will need to write the differential equations. A step-by-step description of the recommended procedure, called the *phasor method*, is as follows: **Phasor method**

(1) Construct an equivalent phasor circuit by replacing all inductors L and capaci-tors C with their impedances $j\omega L$ and $\frac{1}{j\omega C}$, calculated at the source frequency ω, and replacing all the signals (input signals as well as unknown responses) with their phasors.

(2) Construct phasor KVL and KCL equations for the equivalent circuit, using the phasor V–I relation $V = ZI$.

(3) Solve the equations for the unknown phasors.

(4) Translate each phasor to its co-sinusoid at the source frequency ω.

This method will work in all cases when all the sources in the circuit have the same frequency. (The procedure for sources with different frequencies will be described in Chapter 5.)

Because the voltage–current relations $v(t) = R i(t)$ and $V = ZI$ have the same form, step 2 of the phasor method produces a set of algebraic equations (instead of differential equations), just as in resistive circuit analysis. As a consequence, phasor circuits inherit all of the properties of resistive DC circuits, so that the various analysis strategies discussed in Chapter 2 can be applied: series and parallel combinations, voltage and current division, source transformations, the superposition method, Thevenin and Norton equivalents, and the node-voltage and loop-current methods. The next section consists of a sequence of examples where we use the phasor method to solve steady-state circuit problems. The section is titled "Sinusoidal Steady-State Analysis" because the term *sinusoidal steady state* commonly is used to refer to the steady-state response of dissipative LTI systems with co-sinusoidal inputs (as already indicated in the introduction to this chapter).

4.2   Sinusoidal Steady-State Analysis

4.2.1   Impedance combinations and voltage and current division

Impedance combinations: Impedances in series or parallel can be combined to obtain equivalent impedances. Series and parallel equivalents of impedances Z_1 and Z_2 are

$$Z_s = Z_1 + Z_2 \quad \text{(series equivalent)}$$

and

$$Z_p = \frac{Z_1 Z_2}{Z_1 + Z_2} \quad \text{(parallel equivalent)}.$$

For instance, for $Z_1 = 6 \ \Omega$ and $Z_2 = j8 \ \Omega$,

$$Z_s = 6 + j8 \ \Omega$$

and

$$Z_p = \frac{6(j8)}{6 + j8} = \frac{j48}{6 + j8} = \frac{j48(6 - j8)}{100} = 3.84 + j2.88 \ \Omega.$$

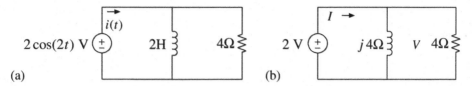

Figure 4.8 (a) An *RL* circuit with a parallel inductor and resistor, and (b) the equivalent phasor circuit.

Example 4.12

In the circuit shown in Figure 4.8a, the source voltage

$$v(t) = 2\cos(2t) \text{ V}$$

has phasor

$$V = 2\,\text{V}.$$

Determine the steady-state current $i(t)$ in the circuit.

Solution   Figure 4.8b shows the equivalent phasor circuit with impedances

$$Z_1 = j(2\,\frac{\text{rad}}{\text{s}})(2\,\text{H}) = j4\,\Omega$$

and

$$Z_2 = 4\,\Omega$$

for the inductor and resistor, respectively. The parallel equivalent of Z_1 and Z_2 is

$$Z_p = \frac{Z_1 Z_2}{Z_1 + Z_2} = \frac{(j4)(4)}{j4 + 4} = \frac{j4}{1 + j1}\,\Omega.$$

Therefore,

$$I = \frac{V}{Z_p} = \frac{2(1 + j1)}{j4} = \frac{1 - j}{2} = \frac{1}{\sqrt{2}} e^{-j\frac{\pi}{4}} \text{ A}$$

and

$$i(t) = \frac{1}{\sqrt{2}} \cos(2t - \frac{\pi}{4}) \text{ A}.$$

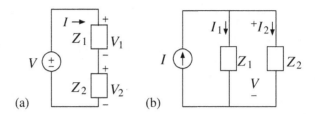

Figure 4.9 (a) Phasor voltage division, and (b) current division.

Voltage and current division: The equations for voltage and current division in phasor circuits (see Figure 4.9) have the forms

$$V_1 = V\frac{Z_1}{Z_1 + Z_2}, \quad V_2 = V\frac{Z_2}{Z_1 + Z_2},$$

and

$$I_1 = I\frac{Z_2}{Z_1 + Z_2}, \quad I_2 = I\frac{Z_1}{Z_1 + Z_2},$$

respectively.

Example 4.13

In the circuit shown in Figure 4.10a, calculate the steady-state voltage $v(t)$.

Solution   In the equivalent phasor circuit shown in Figure 4.10b,

$$\frac{1}{j(1\,\frac{\text{rad}}{\text{s}})(1\,\text{F})} = -j\,\Omega$$

denotes the impedance of the 1 F capacitor. Applying voltage division in the phasor circuit,

$$V = 1\,\text{V}\frac{1}{-j+1} = \frac{1}{1-j}\,V = \frac{1}{\sqrt{2}}e^{j\frac{\pi}{4}}\,\text{V}.$$

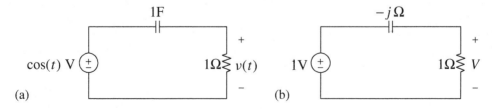

Figure 4.10 (a) An *RC* circuit in sinusoidal steady-state, and (b) the equivalent phasor circuit.

Hence,

$$v(t) = \frac{1}{\sqrt{2}} \cos(t + \frac{\pi}{4}) \, \text{V}.$$

Example 4.14

The phasor equivalent of Figure 4.11a is shown in Figure 4.11b. Determine the steady-state current $i(t)$.

Solution   The parallel equivalent impedance of the inductor and capacitor in Figure 4.11b is

$$Z_p = \frac{(j8)(-j4)}{j8 + (-j4)} = \frac{32}{j4} = -j8 \, \Omega.$$

Therefore, current division gives

$$I = 2e^{-j\frac{\pi}{3}} \frac{-j8}{8 - j8} = 2\angle - 60° \frac{1\angle - 90°}{\sqrt{2}\angle - 45°}$$

$$= 2\angle - 60° \left(\frac{1}{\sqrt{2}} \angle - 45° \right) = \sqrt{2}\angle - 105° \, \text{A}.$$

So,

$$i(t) = \sqrt{2} \cos(4t - 105°) \, \text{A}.$$

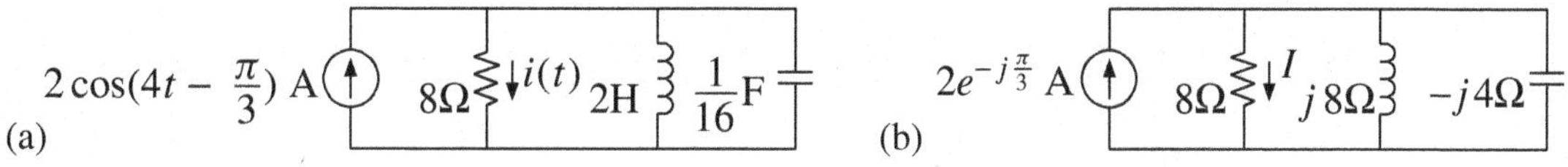

Figure 4.11   (a) A circuit with a co-sinusoidal current source, and (b) the equivalent phasor circuit.

4.2.2   Source transformations and superposition method

Source transformations: Source transformations can be used in phasor analysis just as in resistive circuit analysis. The phasor networks shown in Figure 4.12 are equivalent when

$$V_s = Z_s I_s \iff I_s = \frac{V_s}{Z_s}.$$

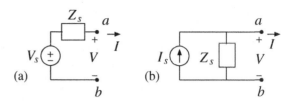

Figure 4.12 Equivalent sinusoidal steady-state networks for $V_s = Z_s I_s$.

Example 4.15
Find an expression for phasor V in Figure 4.13a in terms of source phasors V_s and I_s.

Solution   In Figures 4.13b through 4.13e we demonstrate the simplification of the given phasor network via source transformations and element combinations. The final network is the Thevenin equivalent.[4] Clearly,

$$V = (\sqrt{2}\angle -45°\ \Omega)I_s + (\frac{1}{\sqrt{2}}\angle -45°)V_s.$$

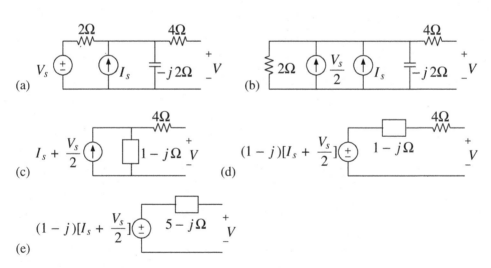

Figure 4.13 Network simplification with source transformations (a→b, c→d) and element combinations (b→c, d→e). Network (e) is the Thevenin equivalent of network (a).

---

[4]With no loss of generality, the terminal voltage phasor of any linear network in sinusoidal steady-state is $V = V_T - Z_T I$, where I is the terminal current and V_T is the open-circuit voltage of the network. Hence, a linear network in sinusoidal steady-state can be represented by its Thevenin equivalent with a Thevenin voltage V_T and impedance $Z_T = \frac{V_T}{I_N}$, where I_N is the short-circuit current phasor of the network (in exact analogy with resistive networks).

Superposition method: In Example 4.15, the voltage phasor V was shown to be a weighted superposition of independent source phasors V_s and I_s in the circuit. Notice that the weighting constants are complex quantities.

Using the superposition method, we can express any signal phasor as a weighted superposition of *all* the independent source phasors in any linear circuit. The superposed terms represent individual contributions of source elements to the signal phasor. We determine the contribution of each source after suppressing all other sources in the circuit (shorting the voltage sources and opening the current sources).

Example 4.16
Determine the Thevenin equivalent of the network shown in Figure 4.14a, using the superposition method.

Solution   Since the network terminals are open, the source current I_s flows down the $j4\,\Omega$ inductor, generating a voltage drop of $j4I_s$ from top to bottom of the element. Therefore, when the voltage source is suppressed (i.e., replaced by a short), the output voltage of the network is $j4I_s$. When the current source is suppressed, the inductor carries no current and the output voltage is simply V_s. Superposition of these contributions yields

$$V = V_s + j4I_s,$$

which is the Thevenin voltage of the network. The Thevenin impedance of the network is the equivalent network impedance $2 + j4\,\Omega$ when both sources are suppressed. The Thevenin equivalent is shown in Figure 4.14b.

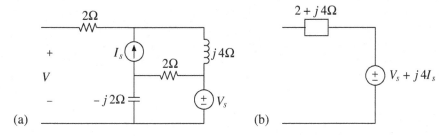

(a)   (b)

Figure 4.14   (a) A phasor network with two independent sources and (b) its Thevenin equivalent.

4.2.3   Node-voltage and loop-current methods

To determine the Thevenin and Norton equivalents of the network shown in Figure 4.15a, we will calculate the open-circuit voltage and short-circuit current in Examples 4.17 and 4.18 below, using the node-voltage and loop-current methods.

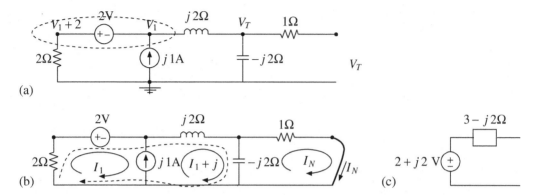

(a)

(b)

(c)

Figure 4.15 (a) A phasor network with an open-circuit voltage phasor V_T, (b) the same network terminated with an external short carrying a phasor current I_N, and (c) the Thevenin equivalent of the same network.

Example 4.17

The network shown in Figure 4.15a already has been marked in preparation for node-voltage analysis. Determine V_T and V_1, using the node-voltage method.

Solution   The KCL equation for the super-node on the left is

$$\frac{V_1 + 2}{2} + \frac{V_1 - V_T}{j2} = j1,$$

while, for the remaining node, the KCL equation is

$$\frac{V_T - V_1}{j2} + \frac{V_T}{-j2} = 0.$$

Note that the second equation implies that $V_1 = 0$. Hence, the first equation simplifies to

$$1 - \frac{V_T}{j2} = j1 \Rightarrow V_T = 2 + j2 \, \text{V}.$$

Example 4.18

After termination by an external short, the network of Figure 4.15a appears as shown in Figure 4.15b. The diagram in Figure 4.15b has been marked in preparation for loop-current analysis. Determine the Norton current I_N for the network shown in Figure 4.15a by applying the loop-current method to Figure 4.15b

Solution   The KVL equation for the dashed super-loop in Figure 4.15b is

$$2 + j2(I_1 + j) + (-j2)(I_1 + j - I_N) + 2I_1 = 0,$$

while, for the remaining loop, the KVL equation is

$$1I_N + (-j2)(I_N - (I_1 + j)) = 0.$$

The first equation yields

$$I_1 = -1 - jI_N,$$

while the second one simplifies to

$$I_N(1 - j2) = 2(1 - jI_1).$$

Eliminating I_1, we obtain

$$I_N(1 - j2) = 2(1 + j1 - I_N) \Rightarrow I_N = \frac{2 + j2}{3 - j2} \text{ A.}$$

Example 4.19
Determine the Thevenin equivalent of the network shown in Figure 4.15a.

Solution   From the previous Examples 4.17 and 4.18, we know that

$$V_T = 2 + j2 \text{ V} \quad \text{and} \quad I_N = \frac{2 + j2}{3 - j2} \text{ A.}$$

Thus, the Thevenin impedance is

$$Z_T = \frac{V_T}{I_N} = 3 - j2 \ \Omega.$$

The Thevenin equivalent network is shown in Figure 4.15c.

4.3 Average and Available Power

For circuits in sinusoidal steady-state, the absorbed power of a circuit element

$$p(t) = v(t)i(t)$$

is necessarily a function of time and is referred to as *instantaneous power*. The net power absorbed by each element corresponds to the *average value* of instantaneous power $p(t)$. For instance, a 60-W lightbulb absorbs 60 W of net power, while the instantaneous power fluctuates between 0 and 120 W. A 60-W lightbulb also shines brighter than a 45-W lightbulb because, on the average, it converts more joules per second into light than does the 45-W bulb.

The net absorbed power—that is, the average value of $p(t) = v(t)i(t)$ over one oscillation period $T = \frac{2\pi}{\omega}$ for signals $v(t)$ and $i(t)$—can be calculated as

$$P = \frac{1}{T} \int_{t=0}^{T} v(t)i(t)dt = \frac{1}{T} \int_{t=0}^{T} |V|\cos(\omega t + \theta) \times |I|\cos(\omega t + \phi)dt.$$

This integral can be evaluated using quite a lot of algebra, but it turns out that P can be computed far more easily, directly from the signal phasors V and I, as shown next.

4.3.1 Average power

We next derive the phasor formula for the average absorbed power P by expressing the instantaneous power $p(t) = v(t)i(t)$ as the sum of a constant term and a zero-average time-varying term. Since $v(t)$ and $i(t)$ are co-sinusoids, the instantaneous power works out to be

$$
\begin{aligned}
p(t) = v(t)i(t) &= \text{Re}\{Ve^{j\omega t}\}\text{Re}\{Ie^{j\omega t}\} \\
&= (\frac{Ve^{j\omega t} + V^*e^{-j\omega t}}{2})(\frac{Ie^{j\omega t} + I^*e^{-j\omega t}}{2}) \\
&= \frac{VIe^{j2\omega t} + VI^* + V^*I + V^*I^*e^{-j2\omega t}}{4} \\
&= \frac{VI^* + V^*I}{4} + \frac{VIe^{j2\omega t} + V^*I^*e^{-j2\omega t}}{4} \\
&= \frac{1}{2}\text{Re}\{VI^*\} + \frac{1}{2}\text{Re}\{VIe^{j2\omega t}\},
\end{aligned}
$$

where we have made use of the identity $\text{Re}\{C\} = \frac{C+C^*}{2}$ exactly four times. The last line expresses $p(t)$ as a superposition of a constant (the first term) and a co-sinusoid of frequency 2ω (the second term). Because the average value (integrated across the interval from 0 to T) of the co-sinusoid is zero, the average value of $p(t)$ is just the first term

Net absorbed power P

$$P = \frac{1}{2}\text{Re}\{VI^*\},$$

which is the net power absorbed by an element having voltage and current phasors V and I.

For a resistor,

$$V = RI$$

so that

$$P = \frac{1}{2}\text{Re}\{(RI)I^*\} = \frac{R|I|^2}{2} = \frac{|V|^2}{2R},$$

where $|V|$ and $|I|$ are the amplitudes of the resistor voltage and current. The same result also can be expressed as

$$P = RI_{\text{rms}}^2 = \frac{V_{\text{rms}}^2}{R},$$

where $I_{\text{rms}} \equiv \frac{|I|}{\sqrt{2}}$ and $V_{\text{rms}} \equiv \frac{|V|}{\sqrt{2}}$ are known as the rms, or *effective*, amplitudes.

Example 4.20
A 60-W lightbulb is designed to absorb 60 W of net power when utilized with $V_{\text{rms}} = 120$ V available from a wall outlet. Estimate the resistance of a 60-W lightbulb. What is the peak value of the sinusoidal voltage waveform at a wall outlet?

Solution   Using one of the preceding formulas for P, we find that

$$R = \frac{V_{\text{rms}}^2}{P} = \frac{120^2}{60} = 240 \ \Omega.$$

The peak value of the voltage waveform at a wall outlet is approximately

$$|V| = \sqrt{2}V_{\text{rms}} = 169.7 \text{ V}.$$

(The power company does not always deliver exactly $V_{\text{rms}} = 120$ V.)

Example 4.21
A signal

$$f(t) = A\cos(\omega t + \theta)$$

is applied to a 1 Ω resistor. What is the average power delivered by $f(t)$ to the resistor?

Solution   First, we note that, since $R = 1 \ \Omega$,

$$v(t) = i(t) = f(t)$$

in this problem; that is, $f(t)$ stands for both the element voltage $v(t)$ and element current $i(t)$. The corresponding phasors are all equal—$V = I = F = Ae^{j\theta}$—and the average power is

$$P = \frac{|V|^2}{2R} = \frac{1}{2}A^2.$$

Signal
$A\cos(\omega t + \theta)$
transfers
$\frac{1}{2}A^2$
average power to a 1Ω resistor

This is an important result that should be committed to memory: Average power from a co-sinusoid to a 1 ohm resistor is half the square of the signal amplitude.

For ideal inductors and capacitors,

$$V = jXI,$$

where the reactance X (either ωL or $-\frac{1}{\omega C}$) is real; therefore,

$$P = \frac{1}{2}\text{Re}\{(jXI)I^*\} = \frac{X|I|^2}{2}\text{Re}\{j\} = 0.$$

Capacitors and inductors absorb no net power, because they return their instantaneous absorbed power back to the circuit.

Example 4.22

Determine the net power absorbed by each element in the circuit shown in Figure 4.16a.

Solution   Clearly, the inductor will absorb no net power. To calculate P for the resistor and the voltage source, we first calculate the phasor current I for the loop. From Figure 4.16b,

$$I = \frac{1\text{ V}}{j4 + 3\text{ }\Omega} = \frac{1}{5}\angle 53.13° \text{ A}.$$

Therefore, for the resistor,

$$P = \frac{|I|^2 R}{2} = \frac{(\frac{1}{5})^2 3}{2} = \frac{3}{2 \times 25} = 0.06 \text{ W}.$$

Energy conservation requires that the net power absorbed by the source must be $P = -0.06$ W. Indeed, since the phasor voltage drop for the source in

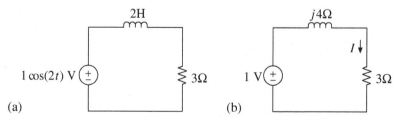

(a)   **(b)**

Figure 4.16   (a) A circuit with a single energy source, and (b) the equivalent phasor circuit.

the direction of I is -1 V, the average absorbed power of the source is

$$P = \frac{1}{2}\text{Re}\{VI^*\} = \frac{1}{2}\text{Re}\{(-1 \text{ V})(\frac{1}{3+j4} \text{ A})^*\}$$

$$= -\frac{1}{2}\text{Re}\{\frac{3+j4}{25}\} = -\frac{1}{2}\frac{3}{25} = -0.06 \text{ W}.$$

4.3.2 Available power and maximum power transfer

What is the *available* net power that can be delivered from a linear network, operating in sinusoidal steady state, to an external load? To answer this question, we will examine the phasor circuit shown in Figure 4.17, where the Thevenin network on the left represents an arbitrary linear network operating in sinusoidal steady state. Our approach will be to find the value of the average power that is delivered to the load and then to find the value of the load that will maximize this average power.

In Figure 4.17,

$$V_L = \frac{V_T Z_L}{Z_T + Z_L}$$

and

$$I_L = \frac{V_T}{Z_T + Z_L}.$$

Therefore, the net power delivered to (and absorbed by) the load Z_L is

$$P_L = \frac{1}{2}\text{Re}\{V_L I_L^*\} = \frac{1}{2}\text{Re}\{\frac{|V_T|^2 Z_L}{|Z_T + Z_L|^2}\} = \frac{|V_T|^2 R_L}{2|Z_T + Z_L|^2},$$

where $R_L \equiv \text{Re}\{Z_L\}$ is the resistive part of the load impedance Z_L.

Clearly, for a given network (i.e., for fixed V_T and Z_T), the value of P_L depends on both the load resistance R_L and the load reactance $X_L \equiv \text{Im}\{Z_L\}$. We wish to determine the maximum possible value for P_L.

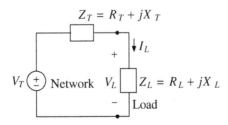

Figure 4.17 A linear phasor network terminated by an external load.

Substituting $Z_T = R_T + jX_T$ and $Z_L = R_L + jX_L$ into the formula for P_L yields

$$P_L = \frac{|V_T|^2 R_L}{2|(R_T + R_L) + j(X_T + X_L)|^2}.$$

For any fixed value of R_L, this expression is maximized when

$$X_L = -X_T,$$

because the denominator of P_L is then reduced to its smallest possible value, namely, $2(R_T + R_L)^2$. Choosing $X_L = -X_T$, the net power formula becomes

$$P_L = \frac{|V_T|^2 R_L}{2(R_T + R_L)^2},$$

which is maximized when

$$R_L = R_T.$$

as we learned in Chapter 2.

In summary then, an external load with resistance $R_L = R_T$ and reactance $X_L = -X_T$, that is, with an impedance

$$Z_L = R_T - jX_T = Z_T^*,$$

Average available power P_a

will extract the full available power from a network having a Thevenin impedance Z_T. We obtain the formula for the available power of a network by evaluating the preceding formula for P_L, with $R_L = R_T$. The result is

$$P_a = \frac{|V_T|^2}{8R_T}.$$

Matched load

So, the available power of a network depends on the magnitude of its Thevenin voltage phasor V_T and only the resistive (real) part R_T of its Thevenin impedance $Z_T = R_T + jX_T$. The available power will be delivered to any load having an impedance $Z_L = Z_T^*$. Such loads are known as *matched loads*.

Example 4.23
Determine the available power of the network shown in Figure 4.18.

Solution   We first use the superposition method to determine the open-circuit voltage V as

$$V = 2 \text{ V} + (2 + j3) \text{ V} = 4 + j3 \text{ V},$$

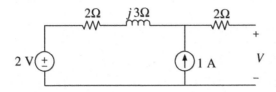

Figure 4.18 A phasor network.

where the first term is the contribution of the voltage source and the second term is the contribution of the current source. Thus, for this network,

$$|V_T|^2 = |4 + j3|^2 = 25 \text{ V}^2.$$

Next, we note that

$$Z_T = 4 + j3 \ \Omega$$

by calculating the equivalent impedance of the network after source suppression. Therefore, for this network, $R_T = 4 \ \Omega$ and the available power is

$$P_a = \frac{|V_T|^2}{8R_T} = \frac{25}{8 \times 4} = 0.78125 \text{ W}.$$

This amount of average power will be transferred to any matched load having an impedance $Z_L = 4 - j3 \ \Omega$.

Example 4.24
What load Z_L is *matched* to the network shown in Figure 4.19 and what is the available power of the network?

Solution $Z_L = Z_T^*$, where Z_T is the Thevenin impedance of the network shown in Figure 4.19. Note that, for all possible loads,

$$I_x = \frac{1 - 2I_x}{1} \ \Rightarrow \ I_x = \frac{1}{3} \text{ A}.$$

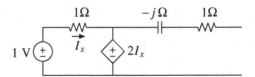

Figure 4.19 A phasor network with a current-controlled voltage source.

Hence, the open-circuit voltage at the network terminals is $V_T = 2I_x = \frac{2}{3}$ V. The short-circuit current is

$$I_N = \frac{2I_x}{1-j} = \frac{\frac{2}{3}}{1-j}\ \text{A}.$$

So, for this network

$$Z_T = \frac{V_T}{I_N} = 1 - j\ \Omega,$$

and the matched load impedance is $Z_L = Z_T^* = 1 + j\ \Omega$. The available power of the network is

$$P_a = \frac{|V_T|^2}{8R_T} = \frac{(\frac{2}{3})^2}{8\cdot 1} = \frac{1}{18}\ \text{W}.$$

4.4   Resonance

Consider the source-free circuits shown in Figures 4.20a and 4.20b. We next examine whether the signals marked $v(t)$ and $i(t)$ in these circuits can be co-sinusoidal waveforms, despite the absence of source elements.

For the RC circuit shown in Figure 4.20a, the phasor KVL equation, expressed in terms of the phasor I of a co-sinusoidal $i(t)$, is

$$(R + \frac{1}{j\omega C})I = 0.$$

Because $R + \frac{1}{j\omega C}$ cannot equal zero, the equation requires that

$$I = 0.$$

Hence, in the RC circuit shown in Figure 4.20a we cannot have co-sinusoidal $i(t)$ and $v(t)$.

By contrast, the phasor KVL equation for the LC circuit shown in Figure 4.20b,

$$(j\omega L + \frac{1}{j\omega C})I = 0,$$

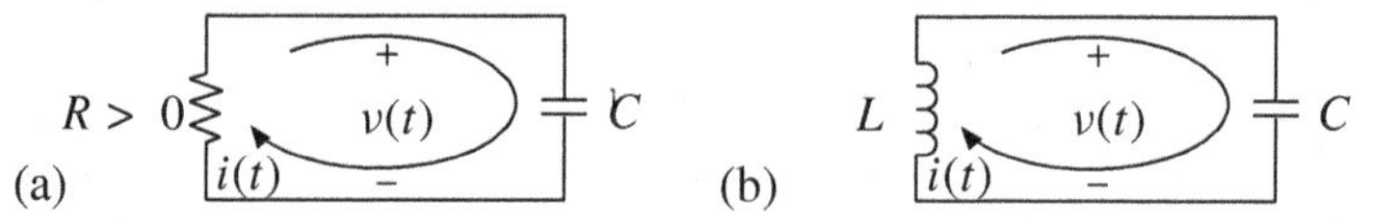

Figure 4.20   (a) A source-free RC circuit, and (b) a source-free LC circuit.

can be satisfied by *any* I, so long as

$$j\omega L + \frac{1}{j\omega C} = 0 \Rightarrow \omega = \frac{1}{\sqrt{LC}} \equiv \omega_o.$$

Thus, in the circuit of Figure 4.20b, co-sinusoidal signals

$$i(t) = \mathrm{Re}\{Ie^{j\omega_o t}\} = |I| \cos(\omega_o t + \theta)$$

and

$$v(t) = \mathrm{Re}\{\frac{I}{j\omega_o C} e^{j\omega_o t}\} = \frac{|I|}{\omega_o C} \sin(\omega_o t + \theta)$$

are possible with arbitrary $|I|$ and $\theta = \angle I$. The oscillation frequency

$$\omega_o = \frac{1}{\sqrt{LC}}$$

Resonant
frequency

is known as the *resonant frequency* of the circuit. The phenomenon itself (i.e., the possible existence of steady-state co-sinusoidal oscillations in a source-free circuit) is known as *resonance*.

Resonance is possible in the LC circuit of Figure 4.20b because the circuit is non-dissipative. As we learned in Section 3.4, circuits with no dissipative elements (i.e., resistors) can exhibit non-transient zero-input response. Resonance in the foregoing LC circuit is an example of such behavior. The inclusion of a series or parallel resistor in the circuit, added in Figure 4.21, introduces dissipation and spoils the possibility of source-free oscillations. We can see this in Figure 4.21a by writing the KVL equation

$$Z_s I = 0,$$

where the series equivalent impedance is

$$Z_s = R + j\omega L + \frac{1}{j\omega C} = R + j(\omega L - \frac{1}{\omega C}).$$

If $R \neq 0$, then the KVL equation can be satisfied only with $I = 0$. Likewise, in Figure 4.21b, the KCL equation is

$$\frac{V}{R} + \frac{V}{j\omega L} + \frac{V}{\frac{1}{j\omega C}} \equiv \frac{V}{Z_p} = 0,$$

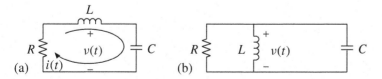

(a)   (b)

Figure 4.21 (a) A source-free series *RLC* circuit, and (b) a source-free parallel *RLC* circuit.

where the parallel equivalent impedance is

$$Z_p = \frac{1}{\frac{1}{R} + j(\omega C - \frac{1}{\omega L})}.$$

If R is not infinite (creating an open circuit), then the denominator of Z_p cannot be zero and, therefore, Z_p cannot be infinite. Thus, the KCL equation can be satisfied only by having $V = 0$. In summary, the presence of a resistor prevents both circuits in Figure 4.21 from supporting unforced co-sinusoidal signals.

Although series and parallel RLC networks cannot exhibit undamped resonant oscillations, the resonant frequency $\omega_o = \frac{1}{\sqrt{LC}}$ remains a significant system parameter in such networks. At frequency $\omega = \omega_o$ the equivalent impedances Z_s and Z_p of the networks (see Figure 4.22) reduce to $Z_s = Z_p = R$. Hence, the series equivalent impedance of L and C in Figure 4.22a is an effective short circuit at $\omega = \omega_o$. Likewise, the parallel equivalent of L and C, shown in Figure 4.22b, is an effective open circuit. Thus, the current response of the series RLC network to an external voltage source, and also the voltage response of the parallel RLC network to an external current source, are maximized at the resonant frequency ω_o. These behaviors of RLC networks are known as *series* and *parallel resonance*, respectively, and are exploited in linear filter circuits (as we will see in Chapters 5 and 11) to obtain frequency sensitive system response.

Series and parallel resonance

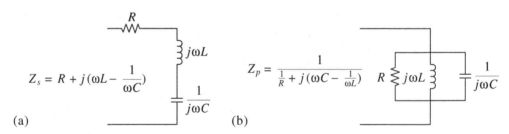

Figure 4.22   (a) Series *RLC* network and its equivalent impedance, and (b) parallel *RLC* network and its equivalent impedance.

Example 4.25
In the series RLC circuit shown in Figure 4.23a, with an external voltage input

$$v(t) = \cos(\omega t),$$

determine the loop current and all of the element voltages if

$$\omega = \omega_o = \frac{1}{\sqrt{LC}}.$$

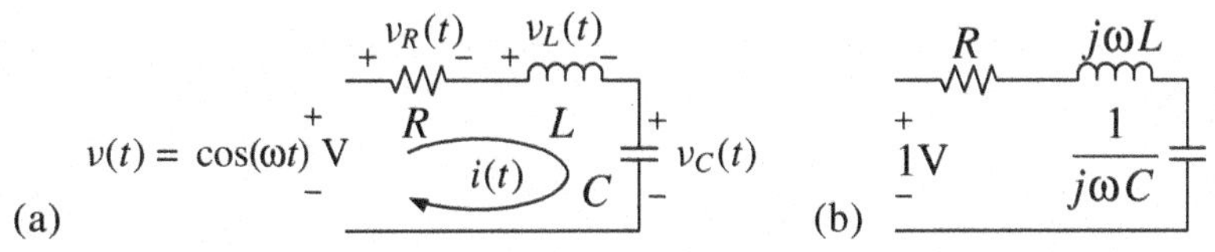

Figure 4.23 (a) A series *RLC* network with a cosine input, and (b) the equivalent phasor network.

Solution Since the series combination of L and C in the circuit is an effective short at $\omega = \omega_o$, the resistor voltage phasor V_R equals the source voltage phasor $V = 1$ V; hence,

$$I = \frac{1}{R}\text{ A},$$

$$V_L = (j\omega_o L)I = j\frac{1}{\sqrt{LC}}L\frac{1}{R} = j\frac{1}{R}\sqrt{\frac{L}{C}}\text{ V},$$

and

$$V_C = (\frac{1}{j\omega_o C})I = -j\frac{\sqrt{LC}}{C}\frac{1}{R} = -j\frac{1}{R}\sqrt{\frac{L}{C}}\text{ V}.$$

Notice that $V_L + V_C = 0$, confirming that the series combination of L and C is an effective short at resonance. Translating these phasors to co-sinusoids, we obtain

$$i(t) = \frac{1}{R}\cos(\omega_o t)\text{ A},$$

$$v_R(t) = \cos(\omega_o t)\text{ V},$$

$$v_L(t) = \frac{1}{R}\sqrt{\frac{L}{C}}\cos(\omega_o t + \frac{\pi}{2})\text{ V},$$

$$v_C(t) = -v_L(t).$$

Figure 4.24 shows plots of the voltage waveforms for the special case with $R = 0.5\ \Omega$, $L = 1$ H, $C = 1$ F, and $\omega = \omega_o = 1\ \frac{\text{rad}}{\text{s}}$. Although the amplitudes of $v_L(t)$ and $v_C(t)$ are greater than the amplitude of the system input $v(t)$, KVL is not violated around the loop, because $v_L(t) + v_C(t) = 0$ (effective short) and $v(t) = v_R(t)$. The large amplitude response of $v_L(t)$ and $v_C(t)$ is a consequence of the behavior of the series RLC network at resonance and the relatively small value chosen for R.

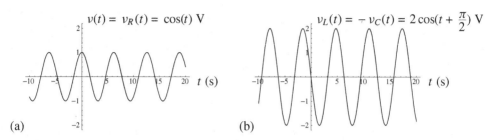

Figure 4.24 (a) Voltage waveforms $v(t) = v_R(t)$, and (b) $v_L(t) = -v_c(t)$ for the resonant system examined in Example 4.25, for the special case with $R = 0.5\,\Omega$, $L = 1\,\text{H}$, $C = 1\,\text{F}$, and $\omega = \omega_0 = 1\,\frac{\text{rad}}{\text{s}}$. Notice that the amplitudes of the inductor and capacitor signals in (b) are larger than the amplitude of the input signal $v(t)$ in (a).

EXERCISES

4.1   Determine the phasor F of the following co-sinusoidal functions $f(t)$:

  (a) $f(t) = 2\cos(2t + \frac{\pi}{3})$.

  (b) $f(t) = A\sin(\omega t)$.

  (c) $f(t) = -5\sin(\pi t)$.

4.2   Find the cosine function $f(t)$, with frequency $\omega = 2\,\frac{\text{rad}}{\text{s}}$, corresponding to the following phasors:

  (a) $F = j2$.

  (b) $F = 3e^{-j\frac{\pi}{6}}$.

  (c) $F = j2 + 3e^{-j\frac{\pi}{6}}$.

4.3   Use the phasor method to determine the amplitude and phase shift (in rad) of the following signals when written as cosines:

  (a) $f(t) = 3\cos(4t) - 4\sin(4t)$.

  (b) $g(t) = 2(\cos(\omega t) + \cos(\omega t + \pi/4))$.

4.4   A circuit component is conducting a current $i(t) = 2\cos(2\pi t + \frac{\pi}{3})\,\text{A}$, and its impedance is $Z = 1 + j\,\Omega$. Plot $i(t)$ and the voltage drop $v(t)$ in the current direction as a function of time for $0 \le t \le 2\,\text{s}$.

4.5   **(a)** Calculate the series equivalent impedance of the following network for
$\omega = 1$ rad/s in rectangular and polar forms and determine the steady-
state current $i(t)$, given that $v(t) = 2\cos(t)$ V:

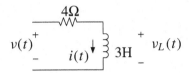

(b) What is the phasor of the inductor voltage $v_L(t)$ in this network, given
that $v(t) = 2\cos(t)$ V?

4.6   Consider the following circuit:

Determine the steady-state current $i(t)$, using phasor current division.

4.7   In the following circuit determine the node-voltage phasors V_1, V_2, and V_3
and express them in polar form:

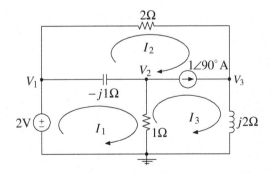

4.8   In the circuit shown for Problem 4.7, determine the loop-current phasors I_1,
I_2, and I_3 and express them in polar form.

4.9   Use the phasor method to determine $v_1(t)$ in the following circuit:

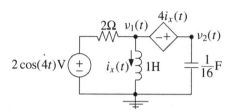

4.10   In the following circuit determine the phasor V and express it in polar form:

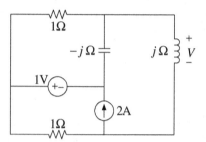

4.11   Use the phasor method to determine the steady-state voltage $v(t)$ in the following op-amp circuit:

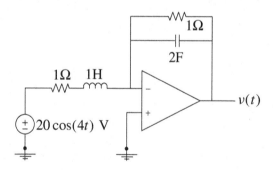

4.12   Use the following network to answer (a) through (d):

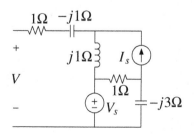

    (a) Determine the phasor V when $I_s = 0$.

    (b) Determine the phasor V when $V_s = 0$.

    (c) Determine V when $V_s = 4$ V and $I_s = -2$ A, and calculate the average power absorbed in the resistors.

    (d) What are the Thevenin equivalent and the available average power of the network when $V_s = 4$ V and $I_s = -2$ A?

4.13   Determine the impedance Z_L of a load that is matched to the following network at terminals a and b, and determine the net power absorbed by the matched load.

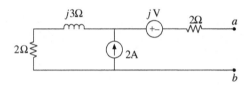

4.14   **(a)** Calculate the equivalent impedance of the following network for (i) $\omega = 5$ krad/s, (ii) $\omega = 25$ krad/s, and (iii) $\omega = 125$ krad/s:

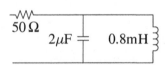

    (b) Assuming a cosine voltage input to the network, with a fixed amplitude and variable frequency ω, at which value of ω is the amplitude of the capacitor voltage maximized? At the same frequency, what will be the amplitude of the resistor current?

5

Frequency Response $H(\omega)$ *of LTI Systems*

Frequency response $H(\omega)$ and properties; LTI system response to co-sinusoids and multi-frequency inputs; resonant and non-dissipative systems

Figure 5.1a shows an LTI system with input signal $v_i(t)$ and output $v_o(t)$. In Chapter 4 we learned how to calculate the steady-state output of such systems when the input is a co-sinusoid (just a simple phasor calculation in Figure 5.1b, for instance). In Chapter 7 we will learn how to calculate the zero-state system response to *any* input signal of practical interest (e.g., a rectangular or triangular pulse, a talk show, a song, a lecture) by using the following facts:

(1)   All practical signals that can be generated in the lab or in a radio station can be expressed as a superposition of co-sinusoids with different frequencies, phases, and amplitudes.

(2)   The output of an LTI system is the superposition of the individual responses caused by the co-sinusoidal components of the input signal.

In this chapter we lay down the conceptual path from Chapter 4 (co-sinusoids) to Chapter 7 (arbitrary signals), and we do some practical calculations concerning dissipative LTI circuits and systems with multifrequency inputs. Section 5.1 introduces the concept of *frequency response $H(\omega)$* of an LTI system and shows how to determine $H(\omega)$ for linear circuits and ODEs. We discuss general properties of the frequency response $H(\omega)$ in Section 5.2. Sections 5.3 and 5.4 describe the applications of $H(\omega)$ in single- and multi-frequency system response calculations. Finally, in Section 5.5 we revisit the resonance phenomenon first encountered in Section 4.4.

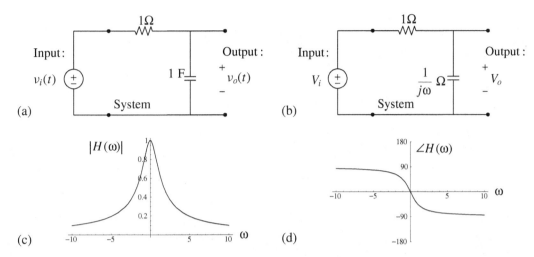

Figure 5.1 (a) A single-input LTI system, (b) its phasor representation, (c) $|H(\omega)|$ vs ω, and (d) $\angle H(\omega)$ (in degrees) vs ω.

5.1 The Frequency Response $H(\omega)$ of LTI Systems

We learned in Chapter 4 that a dissipative LTI circuit with a co-sinusoidal source (independent voltage or current source), namely,

$$f(t) = \text{Re}\{Fe^{j\omega t}\} = |F|\cos(\omega t + \angle F),$$

produces a steady-state output response $y(t)$ (a voltage or current)

$$y(t) = \text{Re}\{Ye^{j\omega t}\} = |Y|\cos(\omega t + \angle Y),$$

where the output phasor

$$Y = |Y|\angle Y$$

is proportional to the input phasor

$$F = |F|\angle F.$$

The proportionality between Y and F depends, in general, on the source frequency ω, because the impedances in the circuit depend on ω. Hence, in a single-source circuit the relationship between source and response phasors F and Y has the form

$$Y = H(\omega)F,$$

where the function $H(\omega)$, with variable ω, is said to be the *frequency response* of the circuit.

Frequency response $H(\omega)$

The next four examples illustrate how the frequency response $H(\omega)$ can be determined in LTI circuits and systems. We also introduce the concepts of amplitude response $|H(\omega)|$ and phase response $\angle H(\omega)$.

Example 5.1

For the system shown in Figure 5.1a, the input is voltage signal

$$f(t) = v_i(t)$$

and the output is voltage signal

$$y(t) = v_o(t).$$

Determine the frequency response of the system

$$H(\omega) = \frac{Y}{F},$$

where F and Y are the input and output signal phasors when the input is specified as a co-sinusoid with frequency ω.

Solution   From the equivalent phasor circuit shown in Figure 5.1b, we obtain, using voltage division,

$$V_o = \frac{\frac{1}{j\omega}}{1 + \frac{1}{j\omega}} V_i = \frac{1}{1 + j\omega} V_i.$$

Since $F = V_i$ and $Y = V_o$, it follows that $Y = \frac{1}{1+j\omega} F$, so

$$H(\omega) = \frac{Y}{F} = \frac{1}{1 + j\omega}.$$

Because $\frac{1}{1+j\omega} = \frac{1}{\sqrt{1+\omega^2}} \angle - \tan^{-1}(\omega)$, we also can write

$$H(\omega) = |H(\omega)| \angle H(\omega),$$

where

$$|H(\omega)| \equiv \frac{1}{\sqrt{1 + \omega^2}}$$

and

$$\angle H(\omega) \equiv - \tan^{-1}(\omega)$$

are known as the *amplitude* and *phase responses*, respectively. The variations of $|H(\omega)|$ and $\angle H(\omega)$ with frequency ω are plotted in Figures 5.1c and 5.1d, respectively.

The plot of $|H(\omega)|$ in Figure 5.1c shows how the amplitude of the output signal depends on the frequency of the input signal. For example, an input signal with frequency near zero is passed with nearly unity scaling of the amplitude (amplitude of the output will be nearly the same as the amplitude of the input), whereas inputs with high frequencies will be greatly attenuated (amplitude of the output will be nearly zero). As a consequence of this behavior, the circuit in Figure 5.1a is referred to as a *low-pass filter*. The plot of $\angle H(\omega)$ shows how the phase of the output signal depends on the frequency of the input signal. For an input frequency that is near zero, the phase of the output will be nearly the same as that of the input. For very large frequencies, the phase of the output will be retarded by approximately $90°$. We will study this example further in Section 5.3.

Amplitude and phase response $|H(\omega)|$ and $\angle H(\omega)$

Example 5.2

For the system shown in Figure 5.2a determine the frequency response $H(\omega)$.

Solution   From the phasor circuit in Figure 5.2b,

$$V_o = \frac{1}{\frac{1}{j\omega} + 1} V_i = \frac{j\omega}{1 + j\omega} V_i.$$

Therefore,

$$H(\omega) = \frac{V_o}{V_i} = \frac{j\omega}{1 + j\omega}$$

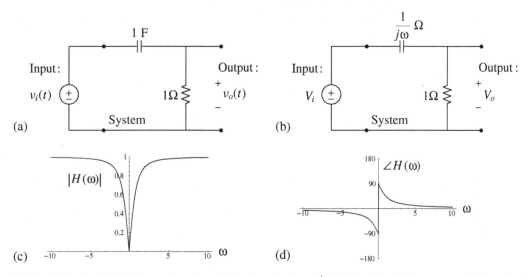

Figure 5.2 (a) A simple high-pass system, (b) its phasor representation, (c) $|H(\omega)|$ vs ω, and (d) $\angle H(\omega)$ (in degrees) vs ω.

for this circuit. Note that

$$\frac{j\omega}{1+j\omega} = \frac{j\omega}{1+j\omega}\frac{1-j\omega}{1-j\omega} = \frac{\omega(\omega+j)}{1+\omega^2} = \frac{|\omega||\omega+j|}{1+\omega^2}\angle\tan^{-1}(\frac{1}{\omega}).$$

Therefore,

$$|H(\omega)| = \frac{|\omega|}{\sqrt{1+\omega^2}}$$

and

$$\angle H(\omega) = \tan^{-1}(\frac{1}{\omega}).$$

The variations of $|H(\omega)|$ and $\angle H(\omega)$ with frequency ω are plotted in Figures 5.2c and 5.2d, respectively.

We see in Figure 5.2c that an input signal with frequency near zero will be almost completely attenuated (amplitude of the output will be nearly zero), whereas inputs with high frequencies will be passed without attenuation. As a consequence, the circuit in Figure 5.2a is referred to as a *high-pass filter*. The plot of $\angle H(\omega)$ shows that for a positive input frequency that is nearly zero, the phase of the output will be advanced by approximately 90°. For very large frequencies, the phase of the output will be nearly the same as that of the input. We will study this example further in Section 5.3

Example 5.3
For the system shown in Figure 5.3a, the input is current signal $f(t)$ and the output is voltage signal $y(t)$. Determine the frequency response of the system $H(\omega) = \frac{Y}{F}$.

Solution   Using the phasor circuit, we find that

$$Y = Z_p F,$$

where Z_p is the parallel equivalent impedance,

$$Z_p = \frac{1}{\frac{1}{1\,\Omega} + \frac{1}{j\omega\,\Omega} + j\omega\,\Omega^{-1}} = \frac{j\omega}{1-\omega^2+j\omega}\,\Omega.$$

Therefore,

$$H(\omega) = \frac{Y}{F} = \frac{j\omega}{1-\omega^2+j\omega}\,\Omega.$$

The amplitude and phase responses are

$$|H(\omega)| = \frac{|\omega|}{\sqrt{(1-\omega^2)^2+\omega^2}}$$

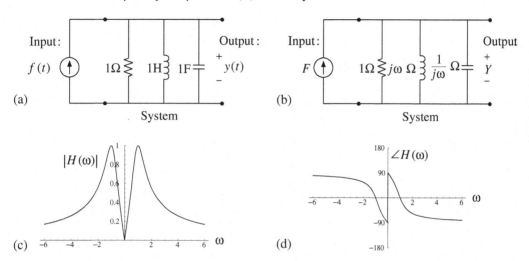

Figure 5.3 (a) A simple band-pass system, (b) its phasor representation, (c) $|H(\omega)|$ vs ω, and (d) $\angle H(\omega)$ (in degrees) vs ω.

and

$$\angle H(\omega) = \tan^{-1}(\frac{1 - \omega^2}{\omega}),$$

which are plotted in Figures 5.3c and 5.3d.

Figure 5.3c shows that both low and high frequencies are attenuated, whereas some frequencies lying between the lowest and highest are passed with significant amplitude. Thus, the circuit in Figures 5.3a is called a *band-pass filter*. We will offer further comments on this example in Section 5.3.

Example 5.4
A linear system with some input $f(t)$ and output $y(t)$ is described by the ODE

$$\frac{dy}{dt} + 4y(t) = \frac{df}{dt} + 2f(t).$$

Determine the frequency response

$$H(\omega) = \frac{Y}{F}$$

of the system. Also, identify the amplitude response $|H(\omega)|$ and phase response $\angle H(\omega)$.

Solution   Using the derivative and superposition rules for phasors, we convert the ODE into its algebraic phasor form

$$j\omega Y + 4Y = j\omega F + 2F,$$

which implies that

$$(4 + j\omega)Y = (2 + j\omega)F.$$

Hence,

$$H(\omega) = \frac{Y}{F} = \frac{2 + j\omega}{4 + j\omega}.$$

The amplitude and phase response of the system are

$$|H(\omega)| = \frac{\sqrt{4 + \omega^2}}{\sqrt{16 + \omega^2}}$$

and

$$\angle H(\omega) = \tan^{-1}(\frac{\omega}{2}) - \tan^{-1}(\frac{\omega}{4}),$$

respectively.

5.2   Properties of Frequency Response $H(\omega)$ of LTI Circuits

Table 5.1 lists some of the general properties of the frequency response $H(\omega)$ of LTI circuits introduced in the previous section. The reason for the conjugate symmetry condition

$$H(-\omega) = H^*(\omega)$$

	Description	Property
1	Conjugate symmetry	$H(-\omega) = H^*(\omega)$
2	Even amplitude response	$|H(-\omega)| = |H(\omega)|$
3	Odd phase response	$\angle H(-\omega) = -\angle H(\omega)$
4	Real DC response	$H(0) = H^*(0)$ is real valued
5	Steady-state response to $e^{j\omega t}$	$e^{j\omega t} \longrightarrow \boxed{\text{LTI}} \longrightarrow H(\omega)e^{j\omega t}$

Table 5.1   Properties of the frequency response $H(\omega)$ of LTI circuits.

(property 1) can be traced back to the fact that capacitor and inductor impedances $\frac{1}{j\omega C}$ and $j\omega L$ satisfy the same property—for example,

$$-j\omega L = (j\omega L)^*.$$

Because the ω dependence enters $H(\omega)$ via only the capacitor and inductor impedances, the frequency response $H(\omega)$ of an LTI circuit will always be conjugate symmetric. Linear ODEs with real-valued constant coefficients, which describe such circuits, will also have conjugate symmetric frequency response functions.[1]

One consequence of $H(-\omega) = H^*(\omega)$ is that

$$|H(-\omega)| = |H(\omega)|;$$

that is, the amplitude response $|H(\omega)|$ is an *even function* of frequency ω (property 2). A second consequence is

$$\angle H(-\omega) = -\angle H(\omega),$$

which indicates that the phase response is an *odd function* of ω (property 3). Notice that the amplitude and phase response curves shown in Figures 5.1 through 5.3 exhibit the even and odd properties of $|H(\omega)|$ and $\angle H(\omega)$ just mentioned. Notice also that $H(0)$ is real valued in each case,[2] consistent with property 4 in Table 5.1.

Complex-valued functions

$$e^{j\omega t}$$

and

$$H(\omega)e^{j\omega t} = |H(\omega)|e^{j(\omega t + \angle H(\omega))}$$

can be re-expressed in *pair form* (see Appendix A, Section 6) as

$$(\cos(\omega t), \ \sin(\omega t))$$

and

$$|H(\omega)|(\cos(\omega t + \angle H(\omega)), \ \sin(\omega t + \angle H(\omega))),$$

respectively. Therefore, property 5,

$$e^{j\omega t} \longrightarrow \boxed{\text{LTI}} \longmapsto H(\omega)e^{j\omega t},$$

---

[1]It is possible to define LTI systems with a frequency response for which conjugate symmetry is not true—for instance, a linear ODE with complex-valued coefficients.

[2]$H(-\omega) = H^*(\omega)$ implies that $H(0) = H^*(0)$, which indicates that the DC response $H(0)$ must be real, since only real numbers can equal their conjugates.

in Table 5.1 can be viewed as *shorthand* for the fact that in sinusoidal steady-state

$$\cos(\omega t) \longrightarrow \boxed{\text{LTI}} \longrightarrow |H(\omega)| \cos(\omega t + \angle H(\omega)),$$

as well as

$$\sin(\omega t) \longrightarrow \boxed{\text{LTI}} \longrightarrow |H(\omega)| \sin(\omega t + \angle H(\omega)).$$

These in turn can be readily inferred from the phasor relation

$$Y = H(\omega)F,$$

with $F = 1$ and $-j$, corresponding to input functions $f(t) = \cos(\omega t)$ and $\sin(\omega t)$, respectively.

Complex exponentials are *eigen-functions* of LTI circuits and systems

Because the steady-state output due to input $e^{j\omega t}$ is the same as the input signal, except scaled by a constant $H(\omega)$, the complex exponential input $e^{j\omega t}$ is sometimes called an "eigenfunction" of an LTI system. Also the corresponding "eigenvalues" constitute the frequency response $H(\omega)$ of the system.

Property 5 also is handy when dealing with real-valued input signals such as $\cos(\omega t)$, $\sin(\omega t)$, and their weighted linear superpositions, because all such signals can be expressed in terms of $e^{j\omega t}$ with variable ω and its conjugate, as for example in,

$$\cos(\omega t) = \frac{e^{j\omega t} + e^{-j\omega t}}{2}.$$

Thus, property 5 and the conjugate symmetry of $H(\omega)$ can be used directly to infer the steady-state response of LTI circuits to various types of real-valued inputs. Starting in the next section, we will make use of these properties and the superposition principle to describe the steady-state response of LTI circuits to sine and cosine inputs with arbitrary amplitudes and phase shifts (in Section 5.3), multifrequency sums of co-sinusoids (in Section 5.4), arbitrary periodic signals (Chapter 6), and nearly arbitrary aperiodic signals (Chapter 7).

5.3   LTI System Response to Co-Sinusoidal Inputs

We have just seen that in steady-state

$$\cos(\omega t) \longrightarrow \boxed{\text{LTI}} \longrightarrow |H(\omega)| \cos(\omega t + \angle H(\omega))$$

and

$$\sin(\omega t) \longrightarrow \boxed{\text{LTI}} \longrightarrow |H(\omega)| \sin(\omega t + \angle H(\omega)).$$

Since linearity implies that amplitude-scaled inputs cause similarly scaled outputs and time-invariance means that delayed inputs cause equally delayed outputs, we can infer that

$$|F| \cos(\omega t + \theta) \longrightarrow \boxed{\text{LTI}} \longrightarrow |H(\omega)||F| \cos(\omega t + \theta + \angle H(\omega))$$

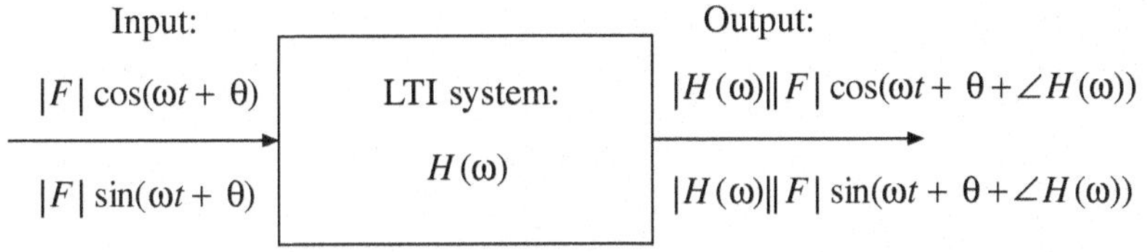

Input: Output:

$|F|\cos(\omega t + \theta)$ LTI system: $|H(\omega)\|F|\cos(\omega t + \theta + \angle H(\omega))$

$H(\omega)$

$|F|\sin(\omega t + \theta)$ $|H(\omega)\|F|\sin(\omega t + \theta + \angle H(\omega))$

Figure 5.4 Steady-state response of LTI systems $H(\omega)$ to cosine and sine inputs, with arbitrary amplitudes $|F|$ and phase shifts θ.

and

$$|F|\sin(\omega t + \theta) \longrightarrow \boxed{\text{LTI}} \longrightarrow |H(\omega)||F|\sin(\omega t + \theta + \angle H(\omega)).$$

These steady-state input–output relations for LTI systems, summarized in Figure 5.4, indicate that LTI systems convert their *co-sinusoidal inputs* of frequency ω into *co-sinusoidal outputs* having the same frequency and the following *amplitude* and *phase* parameters:

(1) Output amplitude = Input amplitude *multiplied by* $|H(\omega)|$, and
(2) Output phase = Input phase *plus* $\angle H(\omega)$.

Therefore, knowledge of the frequency response $H(\omega)$ is sufficient to determine how an LTI system responds to co-sinusoidal signals and their superpositions in steady state. Furthermore, if the steady-state response of a system to a co-sinusoidal input is not a co-sinusoid of the same frequency, then the system cannot be LTI.

$H(\omega)$ **represents the steady-state behavior of LTI circuits**

Example 5.5
Return to Example 5.1 (see Figure 5.1a), where we found the system frequency response to be

$$H(\omega) = \frac{1}{1 + j\omega} = \frac{1}{\sqrt{1 + \omega^2}} \angle - \tan^{-1}(\omega).$$

Consider two different inputs,

$$f_1(t) = 1\cos(0.5t) \text{ V}$$

and

$$f_2(t) = 1\cos(2t) \text{ V}.$$

Determine the steady-state system responses $y_1(t)$ and $y_2(t)$ to $f_1(t)$ and $f_2(t)$.

Solution   Applying the input–output relation shown in Figure 5.4, we note that

$$y_1(t) = |H(0.5)|1\cos(0.5t + \angle H(0.5)) \text{ V},$$

where $|H(0.5)|$ and $\angle H(0.5)$ are the amplitude and phase response evaluated at the frequency $\omega = 0.5$ rad/s of the input $f_1(t)$. Since

$$|H(0.5)| = \frac{1}{\sqrt{1 + 0.5^2}} = 0.894$$

and

$$\angle H(0.5) = -\tan^{-1}(0.5) = -26.56°,$$

it follows that

$$y_1(t) = 0.894 \cos(0.5t - 26.56°) \text{ V}.$$

Likewise,

$$|H(2)| = \frac{1}{\sqrt{1 + 2^2}} = 0.447$$

and

$$\angle H(2) = -\tan^{-1}(2) = -63.43°,$$

and so

$$y_2(t) = |H(2)|1 \cos(2t + \angle H(2)) \text{ V} = 0.447 \cos(2t - 63.43°) \text{ V}.$$

A summary of these results is presented in Figure 5.5. Study the plots carefully to better understand that system

$$H(\omega) = \frac{1}{1 + j\omega}$$

Low-pass filter

is a low-pass filter.

Example 5.6
Return to Example 5.2 (see Figure 5.2a), where we found the system frequency response to be

$$H(\omega) = \frac{j\omega}{1 + j\omega} = \frac{|\omega|}{\sqrt{1 + \omega^2}} \angle \tan^{-1}(\frac{1}{\omega}).$$

Consider two different inputs

$$f_1(t) = 2 \sin(0.5t) \text{ V}$$

and

$$f_2(t) = 1 \cos(2t + 45°) \text{ V}.$$

Determine the steady-state system responses $y_1(t)$ and $y_2(t)$ to $f_1(t)$ and $f_2(t)$.

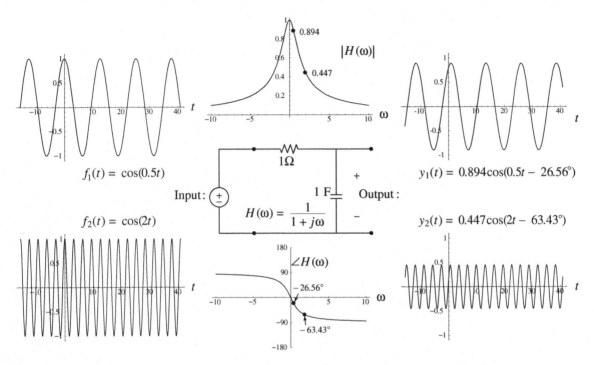

$f_1(t) = \cos(0.5t)$

Input: $\pm$

$f_2(t) = \cos(2t)$

$1\,\Omega$

$1\,F$

$H(\omega) = \dfrac{1}{1 + j\omega}$

Output:

$y_1(t) = 0.894\cos(0.5t - 26.56°)$

$y_2(t) = 0.447\cos(2t - 63.43°)$

Figure 5.5   A summary plot of the responses of the system $H(\omega) = \frac{1}{1+j\omega}$ to co-sinusoidal inputs $f_1(t)$ and $f_2(t)$ examined in Example 5.5. Note that the higher frequency input (bottom left signal) is attenuated more strongly than the lower frequency input (upper left). The system is therefore a low-pass filter.

Solution   Once again, using the same input–output relation based on frequency response, we obtain

$$y_1(t) = |H(0.5)|2\sin(0.5t + \angle H(0.5))\ \text{V}.$$

Since

$$|H(0.5)| = \frac{|0.5|}{\sqrt{1 + 0.5^2}} = 0.447$$

and

$$\angle H(0.5) = \tan^{-1}\left(\frac{1}{0.5}\right) = 63.43°,$$

it follows that

$$y_1(t) = 0.894\sin(0.5t + 63.43°)\ \text{V}.$$

Likewise,

$$|H(2)| = \frac{|2|}{\sqrt{1 + 2^2}} = 0.894$$

and

$$\angle H(2) = \tan^{-1}(\frac{1}{2}) = 26.56°,$$

and, therefore,

$$y_2(t) = |H(2)|1 \cos(2t + 45° + \angle H(2)) = 0.894 \cos(2t + 71.56°) \text{ V}.$$

A summary of these results is presented in Figure 5.6. Examine the plots carefully to better understand that system

$$H(\omega) = \frac{j\omega}{1 + j\omega}$$

High-pass filter   is a high-pass filter.

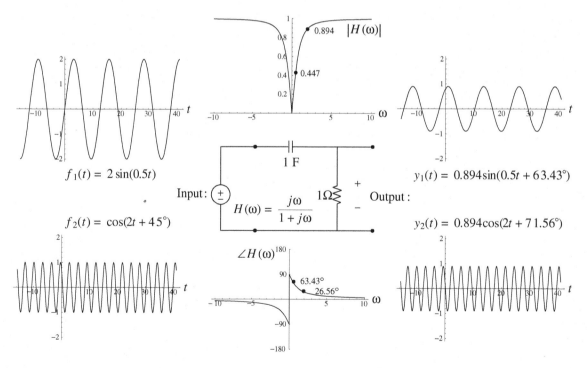

Figure 5.6   A summary plot of the responses of the system $H(\omega) = \frac{j\omega}{1+j\omega}$ to co-sinusoidal inputs $f_1(t)$ and $f_2(t)$ examined in Example 5.6. Note that the low-frequency input (upper left signal) is attenuated more strongly than the high-frequency input (bottom left). The system is therefore a high-pass filter.

While the preceding Examples 5.5 and 5.6 illustrate that systems

$$H(\omega) = \frac{1}{1 + j\omega}$$

and

$$H(\omega) = \frac{j\omega}{1 + j\omega}$$

function as low-pass and high-pass filters, respectively, recall that we found the system

$$H(\omega) = \frac{j\omega}{1 - \omega^2 + j\omega}$$

Band-pass filter

of Example 5.3 in Section 5.1 to be a band-pass filter. Returning to Example 5.3, we note the amplitude response curve $|H(\omega)|$ shown in Figure 5.3c peaks at $\omega = \pm1$ rad/s and vanishes as $\omega \to 0$ and $\omega \to \pm\infty$. Therefore, only those co-sinusoidal inputs with frequencies ω in the vicinity of 1 rad/s pass through the system with relatively small attenuation. This occurs because 1 rad/s is the resonant frequency of the parallel LC combination in the circuit. At resonance, the parallel LC combination is an effective open circuit and all the source current is routed through the resistor to generate a peak response. Conversely, in the limits as $\omega \to 0$ and $\omega \to \pm\infty$, respectively, the inductor and the capacitor behave as effective shorts, and force the output voltage to zero.

Example 5.7
An LTI system $H(\omega)$ that is known to be a low-pass filter converts its input

$$f(t) = 2\sin(12t)$$

into a steady-state output

$$y(t) = \sqrt{2}\sin(12t + \theta)$$

for some real valued constant θ. Determine $H(12)$ and also compare the *average power* that the input $f(t)$ and output $y(t)$ would deliver to a 1 Ω resistor.

Solution   First, we note that

$$|H(12)| = \frac{|Y|}{|F|} = \frac{\sqrt{2}}{2} = \frac{1}{\sqrt{2}}.$$

Thus, according to the available information,

$$H(12) = \frac{1}{\sqrt{2}} e^{j\theta}.$$

For co-sinusoids, the average power into a 1 Ω load is obtained as one-half the square of the signal amplitude. (See Example 4.21 in Section 4.3.1.) Thus. the average power per Ω for the input $f(t)$ is

$$P_f = \frac{1}{2}|F|^2 = 2,$$

while it is

$$P_y = \frac{1}{2}|Y|^2 = 1$$

for the output $y(t)$. Thus,

$$\frac{P_y}{P_f} = \frac{1}{2},$$

Half-power frequency of a low-pass filter

which makes $\omega = 12$ rad/s the *half-power frequency* of filter $H(\omega)$.

Note that for any ω and any LTI system, the power ratio $\frac{P_y}{P_f}$ can be obtained as $|H(\omega)|^2$, the square of the system amplitude response.

Example 5.8
A system converts its input

$$f(t) = 5\sin(12t)$$

into a steady-state output

$$y(t) = 25\sin(12t - 45°) + 2.5\sin(24t - 90°).$$

Is the system LTI?

Solution   No, the system is not LTI because, if it were, then the

$$2.5\sin(24t - 90°)$$

component of the output $y(t)$ would not be present. LTI systems cannot create new frequencies that are not present in their inputs.

DC response: For $\omega = 0$, the input–output relation

$$e^{j\omega t} \longrightarrow \boxed{\text{LTI}} \longrightarrow H(\omega)e^{j\omega t}$$

reduces to

$$1 \longrightarrow \boxed{\text{LTI}} \longrightarrow H(0).$$

Linearity then implies that for an arbitrary DC input $f(t) = F_o$, the relation is

$$F_o \longrightarrow \boxed{\text{LTI}} \longrightarrow H(0)F_o,$$

where the response $H(0)F_o$ is real valued. (Recall from property 4 in Table 5.1 that $H(0)$ is real valued.)

Example 5.9
What is the steady-state response of the system

$$H(\omega) = \frac{2 + j\omega}{4 + j\omega}$$

to a DC input

$$f(t) = 5?$$

Solution   Since

$$H(0) = \frac{2 + j0}{4 + j0} = 0.5,$$

the steady-state response must be

$$y(t) = H(0)5 = 2.5.$$

Measuring $H(\omega)$ in the lab and dB representation: The steady-state input–output relation for LTI systems, depicted in Figure 5.4, suggests that the frequency response

$$H(\omega) = |H(\omega)|e^{j\angle H(\omega)}$$

of a circuit can be determined in the lab via the following procedure:

(1) Using a variable-frequency signal generator, produce a signal

$$f(t) = \cos(\omega t)$$

and display it on an oscilloscope.

(2) For each setting of the input frequency ω, observe and record the amplitude $|H(\omega)|$ and phase shift $\angle H(\omega)$ from the measured circuit response

$$y(t) = |Y|\cos(\omega t + \angle Y) = |H(\omega)|\cos(\omega t + \angle H(\omega)).$$

The amplitude and phase shift data, $|H(\omega)|$ and $\angle H(\omega)$, collected over a wide range of frequencies ω, then can be displayed as amplitude and phase response plots for the system.

This method generally will reveal if the system is not LTI (in which case the output $y(t)$ corresponding to the cosine input $f(t) = \cos(\omega t)$ usually is not a pure co-sinusoid at frequency ω) or whether the system is non-dissipative (in which case the system output may contain non-decaying components even after the input is turned off). In either case, it will not be possible to infer $|H(\omega)|$ and $\angle H(\omega)$. For such systems, $H(\omega)$ is not a meaningful system characterization. (See Section 5.5 for further discussion on non-dissipative systems.) For the frequent case with dissipative LTI circuits and systems, however, the foregoing method provides a direct experimental means for determining $|H(\omega)|$ and $\angle H(\omega)$.

More information often is revealed about $|H(\omega)|$ if we plot it on a logarithmic scale rather than on a regular, or *linear* scale. That, in effect, is the idea behind the *decibel* definition[3]

$$|H(\omega)|_{\text{dB}} \equiv 10 \log |H(\omega)|^2 = 20 \log |H(\omega)|,$$

Decibel amplitude response $|H(\omega)|_{\text{dB}}$

commonly used in describing the amplitude response function $|H(\omega)|$.

A plot of $|H(\omega)|_{\text{dB}}$ makes it possible to see exceedingly small values of $|H(\omega)|$. This is important when we graph the response of high-quality low-pass, high-pass, and band-pass filters, where we may wish to have the response in the stop band (the frequency band where the signal components are to be blocked) attenuated by a factor of 1000 or more. This situation and others lead to a frequency response magnitude $|H(\omega)|$ that has a very wide dynamic range best viewed on a logarithmic plot. Both linear and decibel, or dB, plots are illustrated for two different $|H(\omega)|$ in Figures 5.7a through 5.7d.

Figure 5.7a, for instance, shows a linear plot of the amplitude response of the low-pass filter

$$H(\omega) = \frac{1}{1 + j\omega},$$

while Figure 5.7b is a plot of the dB amplitude response

$$|H(\omega)|_{\text{dB}} = 20 \log \frac{|1|}{|1 + j\omega|} = 20 \log |1| - 20 \log |1 + j\omega| = -20 \log \sqrt{1 + \omega^2}$$

---

[3]A *decibel* (dB) is one-tenth of a *bel* (B), which is the name given to $\log |H(\omega)|^2 = 2 \log |H(\omega)|$ in honor of the lab practices of Alexander Graham Bell, the inventor of the telephone and hearing aid.

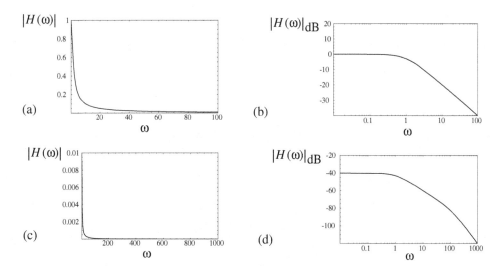

Figure 5.7 Plots of the amplitude response of systems $H(\omega) = \frac{1}{1+j\omega}$ (a), (b) and $H(\omega) = \frac{1}{(1+j\omega)(100+j\omega)}$ (c), (d). Plots (a) and (c) are on a linear scale whereas (b) and (d) are on a decibel, or dB, scale. Note that in the dB plots a logarithmic scale is used for the horizontal axes, according to a common engineering practice.

for the same filter. In the dB plot in Figure 5.7b we also use a logarithmic scale for the horizontal axis representing the frequency variable ω.

Figures 5.7c and 5.7d display the amplitude response of another low-pass filter

$$H(\omega) = \frac{1}{(1 + j\omega)(100 + j\omega)}$$

and the corresponding dB amplitude response

$$|H(\omega)|_{\mathrm{dB}} = 20 \log \frac{|1|}{|1 + j\omega||100 + j\omega|} = -20 \log \sqrt{1 + \omega^2} - 20 \log \sqrt{100^2 + \omega^2}.$$

Clearly, the dB plots shown in Figures 5.7b and 5.7d are more informative than the linear plots shown in Figures 5.7a and 5.7c. For instance, we can see from Figures 5.7b and 5.7d that both filters have similar "flat" amplitude responses for $\omega \ll 1$, a detail that is not as apparent in the linear plots. It is useful to remember that a 20 dB change corresponds to a factor of 10 change in the amplitude $|H(\omega)|$ and a factor of 100 change in $|H(\omega)|^2$. Likewise, a 3 dB change corresponds to a factor of $\sqrt{2}$ variation in $|H(\omega)|$ and a factor of 2 variation in $|H(\omega)|^2$, as summarized in Table 5.2.

Amplitude response	Power response	dB
$\frac{1}{\sqrt{2}}$	$\frac{1}{2}$	-3
1	1	0
$\sqrt{2}$	2	3
2	4	6
$\sqrt{10}$	10	10
10	100	20

Table 5.2 A conversion table for *amplitude* response $|H(\omega)|$, *power* response $|H(\omega)|^2$, and their representation in dB units.

5.4 LTI System Response to Multifrequency Inputs

The steady-state response of LTI systems to multifrequency inputs can be calculated by applying the superposition principle and the input–output relation for co-sinusoids, shown in Figure 5.4.

Consider, for instance, a multifrequency input

$$f(t) = f_1(t) + f_2(t),$$

where

$$f_1(t) = |F_1| \cos(\omega_1 t + \theta_1)$$

and

$$f_2(t) = |F_2| \sin(\omega_2 t + \theta_2).$$

Using superposition, we see that the steady-state response is

$$y(t) = y_1(t) + y_2(t),$$

where (using the input–output relation of Figure 5.4)

$$y_1(t) = |H(\omega_1)||F_1| \cos(\omega_1 t + \theta_1 + \angle H(\omega_1))$$

and

$$y_2(t) = |H(\omega_2)||F_2| \sin(\omega_2 t + \theta_2 + \angle H(\omega_2)).$$

More generally, given a multifrequency input

$$f(t) = \sum_{n=1}^{N} |F_n| \cos(\omega_n t + \theta_n)$$

with arbitrary N, we find that the steady-state output is

$$y(t) = \sum_{n=1}^{N} |H(\omega_n)||F_n| \cos(\omega_n t + \theta_n + \angle H(\omega_n)).$$

This input–output relation for multi-frequency inputs is shown graphically in Figure 5.8 and will be the basis of the calculations presented in the following examples.

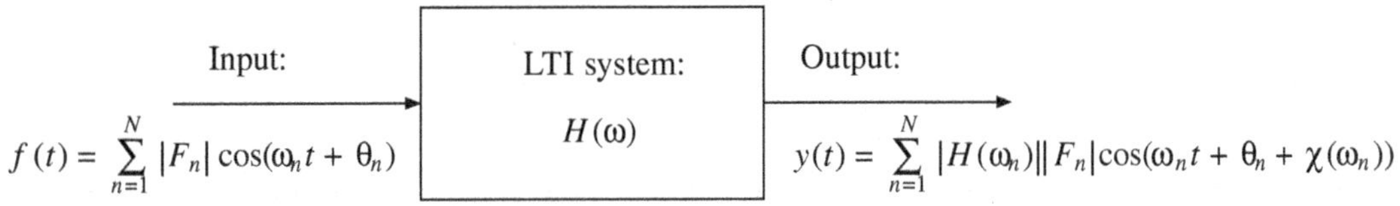

Input:	LTI system:	Output:
$f(t) = \sum_{n=1}^{N} |F_n| \cos(\omega_n t + \theta_n)$	$H(\omega)$	$y(t) = \sum_{n=1}^{N} |H(\omega_n)||F_n| \cos(\omega_n t + \theta_n + \chi(\omega_n))$

Figure 5.8 LTI system response to multifrequency inputs.

Example 5.10

A 1 H inductor current is specified as

$$i(t) = 2 \cos(2t) + 4 \cos(4t) \text{ A.}$$

Determine the inductor voltage $v(t)$, using the v–i relation for the inductor and, confirm that the result is consistent with the input–output relation shown in Figure 5.8.

Solution   Given that $L = 1$ H and $i(t) = 2 \cos(2t) + 4 \cos(4t)$, we obtain

$$v(t) = L \frac{di}{dt} = \frac{d}{dt}(2 \cos(2t) + 4 \cos(4t))$$

$$= -4 \sin(2t) - 16 \sin(4t) = 4 \cos(2t + \frac{\pi}{2}) + 16 \cos(4t + \frac{\pi}{2}) \text{ V.}$$

Now, the frequency response of the same system is

$$H(\omega) = \frac{V}{I} = \frac{j\omega I}{I} = j\omega.$$

Applying the relation in Figure 5.8 with the input signal

$$2 \cos(2t) + 4 \cos(4t),$$

we have

$$y(t) = |H(2)|2\cos(2t + \angle H(2)) + |H(4)|4\cos(4t + \angle H(4)) \text{ V}.$$

This yields the output

$$2 \cdot 2\cos(2t + \frac{\pi}{2}) + 4 \cdot 4\cos(4t + \frac{\pi}{2}) \text{ V},$$

in agreement with the previous result.

Example 5.11

Suppose the input of the low-pass filter

$$H(\omega) = \frac{1}{1 + j\omega}$$

is

$$f(t) = 1\cos(0.5t) + 1\cos(\pi t) \text{ V}.$$

Determine the system output $y(t)$.

Solution   Using the relation given in Figure 5.8, we have

$$y(t) = |H(0.5)|1\cos(0.5t + \angle H(0.5)) + |H(\pi)|1\cos(\pi t + \angle H(\pi)) \text{ V}.$$

Now,

$$|H(0.5)| = \frac{1}{|1 + j0.5|} \approx 0.894$$

and

$$\angle H(0.5) = \angle\frac{1}{1 + j0.5} \approx -26.56°.$$

Likewise,

$$|H(\pi)| = \frac{1}{|1 + j\pi|} \approx 0.303$$

and

$$\angle H(\pi) = \angle\frac{1}{1 + j\pi} \approx -72.34°.$$

Therefore,

$$y(t) \approx 0.894\cos(0.5t - 26.56°) + 0.303\cos(\pi t - 72.34°) \text{ V}.$$

The input and output signals of Example 5.11 are plotted in Figure 5.9a. Notice that the output signal $y(t)$ is a smoothed version of the input, because the low-pass filter $H(\omega)$ has attenuated the high-frequency content that corresponds to rapid signal variation.

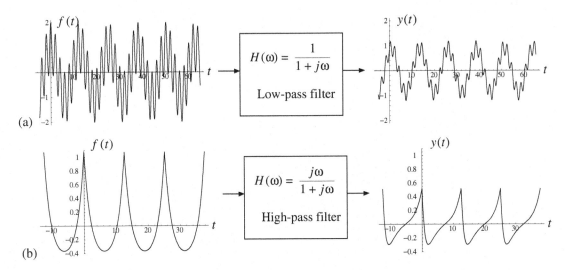

Figure 5.9 Input and output signals $f(t)$ and $y(t)$ for (a) the low-pass filter examined in Example 5.11, and (b) the high-pass filter examined in Example 5.12. Note that both signals in (b) are periodic and have the same period $T_o = 4\pi \approx 12.57$ s.

Example 5.12

Suppose the input of the high-pass filter

$$H(\omega) = \frac{j\omega}{1 + j\omega} = \frac{|\omega|}{\sqrt{1 + \omega^2}} \angle \tan^{-1}(\frac{1}{\omega})$$

is

$$f(t) = \sum_{n=1}^{\infty} \frac{1}{1 + n^2} \cos(0.5nt) \text{ V}.$$

Determine the system output $y(t)$.

Solution   Using the relation given in Figure 5.8, with

$$\omega_n = 0.5n \text{ rad/s}$$

for the specified input, we write

$$y(t) = \sum_{n=1}^{\infty} |H(0.5n)| \frac{1}{1 + n^2} \cos(0.5nt + \angle H(0.5n)) \text{ V}.$$

Now, with the specified amplitude and phase response, we have

$$|H(0.5n)| = \frac{|0.5n|}{\sqrt{1+0.25n^2}}$$

and

$$\angle H(0.5n) = \tan^{-1}\left(\frac{1}{0.5n}\right).$$

Thus,

$$y(t) = \sum_{n=1}^{\infty} \frac{|0.5n|}{\sqrt{1+0.25n^2}} \frac{1}{1+n^2} \cos\left(0.5nt + \tan^{-1}\left(\frac{1}{0.5n}\right)\right) \text{ V}.$$

The input and output signals $f(t)$ and $y(t)$ are plotted[4] in Figure 5.9b. Both $f(t)$ and $y(t)$ are periodic signals with period $T_o = 4\pi$ s. Can you explain why? If not, Chapter 6 will provide the answer.

Glancing at Figure 5.9, try to appreciate that we have just taken a major step toward handling arbitrary inputs in LTI systems. We don't even have names for the complicated input and output signals shown in Figure 5.9!

Example 5.13
What is the steady-state response $y(t)$ of the LTI circuit shown in Figure 5.10a to the input

$$f(t) = 5 + \sin(2t)?$$

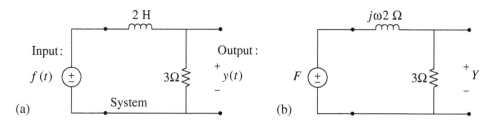

Figure 5.10 (a) An LTI circuit and (b) its phasor equivalent.

---

[4]The plotted curves actually correspond to the sum of the first 100 terms ($n = 1$ to 100) in the expressions for $f(t)$ and $y(t)$. Similar curves calculated with many more terms are virtually indistinguishable from those shown in Figure 5.9b. Thus, the first 100 terms provide a sufficiently accurate representation.

Solution   Applying the phasor method to the phasor equivalent circuit shown in Figure 5.10b, we first deduce that (using voltage division)

$$Y = \frac{3}{3 + j\omega 2} F,$$

which implies that

$$\frac{Y}{F} = H(\omega) = \frac{3}{3 + j\omega 2}.$$

Since the DC response $H(0) = 1$ and

$$H(2) = \frac{3}{3 + j4} = \frac{3}{5\angle 53.13°} = 0.6\angle - 53.13°,$$

the steady-state response to input

$$f(t) = 5 + \sin(2t)$$

is

$$y(t) = |H(0)|5 + |H(2)|\sin(2t + \angle H(2)) = 5 + 0.6\sin(2t - 53.13°).$$

5.5 Resonant and Non-Dissipative Systems

The input–output relation

$$e^{j\omega t} \longrightarrow \boxed{\text{LTI}} \longrightarrow H(\omega)e^{j\omega t},$$

applicable to finding the steady-state response of LTI systems, requires that the system be dissipative. The reason for this restriction is that in non-dissipative systems the steady-state response may contain additional undamped and possibly unbounded terms. For instance, in resonant circuits examined in Section 4.4, we saw that the steady-state output may contain unforced oscillations at a resonance frequency ω_o— for example, $\omega_o = \frac{1}{\sqrt{LC}}$. For such circuits and systems, an $H(\omega)$-based description of the steady-state response is necessarily incomplete.

Consider, for instance, the frequency response of the series RLC circuit shown in Figure 4.23a:

$$H(\omega) = \frac{I}{V} = \frac{1}{R + j\omega L + \frac{1}{j\omega C}} = \frac{j\omega C}{(1 - \omega^2 LC) + j\omega RC}.$$

In the limit, as $R \to 0$,

$$H(\omega) \to \frac{j\omega C}{1 - \omega^2 LC}$$

and the circuit becomes non-dissipative. For $R = 0$, the response to inputs $e^{j\omega t}$ can no longer be described as

$$\frac{j\omega C}{1 - \omega^2 LC} e^{j\omega}.$$

Notice that for $R = 0$, as $\omega \to \frac{1}{\sqrt{LC}}$, we have $|H(\omega)| \to \infty$, indicating that non-dissipative systems can generate unbounded outputs with bounded inputs (an instability phenomenon that we will examine in detail in Chapter 10).

In summary, frequency-response-based analysis methods are extremely powerful and widely used. We will continue to develop these techniques in Chapters 6 through 8. However, the concept of frequency response offers an incomplete and inadequate description of non-dissipative and non-LTI systems. Therefore, we must be careful to apply these techniques only to dissipative LTI systems. Beginning in Chapter 9, we will develop alternative analysis methods that are appropriate for non-dissipative LTI systems

EXERCISES

5.1   Determine the frequency response $H(\omega) = \frac{Y}{F}$ of the circuit shown and sketch $|H(\omega)|$ versus $\omega \geq 0$. In the diagram, $f(t)$ and $y(t)$ denote the input and output signals of the circuit.

$$
\begin{array}{ccc}
 & 1\Omega \quad 0.2\text{H} & \\
 & \underline{\text{—}\!\!\!\bigvee\!\!\!\bigvee\!\!\!\text{—}\!\!\!\text{mmm}} & \\
+ & & + \\
f(t) & \quad 0.05\text{F} & y(t) \\
- & & -
\end{array}
$$

5.2   In the following circuit, determine the frequency response $H(\omega) = \frac{Y}{F}$ and $H(0)$:

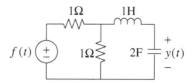

5.3 Determine the frequency response $H(\omega) = \frac{V}{I_s}$ of the circuit in Exercise Problem 3.10 in Chapter 3. Note that $H(\omega)$ can be obtained with the use of the phasor domain circuit as well as the ODE for $v(t)$ given in Problem 3.10.

5.4 Determine the frequency response $H(\omega) = \frac{V}{V_s}$ of the circuit in Exercise Problem 3.17 in Chapter 3. Sketch $|H(\omega)|$ versus $\omega \geq 0$.

5.5 A linear system with input $f(t)$ and output $y(t)$ is described by the ODE

$$\frac{d^2y}{dt^2} + 4\frac{dy}{dt} + 4y(t) = \frac{df}{dt}.$$

Determine the frequency response $H(\omega) = \frac{Y}{F}$ of the system.

5.6 Determine the amplitude response $|H(\omega)|$ and phase response $\angle H(\omega)$ of the system in Problem 5.5. Also, plot $\angle H(\omega)$ versus ω for $-10 < \omega < 10$.

5.7 A linear circuit with input $f(t)$ and output $y(t)$ is described by the frequency response $\frac{Y}{F} = H(\omega) = \frac{j\omega}{4+j\omega}$. Determine the following:

(a) Amplitude of $y(t)$ when $f(t) = 5\cos(3t + \frac{\pi}{4})$ V.

(b) Output $y(t)$ when the input is $f(t) = 8 + 2\sin(4t)$ V.

5.8 A linear system has the frequency response

$$H(\omega) = \frac{1}{(j\omega + 1)(j\omega + 2)} \frac{A}{V}.$$

Determine the system steady-state output $y(t)$ with the following inputs:

(a) $f(t) = 4$ V DC.

(b) $f(t) = 2\cos(2t)$ V.

(c) $f(t) = \cos(2t - 10°) + 2\sin(4t)$ V.

5.9 Repeat Problem 5.8 for a linear system described by the ODE

$$\frac{dy}{dt} + y(t) = 4f(t).$$

5.10 In the circuit of Problem 5.2, the input is $f(t) = 4 + \cos(2t)$. Determine the steady-state output $y(t)$ of the circuit.

5.11 Given an input $f(t) = 5 + 4e^{j2t} + 4e^{-j2t}$ and $H(\omega) = \frac{1+j\omega}{2+j\omega}$, determine the steady-state response $y(t)$ of the system $H(\omega)$ and express it as a real valued signal. Hint: Use the rule $e^{j\omega t} \longrightarrow \boxed{\text{LTI}} \longrightarrow H(\omega)e^{j\omega t}$ and superposition.

5.12 Repeat Problem 5.11 for the input $f(t) = 2e^{-j2t} + (2 + j2)e^{-jt} + (2 - j2)e^{jt} + 2e^{j2t}$.

5.13 Determine whether each of the given steady-state input–output pairs is consistent with the properties of $H(\omega)$ discussed in Section 5.2. Explain your reasoning.

(a) $\cos(25t) \longrightarrow \boxed{\text{System}} \longrightarrow 99.5\sin(25t - \sqrt{\pi})$.

(b) $2\cos(4t) \longrightarrow \boxed{\text{System}} \longrightarrow 1 + 4\cos(4t)$.

(c) $4 \longrightarrow \boxed{\text{System}} \longrightarrow -8$.

(d) $4 \longrightarrow \boxed{\text{System}} \longrightarrow j8$.

(e) $4 \longrightarrow \boxed{\text{System}} \longrightarrow 4\cos(3t)$.

(f) $\sin(\pi t) \longrightarrow \boxed{\text{System}} \longrightarrow \cos(\pi t) + 0.1\sin(\pi t)$.

(g) $\sin(\pi t) \longrightarrow \boxed{\text{System}} \longrightarrow \cos(\pi t) + 0.1\sin(2\pi t)$.

(h) $\sin(\pi t) \longrightarrow \boxed{\text{System}} \longrightarrow \sin^2(\pi t)$.

6

Fourier Series and LTI System Response to Periodic Signals

Suppose a weighted linear superposition of co-sinusoids and/or complex exponentials $e^{j\omega t}$ has harmonically related frequencies ω, meaning that each frequency ω divided by any other is a ratio of integers. Then the resulting sum is a periodic signal, such as those shown in Figure 6.1.

In this chapter we will learn how to express arbitrary periodic signals as such sums and then use the input–output rules for LTI systems established in Chapter 5 to determine the steady-state response of such systems to periodic inputs. In Section 6.1 we will discuss *series representations* of periodic signals—weighted sums of co-sinusoids or complex exponentials, known as *Fourier series*—and introduce the terminology used in the description of periodic signals. Section 6.2 describes the conditions under which such Fourier series representations of periodic signals can exist and how they can be determined. Finally, in Section 6.3 we will examine the response of linear as well as nonlinear systems to periodic inputs. Also, the concepts of average signal power and power spectrum will be introduced.

Periodic signals and Fourier series; average power and power spectrum; linear system response to periodic inputs

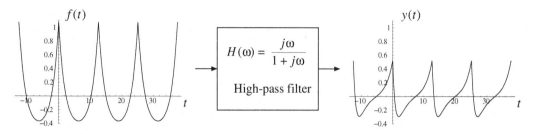

Figure 6.1 A replica of Figure 5.9b, illustrating that LTI systems respond to periodic inputs with periodic outputs.

6.1 Periodic Signals

The signals shown in Figure 6.1 are periodic. In general, a signal $f(t)$ is said to be periodic if there exists some delay t_o such that

$$f(t - t_o) = f(t)$$

for all values of t.[1] A consequence of this basic definition is that

$$f(t - kt_o) = f(t)$$

Signal period

for all integers k. Thus, when periodic signals are delayed by integer multiples of some time interval t_o, they are indistinguishable from their undelayed forms. So, periodic signals consist of replicas, repeated in time. The smallest nonzero value of t_o that satisfies the condition $f(t - t_o) = f(t)$ is said to be the *period* and is denoted by T. For the signals shown in Figure 6.1, the period $T = 4\pi$ s.

Signals $\cos(\omega t)$, $\sin(\omega t)$, and

$$e^{j\omega t} = \cos(\omega t) + j\sin(\omega t)$$

are the simplest examples of periodic signals, having the period $T = \frac{2\pi}{\omega}$ and frequency ω. Consider now the family of periodic signals

$$e^{jn\omega_o t} = \cos(n\omega_o t) + j\sin(n\omega_o t),$$

where n denotes any integer. The period of

$$e^{jn\omega_o t}$$

---

[1] Alternatively if its graph is an endless repetition of the same pattern over and over again, just like the graph in Figure 6.1b.

is $T = \frac{2\pi}{n\omega_o}$, because $\frac{2\pi}{n\omega_o}$ is the smallest nonzero delay t_o that satisfies[2] the constraint

$$e^{jn\omega_o(t-t_o)} = e^{jn\omega_o t}.$$

A weighted linear superposition of signals $e^{jn\omega_o t}$,

$$f(t) \equiv \sum_{n=-\infty}^{\infty} F_n e^{jn\omega_o t},$$

**Exponential
Fourier
series**

where the F_n are constant coefficients, is also periodic with a period $T = \frac{2\pi}{\omega_o}$, which is the smallest nonzero delay t_o that satisfies[3] $f(t - t_o) = f(t)$.

The periodic function $f(t)$, defined above, also can be expressed in terms of a weighted superposition of $\cos(n\omega_o t)$ and $\sin(n\omega_o t)$ signals, or even in terms of $\cos(n\omega_o t + \theta_n)$, when $f(t)$ is real. This gives rise to three different, but related, series representations for periodic signals, as indicated in Table 6.1.

The representations of $f(t)$ shown in Table 6.1, using periodic sums, are known as *Fourier series*. The three equivalent versions shown in the table will be referred to as *exponential*, *trigonometric*, and *compact* forms. Given a set of coefficients F_n that specifies an exponential Fourier series representation of $f(t)$, the equivalence of the trigonometric form can be verified as explained next.

$f(t)$, period $T = \frac{2\pi}{\omega_o}$	Form	Coefficients
$\sum_{n=-\infty}^{\infty} F_n e^{jn\omega_o t}$	Exponential	$F_n = \frac{1}{T} \int_T f(t) e^{-jn\omega_o t} dt$
$\frac{a_0}{2} + \sum_{n=1}^{\infty} a_n \cos(n\omega_o t) + b_n \sin(n\omega_o t)$	Trigonometric	$a_n = F_n + F_{-n}$ $b_n = j(F_n - F_{-n})$
$\frac{c_0}{2} + \sum_{n=1}^{\infty} c_n \cos(n\omega_o t + \theta_n)$	Compact for real $f(t)$	$c_n = 2\|F_n\|$ $\theta_n = \angle F_n$

Table 6.1 Summary of different representations of periodic signal $f(t)$ having period $T = \frac{2\pi}{\omega_o}$ and fundamental frequency ω_o. The formula for F_n in the upper right corner will be derived in Section 6.2.

Verification of trigonometric form: The exponential form can be rewritten as

$$f(t) = F_0 + \sum_{m=1}^{\infty} (F_m e^{jm\omega_o t} + F_{-m} e^{-jm\omega_o t}).$$

---

[2]We need $e^{-jn\omega_o t_o} = 1$, which requires $n\omega_o t_o = 0, 2\pi, 4\pi, \cdots$. The smallest nonzero choice for t_o clearly is $\frac{2\pi}{n\omega_o}$, which is the period T.

[3]$f(t - t_o) = \sum_{n=-\infty}^{\infty} F_n e^{jn\omega_o(t-t_o)} = f(t)$ only if t_o satisfies $|n|\omega_o t_o = 2\pi k$ for *every value of* $|n| \geq$ 1. The smallest nonzero t_o that meets this criterion is $t_o = T = \frac{2\pi}{\omega_o}$.

Replacing $e^{\pm jm\omega_0 t}$ with

$$\cos(m\omega_0 t) \pm j\sin(m\omega_0 t),$$

we obtain

$$f(t) = F_0 + \sum_{m=1}^{\infty}(F_m + F_{-m})\cos(m\omega_0 t) + j(F_m - F_{-m})\sin(m\omega_0 t),$$

Trigonometric Fourier series

which is the same as the trigonometric form with

$$a_n \equiv F_n + F_{-n}$$

and

$$b_n \equiv j(F_n - F_{-n}),$$

for $n \geq 0$.

Notice that for a real-valued signal $f(t)$, the coefficients a_n and b_n of the trigonometric Fourier series

$$f(t) = \frac{a_0}{2} + \sum_{n=1}^{\infty} a_n \cos(n\omega_0 t) + b_n \sin(n\omega_0 t)$$

For-real valued $f(t)$ coefficients $F_{-n} = F_n^*$

must be real-valued. This implies that $F_{-n} = F_n^*$; in other words, the exponential series coefficients F_n are conjugate symmetric when $f(t)$ is real-valued.

We next verify the equivalence of the compact form of the Fourier series to the exponential form, for real-valued $f(t)$—because $f(t)$ is real-valued, we use the fact that the F_n coefficients have conjugate symmetry.

Verification of compact form: Expressing the exponential form as

$$f(t) = \sum_{n=-\infty}^{\infty} F_n e^{jn\omega_0 t} = \sum_{n=-\infty}^{\infty} |F_n| e^{j(n\omega_0 t + \angle F_n)}$$

and assuming that $F_{-n} = F_n^*$ so that $|F_{-n}| = |F_n|$ and $\angle F_{-n} = -\angle F_n$, we have

$$f(t) = F_0 + \sum_{m=1}^{\infty} |F_m|(e^{j(m\omega_0 t + \angle F_m)} + e^{-j(m\omega_0 t + \angle F_m)})$$

$$= F_0 + \sum_{n=1}^{\infty} 2|F_n|\cos(n\omega_0 t + \angle F_n).$$

This is the same as the compact trigonometric form for real $f(t)$, shown in Table 6.1, with

$$c_n \equiv 2|F_n| \quad \text{and} \quad \theta_n \equiv \angle F_n,$$

for $n \geq 0$.

The Fourier series $f(t)$ in Table 6.1 are sums of an infinite number of periodic signals with distinct frequencies $\omega = n\omega_o$ and periods $\frac{2\pi}{n\omega_o}$. It is the longest period, corresponding to the lowest frequency and $n = 1$, that defines an interval across which every periodic component repeats. Therefore, the period of the series is $T = \frac{2\pi}{\omega_o}$. The corresponding lowest frequency $\omega_o = \frac{2\pi}{T}$ is referred to as the *fundamental frequency* of the series. We will refer to the component of $f(t)$ with frequency ω_o as the *fundamental*, and the component having frequency $n\omega_o$ as the *nth harmonic*. Finally, $F_0 = \frac{a_0}{2} = \frac{c_0}{2}$ will be referred to as the DC component of $f(t)$. When the DC component is zero, we will refer to $f(t)$ as having *zero mean*.

Fundamental and harmonics

6.2 Fourier Series

6.2.1 Existence of the Fourier series

In 1807 Jean-Baptiste Joseph Fourier made a shocking announcement that *all* periodic signals with periods $T = \frac{2\pi}{\omega_o}$ can be expressed as weighted linear superpositions of an infinite number of cosine and sine functions $\cos(n\omega_o t)$ and $\sin(n\omega_o t)$. The claim was made during a lecture that Fourier was presenting at the Paris Institute to compete for the Grand Prize in mathematics. Fourier did not win the prize, because the jury, including the prominent mathematicians Laplace and Lagrange, did not quite believe the claim. The story, however, has an all-around happy ending: Laplace and Lagrange were shown to be right in their judgment, because some periodic functions cannot be expressed the way Fourier described. On the other hand, it is now known that all periodic signals $f(t)$ that can be generated in the lab[4] can be expressed exactly as Fourier suggested—that is, as the "Fourier series"

$$f(t) = \sum_{n=-\infty}^{\infty} F_n e^{jn\omega_o t},$$

or its equivalents. (See Table 6.1 in the previous section.) The credit for sorting out which periodic functions can be expressed as Fourier series goes to German mathematician Gustave Peter Lejeune Dirichlet.

---

[4]You should realize, of course, that in the lab we can generate only a finite numbers of periods, due to time limitation. For example, for some $f(t)$ with a fundamental frequency of $\frac{\omega_o}{2\pi} = 1$ MHz, the signal would have 3600 million periods in 1 hour. So long as the number of periods generated during an experiment is large enough, it is reasonable to treat the signal as periodic and represent it by a Fourier series.

If a periodic signal $f(t)$ can be expressed in a Fourier series, then the series coefficients F_n, known as *Fourier coefficients*, can be determined by the formula (see Section 6.2.2 for the derivation)

$$F_n = \frac{1}{T} \int_T f(t) e^{-jn\omega_0 t} dt$$

**Absolutely
integrable
$f(t)$**

where $\int_T$ denotes integration over one period (i.e., $\int_{t=0}^T$, or $\int_{t=-T/2}^{T/2}$, or $\int_{t=t'}^{t'+T}$, where t' is arbitrary). Dirichlet recognized that if $f(t)$ is *absolutely integrable* over a period T, that is, if

$$\int_T |f(t)| dt < \infty,$$

then the Fourier coefficients F_n must be bounded,[5] or $|F_n| < \infty$ for all n. With bounded coefficients F_n, the convergence of the Fourier series of $f(t)$ is possible, and, in fact, guaranteed (as proved by Dirichlet), so long as $f(t)$ has only a finite number of minima and maxima, and a finite number of finite-sized discontinuities,[6] within a single period T (i.e., so long as a plot of $f(t)$ over a period T can be drawn on a piece of paper with a pencil having a finite-width tip). These *Dirichlet sufficiency conditions* for the convergence of the Fourier series—namely, that $f(t)$ be absolutely integrable and be plottable—are satisfied by all periodic signals that can be generated in the lab or in a radio station and displayed on an oscilloscope.

**Dirichlet
conditions**

6.2.2   Orthogonal projections and Fourier coefficients

There is a deep mathematical connection between Fourier series and the representation of a vector in n-dimensional space. We will not explore this fully, but instead will simply raise some of the concepts.

A 3-D vector, say,

$$\vec{v} = (3, -2, 5),$$

can be expressed as a weighted sum of three mutually *orthogonal* vectors

$$\vec{u}_1 \equiv (1, 0, 0)$$

$$\vec{u}_2 \equiv (0, 1, 0)$$

$$\vec{u}_3 \equiv (0, 0, 1)$$

---

[5] Notice that

$$|F_n| = \frac{1}{T} \left| \int_T f(t) e^{-jn\omega_0 t} dt \right| \leq \frac{1}{T} \int_T |f(t) e^{-jn\omega_0 t}| dt = \frac{1}{T} \int_T |f(t)| dt,$$

where we first use the triangle inequality—the absolute value of a sum can't be greater than the sum of the absolute values—and next the fact that the magnitude of a product is the product of the magnitudes, and that $|e^{-jn\omega_0 t}| = 1$. So, if $\int_T |f(t)| dt < \infty$, then $|F_n| < \infty$.

[6] At discontinuity points, the Fourier series converges to a value that is midway between the bottom and the top of the discontinuous jump.

as

$$\vec{v} = 3\vec{u}_1 - 2\vec{u}_2 + 5\vec{u}_3.$$

In general, any 3-D vector can be written as

$$\vec{v} = \sum_{n=1}^{3} V_n \vec{u}_n,$$

where

$$V_n = \vec{v} \cdot \vec{u}_n,$$

since the *dot products*[7]

$$\vec{u}_n \cdot \vec{u}_n = 1$$

and

$$\vec{u}_n \cdot \vec{u}_m = 0 \quad \text{for} \quad m \neq n.$$

Note that $\vec{u}_n \cdot \vec{u}_m = 0$, $m \neq n$, is the *orthogonality condition* pertinent to vectors $\vec{u}_1$, $\vec{u}_2$, and $\vec{u}_3$, which can be regarded as *basis vectors* for all 3-D vectors $\vec{v}$. Furthermore, the coefficients V_n of the vector $\vec{v}$ can be regarded as *projections*[8] of $\vec{v}$ along the basis vectors $\vec{u}_n$.

By analogy, a convergent Fourier series

$$f(t) = \sum_{n=-\infty}^{\infty} F_n e^{jn\omega_o t}$$

can be interpreted as an infinite weighted sum of *orthogonal basis functions*

$$e^{jn\omega_o t}, \quad -\infty \leq n \leq \infty,$$

satisfying an orthogonality condition[9]

$$\int_T (e^{jn\omega_o t})(e^{jm\omega_o t})^* dt = 0 \quad \text{for} \quad m \neq n.$$

---

[7]The scalar, or dot, product of two vectors is the sum of the pairwise products of the two sets of vector coordinates.

[8]A projection of one vector onto another is the component of the first vector that lies in the direction of the second vector.

[9]**Verification:** Assuming $m \neq n$, we find that

$$\int_T (e^{jn\omega_o t})(e^{jm\omega_o t})^* dt = \int_{t=0}^{T=\frac{2\pi}{\omega_o}} e^{j(n-m)\omega_o t} dt = \frac{e^{j(n-m)2\pi} - 1}{j(n-m)\omega_o} = \frac{1-1}{j(n-m)\omega_o} = 0.$$

A Fourier coefficient F_m of $f(t)$ is then the *projection* of $f(t)$ along the basis function $e^{jm\omega_0 t}$. To calculate the coefficient we multiply both sides of the series expression with

$$(e^{jm\omega_0 t})^* = e^{-jm\omega_0 t}$$

and integrate the products on each side across a period T. The result, called the *inner product* (instead of dot product) of $f(t)$ with $e^{jm\omega_0 t}$, is

$$\int_T f(t)e^{-jm\omega_0 t}\,dt = \int_T \sum_{n=-\infty}^{\infty} F_n e^{jn\omega_0 t} e^{-jm\omega_0 t}\,dt$$

$$= \sum_{n=-\infty}^{\infty} F_n \int_T (e^{jn\omega_0 t})(e^{jm\omega_0 t})^*\,dt = T F_m,$$

Fourier coefficients for exponential form

provided that the series is uniformly convergent and hence a term-by-term integration of the series is permissible. We then find (after exchanging m with n),

$$F_n = \frac{1}{T}\int_T f(t)e^{-jn\omega_0 t}\,dt,$$

which can be utilized with any periodic $f(t)$ satisfying the Dirichlet conditions to obtain a Fourier series converging to $f(t)$ at all points where $f(t)$ is continuous.

The trigonometric Fourier series

$$f(t) = \frac{a_0}{2} + \sum_{n=1}^{\infty} a_n \cos(n\omega_0 t) + b_n \sin(n\omega_0 t)$$

can be interpreted as a representation of the periodic signal $f(t)$ in terms of an alternative set of orthogonal basis functions consisting of $\cos(n\omega_0 t)$ and $\sin(n\omega_0 t)$, $n \geq 0$. The pertinent Fourier coefficients a_n and b_n can be determined as projections of $f(t)$ along $\cos(n\omega_0 t)$ and $\sin(n\omega_0 t)$. Equivalently, we can use the relations

$$a_n = F_n + F_{-n}$$

and

$$b_n = j(F_n - F_{-n})$$

(see Table 6.1) and also the formula for the exponential Fourier coefficients F_n obtained previously.

Table 6.2 lists the formulae for the Fourier coefficient for all three forms of the Fourier series. Note that the exponential Fourier series requires the calculation of only a single set of coefficients F_n, while two sets, a_n and b_n, are needed for the trigonometric form. Furthermore, the compact-form coefficients c_n and θ_n can be inferred from the exponential-form coefficients F_n in a straightforward way. For these reasons, we will stress mainly the exponential and compact forms of the Fourier series. We also will see that the exponential Fourier series has the most convenient

$f(t)$, period $T = \frac{2\pi}{\omega_o}$,	Fourier coefficients
$\sum_{n=-\infty}^{\infty} F_n e^{jn\omega_o t}$	$F_n = \frac{1}{T} \int_T f(t) e^{-jn\omega_o t} dt$
$\frac{a_0}{2} + \sum_{n=1}^{\infty} a_n \cos(n\omega_o t) + b_n \sin(n\omega_o t)$	$a_n = \frac{2}{T} \int_T f(t) \cos(n\omega_o t) dt$ $b_n = \frac{2}{T} \int_T f(t) \sin(n\omega_o t) dt$
$\frac{c_0}{2} + \sum_{n=1}^{\infty} c_n \cos(n\omega_o t + \theta_n)$ for real $f(t)$	$c_n = 2\|F_n\|$ $\theta_n = \angle F_n$

Table 6.2 Exponential, trigonometric, and compact-form Fourier coefficients for a periodic signal $f(t)$ having period T and fundamental frequency $\omega_o = \frac{2\pi}{T}$.

form for LTI system response calculations. (See Section 6.3.) The trigonometric form is preferable only when $f(t)$ is either an *even function*—that is, when

$$f(-t) = f(t)$$

(in which case $b_n = 0$)—or an *odd function*—that is, when

$$f(-t) = -f(t)$$

(in which case $a_n = 0$).

6.2.3 Periodic and non-periodic sums

Not all sums of co-sinusoids or complex exponentials are periodic; in particular,

$$g(t) = \sum_{k=1}^{\infty} c_k \cos(\omega_k t + \theta_k)$$

is not periodic unless there exists some number ω_o such that all frequencies ω_k are integer multiples of ω_o. Thus, if the sum is periodic, then all possible ratios of the frequencies ω_k are rational numbers. Furthermore, if the sum is periodic, then its fundamental frequency ω_o is defined to be the largest number whose integer multiples match each and every ω_k.

Example 6.1
Signal

$$p(t) = 2\cos(\pi t) + 4\cos(2t)$$

is not periodic, because the ratio of the two frequencies $\frac{\pi}{2}$ is not a rational number.

Example 6.2
Signal

$$q(t) = \cos(4t) + 5\sin(6t) + 2\cos(7t - \frac{\pi}{3})$$

is periodic, because the frequencies 4, 6, and 7 rad/s are each integer multiples of 1 rad/s or, equivalently, the frequency ratios $\frac{4}{6}$, $\frac{4}{7}$, and $\frac{6}{7}$ (and their inverses) all are rational. Furthermore, since 1 rad/s is the largest frequency whose integer multiples can match 4, 6, and 7 rad/s, the fundamental frequency of $q(t)$ is $\omega_o = 1$ rad/s and its period is $T = \frac{2\pi}{\omega_o} = 2\pi$ s.

Because $q(t)$ in Example 6.2 is periodic, it can be expressed as a Fourier series. In fact, the compact trigonometric series for $q(t)$ is simply

$$q(t) = \cos(4t) + 5\cos(6t - \frac{\pi}{2}) + 2\cos(7t - \frac{\pi}{3}).$$

Thus, the parameters in the compact Fourier series are (compare with Table 6.2) $c_4 = 1$, $\theta_4 = 0$, $c_6 = 5$, $\theta_6 = -\frac{\pi}{2}$ rad, $c_7 = 2$, $\theta_7 = -\frac{\pi}{3}$ rad; all other c_n are zero.

Example 6.3
What is the period of

$$f(t) = 1 + \cos(8\pi t) + 7.6\sin(10\pi t)?$$

Solution   Since 2π is the largest number whose integer multiples ($\times 4$ and $\times 5$) match frequencies 8π and 10π, the fundamental frequency of $f(t)$ is $\omega_o = 2\pi$ rad/s. Therefore, the period of $f(t)$ is $T = \frac{2\pi}{\omega_o} = \frac{2\pi}{2\pi \text{ rad/s}} = 1$ s.

Example 6.4
What are the exponential Fourier series coefficients of $f(t)$ in Example 6.3?

Solution   We can rewrite $f(t)$ from Example 6.3 as

$$f(t) = 1e^{j0\cdot2\pi t} + \frac{e^{j4\cdot2\pi t} + e^{j(-4)\cdot2\pi t}}{2} + 7.6\frac{e^{j5\cdot2\pi t} - e^{j(-5)\cdot2\pi t}}{2j}.$$

The right-hand side of this expression is effectively the exponential Fourier series (see Table 6.2) of $f(t)$. Hence, all Fourier coefficients F_n are zero, except for $F_0 = 1$, $F_4 = \frac{1}{2}$, $F_{-4} = \frac{1}{2}$, $F_5 = -j3.8$, and $F_{-5} = j3.8$.

6.2.4   Properties and calculations of Fourier series

Table 6.3 lists some of the properties of periodic functions and their Fourier series. We will verify some of these properties and use a number of them to assist us in Fourier series calculations. The notation $f(t) \leftrightarrow F_n$, $g(t) \leftrightarrow G_n$, $\cdots$ in Table 6.3 indicates that F_n, G_n, $\cdots$ are the respective exponential Fourier coefficients of $f(t)$, $g(t)$, $\cdots$, where the periodic functions $f(t)$, $g(t)$, $\cdots$ are assumed to have the same fundamental frequency $\omega_o = \frac{2\pi}{T}$. The coefficients a_n, b_n, c_n, and θ_n refer to the coefficients in the trigonometric and compact forms of the Fourier series.

	Name:	Condition:	Property:
1	Scaling	Constant K	$Kf(t) \leftrightarrow KF_n$
2	Addition	$f(t) \leftrightarrow F_n$, $g(t) \leftrightarrow G_n$, $\cdots$	$f(t) + g(t) + \cdots \leftrightarrow F_n + G_n + \cdots$
3	Time shift	Delay t_o	$f(t - t_o) \leftrightarrow F_n e^{-jn\omega_o t_o}$
4	Derivative	Continuous $f(t)$	$\frac{df}{dt} \leftrightarrow jn\omega_o F_n$
5	Hermitian	Real $f(t)$	$F_{-n} = F_n^*$
6	Even function	$f(-t) = f(t)$	$f(t) = \frac{a_0}{2} + \sum_{n=1}^{\infty} a_n \cos(n\omega_o t)$
7	Odd function	$f(-t) = -f(t)$	$f(t) = \sum_{n=1}^{\infty} b_n \sin(n\omega_o t)$
8	Average power		$P \equiv \frac{1}{T} \int_T |f(t)|^2 dt = \sum_{n=-\infty}^{\infty} |F_n|^2$

Table 6.3   Properties of Fourier series.

Example 6.5
Find the exponential and compact Fourier series of $f(t) = |\sin(t)|$ shown in Figure 6.2a.

Solution   The period of $\sin(t)$ is 2π, while the period of $f(t) = |\sin(t)|$ is $T = \pi$ s as can be verified from the graph of $f(t)$ shown in Figure 6.2a. Therefore, we identify the fundamental frequency of $f(t)$ as $\omega_o = \frac{2\pi}{T} = 2$ rad/s. We will calculate F_n by using the integration limits of 0 and π, since $f(t)$ can be described by a single equation on the interval $0 < t < \pi$, namely, $f(t) = |\sin(t)| = \sin(t)$. We then have

$$F_n = \frac{1}{T} \int_T f(t) e^{-jn\omega_o t} \, dt = \frac{1}{\pi} \int_0^{\pi} \sin(t) e^{-jn2t} \, dt = \frac{1}{\pi} \int_0^{\pi} \frac{e^{jt} - e^{-jt}}{2j} e^{-jn2t} \, dt$$

$$= \frac{1}{j2\pi} \int_0^{\pi} (e^{j(1-2n)t} - e^{-j(1+2n)t}) \, dt = \frac{1}{j2\pi} \left(\frac{e^{j(1-2n)t}}{j(1-2n)} - \frac{e^{-j(1+2n)t}}{-j(1+2n)} \right) \Big|_0^{\pi}$$

$$= -\frac{1}{2\pi} \left(\frac{e^{j(1-2n)\pi} - 1}{1 - 2n} + \frac{e^{-j(1+2n)\pi} - 1}{1 + 2n} \right).$$

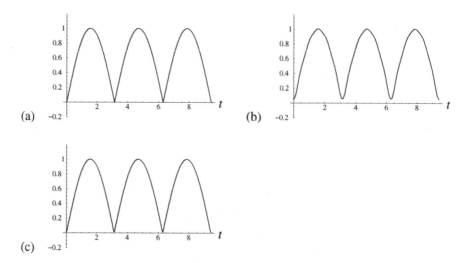

Figure 6.2   (a) A periodic function $f(t) = |\sin(t)|$, (b) plot of Fourier series of $f(t)$ truncated at $n = 5$, and (c) truncated at $n = 20$.

Now,

$$e^{j(1-2n)\pi} = e^{-j(1+2n)\pi} = e^{\pm j\pi} = -1$$

for all integers n, and therefore

$$F_n = \frac{1}{\pi}\left(\frac{1}{1-2n} + \frac{1}{1+2n}\right) = \frac{2}{\pi}\frac{1}{1-4n^2}.$$

The exponential Fourier series of $f(t) = |\sin(t)|$ is then

$$f(t) = \sum_{n=-\infty}^{\infty} \frac{2}{\pi}\frac{1}{1-4n^2}e^{jn2t}.$$

The coefficients for the compact form are, for $n \geq 1$,

$$c_n = 2|F_n| = \frac{4}{\pi}\frac{1}{4n^2-1} = \frac{1}{\pi}\frac{1}{n^2-\frac{1}{4}}$$

and

$$\theta_n = \angle F_n = \pi \text{ rad},$$

where the last line follows because for $n \geq 1$ the F_n are all real and negative, so that their angles all have value π. Also, $F_o = \frac{c_o}{2} = \frac{2}{\pi}$. The compact form of the Fourier series is therefore

$$f(t) = \frac{2}{\pi} + \sum_{n=1}^{\infty} \frac{1}{\pi} \frac{1}{n^2 - \frac{1}{4}} \cos(n2t + \pi).$$

Figures 6.2b and 6.2c show plots of the Fourier series of $f(t)$, but with the sums truncated at $n = 5$ and $n = 20$, respectively (for example, we dropped the sixth and higher-order harmonics from the Fourier series to obtain the curve plotted in Figure 6.2b.) Notice that the curve in Figure 6.2b approximates $f(t) = |\sin(t)|$ very well, except in the neighborhoods where $f(t)$ is nearly zero and abruptly changes direction. The curve in Figure 6.2c, which we obtained by including more terms in the sum (up to the 20th harmonic), clearly gives a finer approximation. Because $f(t)$ is a continuous function, the Fourier series converges to $f(t)$ for all values of t.

Example 6.6

Prove the time-shift property from Table 6.3.

Solution   This property states that

$$f(t) \leftrightarrow F_n \Rightarrow f(t - t_o) \leftrightarrow F_n e^{-jn\omega_o t_o}.$$

To verify it, we first express $f(t)$ in its Fourier series as

$$f(t) = \sum_{n=-\infty}^{\infty} F_n e^{jn\omega_o t}.$$

Replacing t with $t - t_o$ on both sides gives

$$f(t - t_o) = \sum_{n=-\infty}^{\infty} F_n e^{jn\omega_o (t-t_o)} = \sum_{n=-\infty}^{\infty} (F_n e^{-jn\omega_o t_o}) e^{jn\omega_o t}.$$

Hence, the expression in parentheses, $F_n e^{-jn\omega_o t_o}$, is the nth Fourier coefficient for $f(t - t_o)$, proving the time-shift property.

Example 6.7

What are the exponential-form Fourier coefficients G_n of the periodic function

$$g(t) = |\cos(t)|$$

shown in Figure 6.3a? Also, determine the compact-form Fourier series of $g(t)$.

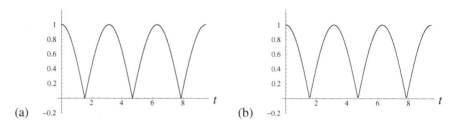

Figure 6.3 (a) Plot of $g(t) = |\cos(t)|$, and (b) plot of its Fourier series truncated at $n = 20$.

Solution   Clearly,

$$g(t) = |\cos(t)| = f(t \pm \frac{\pi}{2}),$$

where $f(t) = |\sin(t)|$ as in Example 6.5. Therefore, using the Fourier coefficients F_n of $f(t)$ from Example 6.5 and the time-shift property from Table 6.3 with $t_o = \frac{\pi}{2}$ s, we obtain the Fourier coefficients G_n of $g(t)$ as

$$G_n = F_n e^{-jn\omega_o t_o} = F_n e^{-jn(2)(\frac{\pi}{2})} = F_n e^{-jn\pi}.$$

Hence, $|G_n| = |F_n|$, and for $n \geq 1$

$$\angle G_n = \angle F_n - n\pi = (1 - n)\pi,$$

since $\angle F_n = \pi$ in that case. The compact form of $g(t)$ is therefore

$$g(t) = \frac{2}{\pi} + \sum_{n=1}^{\infty} \frac{1}{\pi} \frac{1}{n^2 - \frac{1}{4}} \cos(n2t + (1 - n)\pi).$$

Note that the same result also could have been obtained by replacing t with $t - \frac{\pi}{2}$ in the compact form Fourier series for $f(t) = |\sin(t)|$ from Example 6.5. A plot of the series for $g(t)$, truncated at $n = 20$, is shown in Figure 6.3b.

Examples 6.5 and 6.7 illustrated the influence of the angle, or phase coefficient, θ_n on the shape of periodic signals. The amplitude coefficients c_n for $f(t)$ and $g(t)$ are identical, which implies that both functions are constructed with cosines of equal amplitudes. The curves, however, are different, because the phase shifts θ_n of the cosines are different. To further illustrate the impact of θ_n on the shape of a signal waveform, we plot in Figure 6.4 a truncated version of another series,

$$h(t) = \frac{2}{\pi} + \sum_{n=1}^{\infty} \frac{1}{\pi} \frac{1}{n^2 - \frac{1}{4}} \cos(n2t + (1 - n)\frac{\pi}{2}),$$

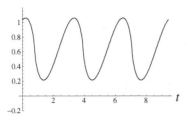

Figure 6.4   Plot of the Fourier series of signal $h(t)$ truncated at $n = 20$.

having the same amplitude coefficients as $f(t)$ and $g(t)$, but different phase coefficients θ_n. Notice that $h(t)$ has a shape that is different from both $f(t)$ and $g(t)$, which is caused by the different Fourier phases. In general, both the Fourier amplitudes and phases affect the shape of a signal.

Example 6.8

Given that (from Example 6.5)

$$f(t) = |\sin(t)| = \sum_{n=-\infty}^{\infty} \frac{2}{\pi} \frac{1}{1 - 4n^2} e^{jn2t},$$

determine the exponential and compact Fourier series of

$$g(t) = |\sin(\frac{1}{2}t)|.$$

Solution   We note that

$$g(t) = f(\frac{t}{2}).$$

Therefore, replacing t in the expression for the exponential Fourier series of $f(t)$ with $\frac{t}{2}$, we obtain the Fourier series of $g(t)$ as

$$g(t) = \sum_{n=-\infty}^{\infty} \frac{2}{\pi} \frac{1}{1 - 4n^2} e^{jnt}.$$

Likewise, the compact form is

$$g(t) = \frac{2}{\pi} + \sum_{n=1}^{\infty} \frac{1}{\pi} \frac{1}{n^2 - \frac{1}{4}} \cos(nt + \pi).$$

Notice that the Fourier series coefficients have not changed. A stretching or squashing of a periodic waveform corresponds to a change in period and fundamental frequency. Comparing $f(t)$ and $g(t)$, the period has increased from π to 2π, and the fundamental frequency has dropped from 2 to 1 rad/s. The waveform $g(t)$ is simply a stretched (by a factor of 2) version of $f(t)$, plotted earlier in Figure 6.2a.

Example 6.9
A periodic signal $f(t)$ with period T is specified as

$$f(t) = e^{-at} \text{ for } 0 \le t < T.$$

A plot of $f(t)$ for $a = 0.5 \text{ s}^{-1}$ and $T = 2$ s is shown in Figure 6.5a. Determine both the exponential and compact Fourier series for $f(t)$, for arbitrary a and T.

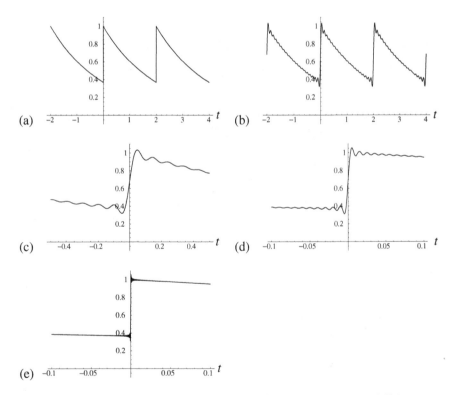

Figure 6.5 (a) A periodic signal $f(t)$; (b) plot of the Fourier series of $f(t)$ truncated at $n = 20$; (c) expanded plot of the same curve near $t = 0$; (d) expanded plot of the same series truncated at $n = 200$; and (e) truncated at $n = 2000$.

Solution   Using the integration limits of 0 to T, we see that

$$F_n = \frac{1}{T} \int_0^T e^{-at} e^{-jn\omega_o t} dt = \frac{1}{T} \int_0^T e^{-(a+jn\omega_o)t} dt = \frac{1}{T} \frac{e^{-(a+jn\omega_o)t}}{-(a+jn\omega_o)} \bigg|_0^T$$

$$= \frac{1 - e^{-(a+jn\omega_o)T}}{(a+jn\omega_o)T} = \frac{1 - e^{-aT}e^{-jn2\pi}}{aT + j2\pi n} = \frac{1 - e^{-aT}}{\sqrt{(aT)^2 + (2\pi n)^2}} e^{-j\tan^{-1}\frac{2\pi n}{aT}}.$$

In the last line we used $\omega_o T = 2\pi$ and $e^{-j2\pi n} = 1$. Thus, the exponential Fourier series is

$$f(t) = \sum_{n=-\infty}^{\infty} \frac{1 - e^{-aT}}{\sqrt{(aT)^2 + (2\pi n)^2}} e^{-j\tan^{-1}\frac{2\pi n}{aT}} e^{jn\omega_o t}.$$

The compact-form coefficients are

$$c_n = 2|F_n| = \frac{2(1 - e^{-aT})}{\sqrt{(aT)^2 + (2\pi n)^2}}$$

and

$$\theta_n = \angle F_n = -\tan^{-1}\frac{2\pi n}{aT}.$$

Therefore, in compact form,

$$f(t) = \frac{1 - e^{-aT}}{aT} + \sum_{n=1}^{\infty} \frac{2(1 - e^{-aT})}{\sqrt{(aT)^2 + (2\pi n)^2}} \cos(n\omega_o t - \tan^{-1}\frac{2\pi n}{aT}).$$

Figure 6.5b displays a plot of the Fourier series for $f(t)$ (assuming that $a = 0.5$ and $T = 2$), truncated at $n = 20$. Note that the plot exhibits small fluctuations about the true $f(t)$ shown in Figure 6.5a. Furthermore, the fluctuations are largest near the points of discontinuity of $f(t)$, at $t = 0, 2, 4$, etc., as shown in more detail in the expanded plot of Figure 6.5c. Including a larger number of higher-order harmonics in the truncated series reduces the widths of the fluctuations, but does not diminish their amplitudes. This is illustrated in Figures 6.5d and 6.5e, which show other expanded plots of the series, but now with the series truncated at $n = 200$ and 2000. In the limit, as the number of terms in the Fourier series becomes infinite, the fluctuation widths vanish as the fluctuations bunch up around the points of discontinuity and the infinite series converges to $f(t)$ at all points, except those where the signal is discontinuous. At points of discontinuity, the series converges to a value that is midway between the bottom and the top of the discontinuous jump. The behavior of a Fourier series near **Gibbs** points of discontinuity, where increasing the number of terms in the series causes **phenomenon** the fluctuations, or "ripples," to be narrower and to bunch up around the points of discontinuity (but with no decrease in amplitude), is known as the *Gibbs phenomenon*, after the American mathematical physicist Willard Josiah Gibbs.

Example 6.10

Determine the exponential and compact-form Fourier series of the square-wave signal

$$p(t) = \begin{cases} 1, & 0 < t < D, \\ 0, & D < t < 1, \end{cases}$$

where $0 < D < 1$ and the signal period is $T = 1$ s. Figure 6.6a shows an example of $p(t)$ with a *duty cycle* of $D = 0.25 = 25\%$.

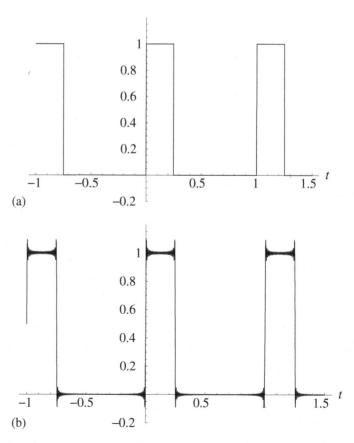

Figure 6.6   (a) A square wave $p(t)$ with 25 percent duty cycle and unity amplitude, and (b) a plot of its Fourier series truncated at $n = 200$.

Solution   Since $T = 1$, the fundamental frequency $\omega_o = \frac{2\pi}{T} = 2\pi$. There-fore, the Fourier coefficients are

$$P_n = \frac{1}{T} \int_0^1 p(t)e^{-jn\omega_o t} dt = \int_0^D e^{-jn2\pi t} dt.$$

For the special case of $n = 0$, we obtain

$$P_0 = D.$$

For $n \neq 0$,

$$P_n = \frac{e^{-jn2\pi t}}{-jn2\pi} \bigg|_0^D = \frac{e^{-jn2\pi D} - 1}{-jn2\pi}$$

$$= e^{-jn\pi D} \frac{e^{-jn\pi D} - e^{jn\pi D}}{-jn2\pi} = \frac{\sin(n\pi D)}{n\pi} e^{-jn\pi D}.$$

This is consistent in the limit $n \to 0$ with P_0 determined above, as can be verifiable by using *l'Hopital's rule*.[10] Thus, in exponential form the Fourier series is

$$p(t) = \sum_{n=-\infty}^{\infty} \frac{\sin(n\pi D)}{n\pi} e^{j(n2\pi t - n\pi D)},$$

and in compact-form is

$$p(t) = D + \sum_{n=1}^{\infty} \frac{2\sin(n\pi D)}{n\pi} \cos(n2\pi t - n\pi D).$$

Figure 6.6b shows a plot of the Fourier series of $p(t)$ for $D = 0.25$, truncated at $n = 200$. Note the Gibbs phenomenon near $t = 0$, $t = 0.25$, etc., where $p(t)$ is discontinuous.

All discontinuous signals having Fourier series coefficients c_n that are propor-tional to $\frac{1}{n}$ (for large n) exhibit the Gibbs phenomenon. Notice that in Examples 6.5 and 6.7, where the signals were continuous, c_n was proportional to $\frac{1}{n^2}$ (for large n) and the Gibbs phenomenon was absent. When c_n decays as $\frac{1}{n^2}$ (or faster), the contribution of higher-order harmonics to the Fourier series is less important than in cases where c_n is proportional to $\frac{1}{n}$, and the Gibbs phenomenon does not occur.

---

[10] Applying l'Hopital's rule yields

$$\lim_{n \to 0} \frac{\sin(n\pi D)}{n\pi} = \lim_{n \to 0} \frac{\frac{d}{dn}\sin(n\pi D)}{\frac{d}{dn} n\pi} = \frac{\pi D \cos(n\pi D)|_{n=0}}{\pi} = D = P_0.$$

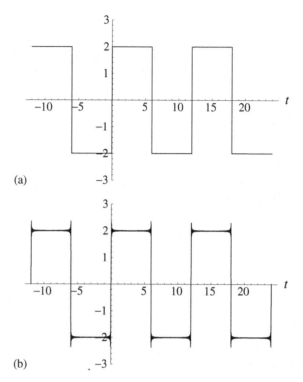

Figure 6.7 (a) A square wave of peak-to-peak amplitude 4 and period 12 s, and (b) its truncated Fourier series plot.

Example 6.11

Let $f(t)$ denote a zero-mean square wave with a period of $T = 12$ s and with a peak-to-peak amplitude of 4, as depicted in Figure 6.7a. Express $f(t)$ as a stretched, scaled, and offset version of $p(t)$ defined in Example 6.10 and then determine the compact-form Fourier series of $f(t)$.

Solution   Consider first $4p(\frac{t}{12})$, which is, for $D = \frac{1}{2}$, a square wave with 50% duty cycle, period $T = 12$ s, and an amplitude of 4. It differs from $f(t)$ shown in Figure 6.7a only by a DC offset of 2. Clearly, then, we can write

$$f(t) = 4p(\frac{t}{12}) - 2$$

for $D = \frac{1}{2}$. Since, for this choice of D, we have, from Example 6.10,

$$p(t) = \frac{1}{2} + \sum_{n=1}^{\infty} \frac{2\sin(n\pi\frac{1}{2})}{n\pi} \cos(n2\pi t - n\pi\frac{1}{2}),$$

it follows, by direct substitution, that

$$f(t) = 4p(\frac{t}{12}) - 2 = \sum_{n=1}^{\infty} \frac{8\sin(n\frac{\pi}{2})}{n\pi} \cos(n\pi\frac{t}{6} - n\frac{\pi}{2})$$

$$= \sum_{n=1 \text{ (odd)}}^{\infty} \frac{8\sin^2(n\frac{\pi}{2})}{n\pi} \sin(n\pi\frac{t}{6})$$

$$= \sum_{n=1 \text{ (odd)}}^{\infty} \frac{8}{n\pi} \sin(n\frac{\pi}{6}t) = \sum_{n=1 \text{ (odd)}}^{\infty} \frac{8}{n\pi} \cos\left(n\frac{\pi}{6}t - \frac{\pi}{2}\right),$$

where we made use of the trig identity

$$\cos(a - b) = \cos a \cos b + \sin a \sin b$$

to reach line 2. Figure 6.7b shows a truncated Fourier series plot of $f(t)$ obtained from the preceding expression.

Example 6.12

Obtain the trigonometric-form Fourier series for $f(t)$ in Figure 6.7a, using the formula for b_n from Table 6.2.

Solution   The period of $f(t)$ is $T = 12$ s and

$$\omega_o = \frac{2\pi}{T} = \frac{\pi}{6} \frac{\text{rad}}{\text{s}}.$$

The function is odd, so by property 7 from Table 6.3, all $a_n = 0$. Using the formula for b_n from Table 6.2,

$$b_n = \frac{2}{12} \int_{-6}^{6} f(t) \sin(n\frac{\pi}{6}t)dt.$$

Since the integrand is even—note that odd $f(t) \times$ odd $\sin(n\frac{\pi}{6}t)$ is an even function—we can evaluate b_n as

$$b_n = \frac{2 \times 2}{12} \int_{0}^{6} 2\sin(n\frac{\pi}{6}t)dt = \frac{8}{12} \left.\frac{\cos(n\frac{\pi}{6}t)}{-n\frac{\pi}{6}}\right|_{0}^{6} = -\frac{4}{n\pi}(\cos(n\pi) - 1).$$

Clearly, for even n, $b_n = 0$, and for odd n, $b_n = \frac{8}{n\pi}$. Thus, the trigonometric Fourier series is

$$f(t) = \sum_{n=1 \text{ (odd)}}^{\infty} \frac{8}{n\pi} \sin(n\frac{\pi}{6}t),$$

as already determined by another method in Example 6.11.

Example 6.13

The function $q(t)$ is periodic with period $T = 4$ s and is specified as

$$q(t) = \begin{cases} 2t, & 0 < t < 2 \text{ s,} \\ 0, & 2 < t < 4 \text{ s.} \end{cases}$$

Determine the compact trigonometric Fourier series of $q(t)$.

Solution   The fundamental frequency is $\omega_o = \frac{2\pi}{T} = \frac{\pi}{2}$. First, for $n = 0$,

$$Q_0 = \frac{1}{4} \int_0^2 2t\,dt = \frac{1}{4} t^2 \big|_0^2 = 1.$$

For all other values of n,

$$Q_n = \frac{1}{4} \int_0^2 2te^{-jn\frac{\pi}{2}t}\,dt = \frac{1}{2} \int_0^2 t \frac{d}{dt}\left[\frac{e^{-jn\frac{\pi}{2}t}}{-jn\frac{\pi}{2}}\right] dt.$$

(The last line of this expression does not hold for $n = 0$, which is why we examined this case separately.) Using *integration by parts*, we obtain

$$Q_n = \frac{1}{2}\left(\frac{te^{-jn\frac{\pi}{2}t}}{-jn\frac{\pi}{2}}\right)\bigg|_0^2 - \frac{1}{2}\int_0^2 \frac{e^{-jn\frac{\pi}{2}t}}{-jn\frac{\pi}{2}}\,dt = \frac{1}{2}\frac{2e^{-jn\pi} - 0}{-jn\frac{\pi}{2}} - \frac{1}{2}\left(\frac{e^{-jn\frac{\pi}{2}t}}{(-jn\frac{\pi}{2})^2}\right)\bigg|_0^2$$

$$= \frac{(-1)^n}{-jn\frac{\pi}{2}} - \frac{1}{2}\frac{(-1)^n - 1}{-n^2\frac{\pi^2}{4}} = \frac{2((-1)^n - 1) + j2\pi n(-1)^n}{\pi^2 n^2}$$

$$= \begin{cases} \frac{j2}{\pi n}, & \text{for even } n > 0, \\ -\frac{4 + j2\pi n}{\pi^2 n^2}, & \text{for odd } n. \end{cases}$$

After finding the magnitudes and phases of Q_n, for even and odd n, the compact Fourier series can be written as

$$q(t) = 1 + \sum_{n=2 \text{ (even)}}^{\infty} \frac{4}{\pi n} \cos\left(n\frac{\pi}{2}t + \frac{\pi}{2}\right) + \sum_{n=1 \text{ (odd)}}^{\infty} \frac{4\sqrt{4 + \pi^2 n^2}}{\pi^2 n^2}$$

$$\times \cos\left(n\frac{\pi}{2}t + \pi + \tan^{-1}\frac{\pi n}{2}\right).$$

A plot of this series, truncated at $n = 20$, is shown in Figure 6.8.

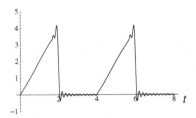

Figure 6.8 Plot of the Fourier series of $q(t)$ truncated at $n = 20$.

Example 6.14
Prove the derivative property from Table 6.3.

Solution   This property states that

$$f(t) \leftrightarrow F_n \implies \frac{df}{dt} \leftrightarrow jn\omega_o F_n$$

when $f(t)$ is a continuous function. In other words, the derivative $f'(t) \equiv \frac{df}{dt}$ will have a Fourier series with Fourier coefficients $jn\omega_o F_n$.
    To verify the property, we differentiate

$$f(t) = \sum_{n=-\infty}^{\infty} F_n e^{jn\omega_o t}$$

to obtain

$$\frac{df}{dt} = \frac{d}{dt} \sum_{n=-\infty}^{\infty} F_n e^{jn\omega_o t} = \sum_{n=-\infty}^{\infty} F_n \frac{d}{dt} e^{jn\omega_o t} = \sum_{n=-\infty}^{\infty} (jn\omega_o F_n) e^{jn\omega_o t}.$$

This is a Fourier series expansion for the function $\frac{df}{dt}$, where the expression in parentheses, $jn\omega_o F_n$, is the nth Fourier coefficient. This proves the derivative property.

Example 6.15
Let

$$g(t) = \frac{df}{dt},$$

where

$$f(t) = |\sin(t)|.$$

Determine the exponential Fourier coefficients of $g(t)$ and its compact trigonometric Fourier series.

Solution   Since $f(t)$ is continuous,

$$G_n = jn\omega_o F_n = jn2F_n,$$

by the derivative property from Table 6.3. Because

$$F_n = \frac{2}{\pi}\frac{1}{1 - 4n^2}$$

from Example 6.5, it follows that

$$G_n = j\frac{4}{\pi}\frac{n}{1 - 4n^2}.$$

The easiest way to obtain the compact trigonometric series for $g(t)$ is to differentiate the compact series for $f(t)$ (from Example 6.5) term by term; the result is

$$g(t) = \sum_{n=1}^{\infty}\frac{1}{\pi}\frac{2n}{n^2 - \frac{1}{4}}\cos(n2t + \frac{3\pi}{2}).$$

A truncated version of this expression is plotted in Figure 6.9.

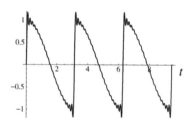

Figure 6.9   A plot of the Fourier series of $g(t) = \frac{d}{dt}|\sin(t)|$ truncated at $n = 20$.

6.3   System Response to Periodic Inputs

6.3.1   LTI system response to periodic inputs

The input–output relation

$$e^{j\omega t} \longrightarrow \boxed{\text{LTI}} \longrightarrow H(\omega)e^{j\omega t}$$

for dissipative LTI systems implies that, with $\omega = n\omega_o$,

$$e^{jn\omega_o t} \longrightarrow \boxed{\text{LTI}} \longrightarrow H(n\omega_o)e^{jn\omega_o t}.$$

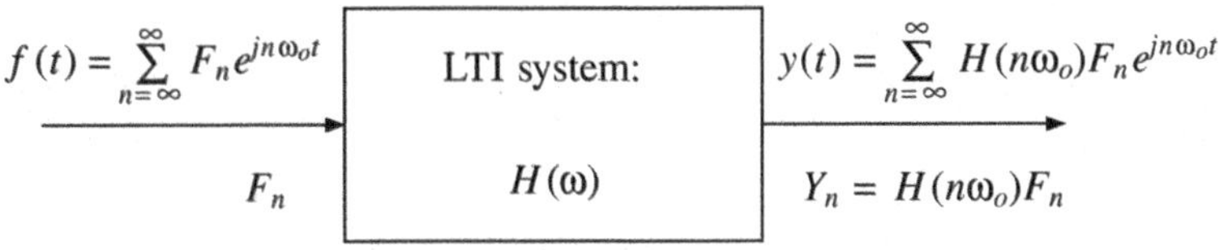

$$f(t) = \sum_{n=\infty}^{\infty} F_n e^{jn\omega_o t} \quad\boxed{\begin{array}{c} \text{LTI system:} \\[1em] H(\omega) \end{array}}\quad y(t) = \sum_{n=\infty}^{\infty} H(n\omega_o)F_n e^{jn\omega_o t}$$

$$F_n \qquad\qquad Y_n = H(n\omega_o)F_n$$

Figure 6.10 Input–output relation for dissipative LTI systems with periodic inputs.

Therefore, using superposition, we get

$$\sum_{n=-\infty}^{\infty} F_n e^{jn\omega_o t} \longrightarrow \boxed{\text{LTI}} \longrightarrow \sum_{n=-\infty}^{\infty} H(n\omega_o)F_n e^{jn\omega_o t}$$

for any set of coefficients F_n. Thus, as illustrated in Figure 6.10, the steady-state response of an LTI system $H(\omega)$ to an arbitrary periodic input

$$f(t) = \sum_{n=-\infty}^{\infty} F_n e^{jn\omega_o t}$$

is the periodic output

$$y(t) = \sum_{n=-\infty}^{\infty} H(n\omega_o)F_n e^{jn\omega_o t}.$$

The input–output relation for periodic signals described in Figure 6.10 indicates that an LTI system $H(\omega)$ simply converts the Fourier coefficients F_n of its periodic input $f(t)$ into the Fourier coefficients

$$Y_n = H(n\omega_o)F_n$$

of its periodic output $y(t)$.

Example 6.16
The input of a linear system

$$H(\omega) = \frac{2 + j\omega}{3 + j\omega}$$

is the periodic function

$$f(t) = \sum_{n=-\infty}^{\infty} \frac{n}{1+n^2} e^{-jn4t}.$$

What are the Fourier coefficients Y_n of the periodic system output $y(t)$?

Solution   The input Fourier coefficients are

$$F_n = \frac{n}{1 + n^2},$$

and the fundamental frequency of the input is $\omega_o = 4 \frac{\text{rad}}{\text{s}}$. Therefore,

$$H(n\omega_o) = H(n4) = \frac{2 + jn4}{3 + jn4}$$

and the Fourier coefficients of the output $y(t)$ are

$$Y_n = H(n\omega_o)F_n = \frac{2 + jn4}{3 + jn4} \frac{n}{1 + n^2}.$$

As a further illustration of the input–output relation in Figure 6.10, consider the three linear systems depicted in Figures 5.1, 5.2, and 5.3. The frequency responses of the systems are

$$H_1(\omega) = \frac{1}{1 + j\omega},$$

$$H_2(\omega) = \frac{j\omega}{1 + j\omega},$$

and

$$H_3(\omega) = \frac{j\omega}{1 - \omega^2 + j\omega}.$$

Suppose that all three systems are excited by the same periodic input

$$f(t) = |\sin(\tfrac{1}{2}t)| = \sum_{n=-\infty}^{\infty} \frac{2}{\pi} \frac{1}{1 - 4n^2} e^{jnt},$$

with fundamental frequency $\omega_o = 1 \frac{\text{rad}}{\text{s}}$. (See Example 6.8 in the previous section.) Let $y_1(t)$, $y_2(t)$, and $y_3(t)$ denote the steady-state response of the systems to this input. Applying the relation in Figure 6.10, we determine that

$$y_1(t) = \sum_{n=-\infty}^{\infty} \frac{1}{1 + jn} \frac{2}{\pi} \frac{1}{1 - 4n^2} e^{jnt},$$

$$y_2(t) = \sum_{n=-\infty}^{\infty} \frac{jn}{1 + jn} \frac{2}{\pi} \frac{1}{1 - 4n^2} e^{jnt},$$

and

$$y_3(t) = \sum_{n=-\infty}^{\infty} \frac{jn}{1 - n^2 + jn} \frac{2}{\pi} \frac{1}{1 - 4n^2} e^{jnt}.$$

These expressions can be readily converted to compact form, and then their truncated plots can be compared with the input signal $|\sin(\frac{1}{2}t)|$ to assess the impact of each system on the input. However, we also can gain some insight by examining how the magnitudes of the Fourier coefficients of input $f(t)$ and responses $y_1(t)$, $y_2(t)$, and $y_3(t)$ compare.

Figures 6.11a through 6.11d display point plots of $|F_n|^2$, $|H_1(n)|^2|F_n|^2$, $|H_2(n)|^2$ $|F_n|^2$, and $|H_3(n)|^2|F_n|^2$ versus harmonic frequency $n\omega_o = n \frac{rad}{s}$, representing the squared magnitude of the Fourier coefficients of $f(t)$, $y_1(t)$, $y_2(t)$, and $y_3(t)$, respectively. (The reason for plotting the squared magnitudes instead of just the magnitudes will be explained in the next section.) Notice that $|H_1(n)|^2|F_n|^2$, representing $y_1(t)$, has been reduced compared with $|F_n|^2$, for $|n| \geq 1$. This shows that system $H_1(\omega)$ attenuates the high-frequency content of the input $f(t)$ in producing $y_1(t)$. As a consequence, the waveform $y_1(t)$, shown in Figure 6.12b, is smoother than the input $f(t)$, shown in Figure 6.12a. This occurs because substantial high-frequency content is necessary to produce rapid variation in a waveform.

By contrast, we see in Figure 6.11c that $H_2(\omega)$ has zeroed out the lowest frequency (DC component) of the input. Hence, the response $y_2(t)$, shown in Figure 6.12c, is a zero-mean signal, but it retains the sharp corners present in $f(t)$, indicating insignificant change in the high-frequency content of the input.

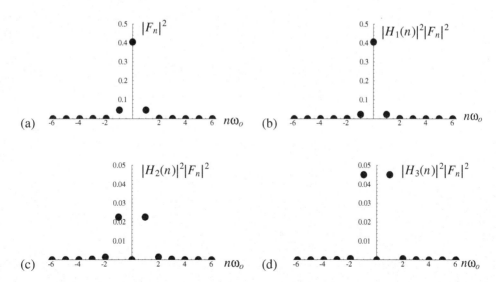

Figure 6.11 (a) $|F_n|^2$ vs $n\omega_o = n$ rad/s, (b) $|H_1(n)|^2|F_n|^2$, (c) $|H_2(n)|^2|F_n|^2$, and (d) $|H_3(n)|^2|F_n|^2$.

Finally, by comparing Figures 6.11c and 6.11d, we should expect that $y_3(t)$ will look more like a sine wave than does $y_2(t)$, because $|H_3(n)|^2|F_n|^2$ is almost entirely dominated by the $n = \pm 1$ harmonic (i.e, the fundamental). Indeed, Figure 6.12d shows that $y_3(t)$ is almost a pure co-sinusoid with frequency 1 rad/s.

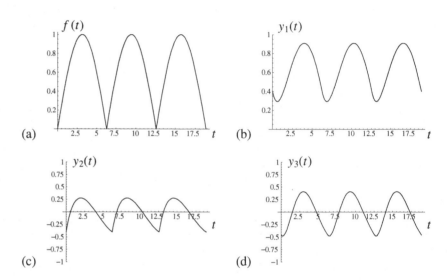

Figure 6.12   (a) $f(t)$, (b) $y_1(t)$, (c) $y_2(t)$, and (d) $y_3(t)$ versus time t (in seconds).

6.3.2   Average power, power spectrum, and Parseval's theorem

The plot of $|F_n|^2$ displayed in Figure 6.11a is known as the *power spectrum* of the periodic signal $f(t)$. The sum of $|F_n|^2$ over all n gives the *average power* of the same signal. The reasoning behind this terminology follows.

Suppose that $f(t)$ denotes a periodic voltage or current signal in a circuit. Then the application of $f(t)$ to a $1\,\Omega$ resistor would lead to an instantaneous power absorption of $f^2(t)\,\mathrm{W}$ from the signal. Therefore, the average power delivered by $f(t)$ to the resistor would be

$$P \equiv \frac{1}{T} \int_T |f(t)|^2 dt.$$

In this integral we have used $|f(t)|^2$, instead of $f^2(t)$, to denote the instantaneous power, because doing so allows us to define instantaneous and average power measures that are suitable for complex-valued $f(t)$. Of course, there is no distinction between $|f(t)|^2$ and $f^2(t)$ for real-valued $f(t)$.

Now, not all $f(t)$ are voltages or currents, nor are they always applied to $1\,\Omega$
Average signal power
resistors. Nevertheless, we still will refer to $|f(t)|^2$ as the *instantaneous signal power* and to P as the *average signal power*. Whatever the true nature of $f(t)$ may be, in practice the cost of generating $f(t)$ will be proportional to P.

The formula for P given above, shows how we can calculate the average signal power by integrating in the time domain, using the waveform $f(t)$ over a period T. Surprisingly, P also can be computed using the Fourier coefficients F_n. This follows by *Parseval's theorem*, which is property 8 in Table 6.3, stated as follows: **Parseval's theorem**

$$P \equiv \frac{1}{T} \int_T |f(t)|^2 dt = \sum_{n=-\infty}^{\infty} |F_n|^2.$$

Thus, the average value of $|f(t)|^2$ over one period *and* the sum of $|F_n|^2$ over all n give the same number,[11] the average signal power P. That is why we interpret $|F_n|^2$ as the power spectrum of $f(t)$—just as the rainbow reveals the spectrum of colors (frequencies) contained in sunlight, $|F_n|^2$ describes how the average power of $f(t)$ is distributed among its different harmonic components at different frequencies $n\omega_o$.   **Power spectrum $|F_n|^2$**

Now, for real-valued signals $f(t)$, we have

$$F_{-n} = F_n^*$$

and

$$c_n = 2|F_n| = 2|F_{-n}|,$$

implying that

$$|F_{-n}|^2 = |F_n|^2 = \frac{c_n^2}{4}, \quad \cdot$$

where c_n is the compact Fourier series coefficient. Therefore, for real $f(t)$,

$$\sum_{n=-\infty}^{\infty} |F_n|^2 = |F_0|^2 + \sum_{m=1}^{\infty} (|F_m|^2 + |F_{-m}|^2) = |F_0|^2 + \sum_{n=1}^{\infty} 2|F_n|^2 = \frac{c_0^2}{4} + \sum_{n=1}^{\infty} \frac{1}{2} c_n^2.$$

So, for real-valued $f(t)$, Parseval's theorem also can be written as   **Parseval's theorem for real valued $f(t)$**

$$P \equiv \frac{1}{T} \int_T |f(t)|^2 dt = \frac{c_0^2}{4} + \sum_{n=1}^{\infty} \frac{1}{2} c_n^2.$$

---

[11] **Proof of Parseval's theorem:** For a periodic $f(t) = \sum_{n=-\infty}^{\infty} F_n e^{jn\omega_o t}$ with period $T = \frac{2\pi}{\omega_o}$,

$$P = \frac{1}{T} \int_T |f(t)|^2 dt = \frac{1}{T} \int_T f(t) f^*(t) dt = \frac{1}{T} \int_T f(t) \left\{ \sum_{n=-\infty}^{\infty} F_n^* e^{-jn\omega_o t} \right\} dt$$

$$= \sum_{n=-\infty}^{\infty} F_n^* \frac{1}{T} \int_T f(t) e^{-jn\omega_o t} dt = \sum_{n=-\infty}^{\infty} F_n^* F_n = \sum_{n=-\infty}^{\infty} |F_n|^2.$$

This formula has a simple, intuitive interpretation. It states that the average power P of a real-valued periodic signal

$$f(t) = \frac{c_0}{2} + \sum_{n=1}^{\infty} c_n \cos(n\omega_o t + \theta_n)$$

is the sum of $\frac{c_0^2}{4}$, representing the DC power, and the terms $\frac{1}{2}c_n^2$, $n > 0$, where each succeeding term represents the average AC power of a harmonic component $c_n \cos(n\omega_o t + \theta_n)$. Recall that, for co-sinusoids, the average power into a 1 Ω resistor is simply one-half the amplitude squared.

Example 6.17

Consider the square-wave signal

$$f(t) = \sum_{n=1 \, (odd)}^{\infty} \frac{8}{n\pi} \cos(n\frac{\pi}{6}t - \frac{\pi}{2}),$$

with period $T = 12$ s plotted in Figure 6.7a. If $f(t)$ is the input of an LTI system with a frequency response

$$H(\omega) = \begin{cases} 1, & \text{for } |\omega| < 2 \text{ rad/s}, \\ 0, & \text{otherwise}, \end{cases}$$

determine the system output $y(t)$ and the average powers P_f and P_y of the input and output signals $f(t)$ and $y(t)$.

Solution   First, we note that the input $f(t)$ consists of co-sinusoids with harmonic frequencies

$$\frac{\pi}{6}, \frac{3\pi}{6}, \frac{5\pi}{6}, \frac{7\pi}{6}, \cdots \frac{\text{rad}}{\text{s}}.$$

Since $H(\omega) = 0$ for $\omega > 2 \frac{\text{rad}}{\text{s}}$, only the fundamental ($n = 1$) and 3rd harmonic ($n = 3$) of $f(t)$ will pass through the specified system. Hence,

$$y(t) = \frac{8}{\pi} \cos(\frac{\pi}{6}t - \frac{\pi}{2}) + \frac{8}{3\pi} \cos(3\frac{\pi}{6}t - \frac{\pi}{2}).$$

Using the second form of Parseval's theorem (applicable to real-valued signals) with $c_1 = \frac{8}{\pi}$ and $c_3 = \frac{8}{3\pi}$, we find the average power in the output $y(t)$ to be

$$P_y = \frac{1}{2}(\frac{8}{\pi})^2 + \frac{1}{2}(\frac{8}{3\pi})^2 \approx 3.602.$$

Calculation of P_f—the average power in the input $f(t)$—is easier in the time domain, using the graph of $f(t)$ shown in Figure 6.7a. We note that the period $T = 12$ s and $|f(t)|^2 = 4$ for $0 < t < 12$ s. Hence,

$$P_f = \frac{1}{T} \int_0^T |f(t)|^2 dt = \frac{1}{12} \int_0^{12} 4 \, dt = 4.$$

Comparison of P_f and P_y shows that only about 10% of the average power in $f(t)$ is contained in the fifth and higher harmonics.

6.3.3 Harmonic distortion

Suppose that

$$y(t) = Af(t) + Bf^2(t)$$

is the actual response of a system to some input $f(t)$, where A and B denote arbitrary constants. Assume that the first term in the expression, which is linear in $f(t)$, is the desired response. The second term, which is nonlinear in $f(t)$, is unintentional, perhaps due to a design error or imperfect electrical components. How might we measure the consequences of the undesired nonlinear term?

As we saw in Chapters 4 and 5, linear systems respond to co-sinusoidal inputs with co-sinusoidal outputs at the same frequency. However, the system defined above will respond to a pure cosine input

$$f(t) = \cos(\omega_o t),$$

with

$$y(t) = A\cos(\omega_o t) + B\cos^2(\omega_o t) = \frac{B}{2} + A\cos(\omega_o t) + \frac{B}{2}\cos(2\omega_o t),$$

since

$$\cos^2 \theta = \frac{1}{2}(1 + \cos(2\theta)).$$

Clearly, the output $y(t)$ is not just the desired pure cosine

$$A\cos(\omega_o t),$$

but also contains a DC term $\frac{B}{2}$ and a second-harmonic term

$$\frac{B}{2}\cos(2\omega_o t).$$

The average power of the output (by Parseval's theorem) is

$$P_y = \frac{B^2}{4} + \frac{A^2}{2} + \frac{B^2}{8},$$

Second-harmonic distortion

where the last two terms represent the average power of the fundamental and the second harmonic, respectively. One common measure of *second-harmonic distortion* in the system output is the ratio of the average power in second-harmonic to the average power in the fundamental:[12]

$$\frac{B^2}{4A^2}.$$

For instance, for $B = 0.1A$, this ratio is 0.25%.

In a more general scenario, a nonlinear system response to a pure cosine input $f(t) = \cos(\omega_o t)$ may have the Fourier series form

$$y(t) = \frac{c_0}{2} + \sum_{n=1}^{\infty} c_n \cos(n\omega_o t + \theta_n),$$

containing many, and possibly an infinite number of, higher-order harmonics. Every term in this response, except for the fundamental, would have been absent if the system were linear. Hence, a sensible and useful measure of the effects of the nonlinearity in this case is the ratio of the total power of the second and higher-order harmonics to the power of the fundamental. This ratio, known as *total harmonic distortion* (THD), can be expressed (by Parseval's theorem) as

Total-harmonic distortion: THD

$$\text{THD} = \frac{\sum_{n=2}^{\infty} \frac{1}{2} c_n^2}{\frac{1}{2} c_1^2} = \frac{\sum_{n=2}^{\infty} c_n^2}{c_1^2} = \frac{P_y - \frac{c_0^2}{4} - \frac{1}{2} c_1^2}{\frac{1}{2} c_1^2},$$

where P_y denotes the average signal power of the output $y(t)$.

Example 6.18

A power plant promises

$$y(t) = \cos(\omega_o t)$$

as a product to its customers, where $\frac{\omega_o}{2\pi} = 60$ Hz. The plant actually delivers the signal

$$y(t) = \cos(\omega_o t) + \frac{1}{9} \cos(3\omega_o t) + \frac{1}{25} \cos(5\omega_o t).$$

What is the THD?

Solution   Clearly,

$$\text{THD} = \frac{(\frac{1}{9})^2 + (\frac{1}{25})^2}{1^2} \approx 1.39\%.$$

---

[12]DC distortion is of less concern because the DC component can be removed by the use of a blocking capacitor or a simple high-pass filter.

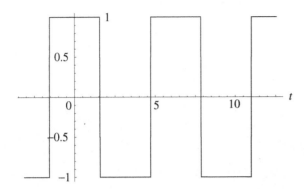

Figure 6.13   A square-wave signal with period $T = 2\pi$ s and a peak-to-peak amplitude of 2.

Example 6.19
The Fourier series of the zero-mean square-wave signal shown in Figure 6.13 is

$$y(t) = \frac{4}{\pi}[\cos(t) - \frac{1}{3}\cos(3t) + \frac{1}{5}\cos(5t) - \frac{1}{7}\cos(7t) + \cdots].$$

Assuming it is desired that $y(t)$ be proportional to $\cos(t)$, what is the THD for this signal?

Solution   Since the signal is zero-mean, $c_0^2 = 0$. Also,

$$\frac{1}{2}c_1^2 = \frac{8}{\pi^2}.$$

Moreover, we can obtain the average power P_y by averaging $y^2(t)$ over a period 2π as

$$P_y = \frac{1}{2\pi}\int_0^{2\pi} 1^2 dt = 1.$$

Therefore, substituting this last expression into the formula for THD, we obtain

$$\text{THD} = \frac{1 - \frac{8}{\pi^2}}{\frac{8}{\pi^2}} = \frac{\pi^2 - 8}{8} \approx 23.4\%.$$

EXERCISES

6.1   Plot the following periodic functions over at least two periods and specify their period T and fundamental frequency $\omega_o = \frac{2\pi}{T}$:

(a) $f(t) = 4 + \cos(3t)$.

(b) $g(t) = 8 + 4e^{-j4t} + 4e^{j4t}$.

(c) $h(t) = 2e^{-j2t} + 2e^{j2t} + 2\cos(4t)$.

6.2   Show that (from Example 6.5)

$$-\frac{1}{2\pi}\left(\frac{e^{j(1-2n)\pi} - 1}{1 - 2n} + \frac{e^{-j(1+2n)\pi} - 1}{1 + 2n}\right) = \frac{2}{\pi}\frac{1}{1 - 4n^2}$$

for all integers n and explain each step of the derivation carefully. You will be expected to perform similar simplifications when you make Fourier coefficient calculations.

6.3   Show that (from Example 6.13 in Section 6.2.4)

$$\frac{1}{2}\frac{2e^{-jn\pi} - 0}{-jn\frac{\pi}{2}} - \frac{1}{2}\left(\frac{e^{-jn\frac{\pi}{2}t}}{(-jn\frac{\pi}{2})^2}\right)\Big|_0^2 = \begin{cases} \frac{j2}{\pi n}, & \text{for even } n > 0, \\ -\frac{4 + j2\pi n}{\pi^2 n^2}, & \text{for odd } n. \end{cases}$$

Explain each step of the derivation carefully.

6.4   The function $f(t)$ is periodic with period $T = 4$ s. Between $t = 0$ and $t = 4$ s the function is described by

$$f(t) = \begin{cases} 1, & 0 < t < 2 \text{ s}, \\ 2, & 2 < t < 4 \text{ s}. \end{cases}$$

(a)   Plot $f(t)$ between $t = -4$ s and $t = 8$ s.

(b)   Determine the exponential Fourier coefficients F_n of $f(t)$ for $n = 0$, $n = 1$, and $n = 2$.

(c)   Using the results of part (b), determine the compact-form Fourier coefficients c_0, c_1, and c_2.

6.5   (a)   Calculate the exponential Fourier series of $f(t)$ plotted below.

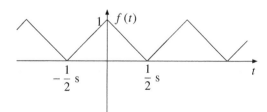

(b) Express the function $g(t)$, shown next, as a scaled and shifted version of the function $f(t)$ from part (a) and determine the Fourier series of $g(t)$ by using the scaling and time-shift properties of the Fourier series. Simplify your result as much as you can.

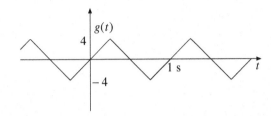

6.6 **(a)** Suppose a periodic signal $f(t)$ is differentiable and $g(t) \equiv \frac{df}{dt}$. Express G_n, the Fourier series coefficients of function $g(t)$, in terms of the Fourier coefficients F_n of $f(t)$.

(b) Determine the exponential Fourier series of the function $h(t)$ plotted below. Hint: If you already have solved Problem 6.5, then note that the derivative of signal $f(t)$ in that problem is related to $h(t)$.

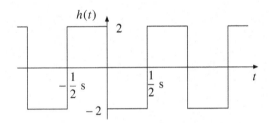

(c) Using the result of part (b), determine the exponential Fourier series of the following signal $s(t)$, assuming that $T = 1$ s:

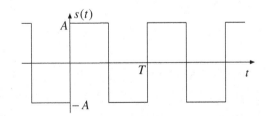

(d) Repeat part (c) for an arbitrary T.

6.7   **(a)** The function

$$z(t) = 6 + 3\cos(4t) + 2\sin(5t)$$

is periodic with period $T = 2\pi$ s and has fundamental frequency $\omega_o = 1$ rad/s. Determine the Fourier series coefficients Z_0, Z_1, Z_2, Z_3, Z_4, and Z_6. What is Z_{-4}?

(b) What is the period of the function

$$q(t) = 5\sin(4t) + 3\cos(6t)$$

and how can the function be expressed in exponential Fourier series form?

(c) Determine the average power P_z and P_q of signals $z(t)$ and $q(t)$, and plot their power spectra versus harmonic frequency $n\omega_o$ in each case.

6.8   For the periodic function $f(t)$ shown here, with exponential Fourier coefficients F_n, determine

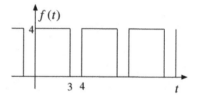

(a) The period T and fundamental frequency ω_o.

(b) The DC level $F_0 = \frac{c_0}{2}$.

(c) F_1, c_1 and θ_1.

(d) F_2, c_2 and θ_2.

(e) The average signal power P_f.

(f) THD.

6.9   The input signal of a linear system with frequency response $H(\omega)$ is

$$f(t) = \frac{1}{2} + \sum_{n=1}^{\infty} \frac{1}{n\pi} \cos(2\pi n t + \frac{\pi}{2}).$$

The input $f(t)$ and frequency response $H(\omega)$ are plotted as:

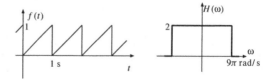

(a) What is the system output $y(t)$ in response to $f(t)$?

(b) Calculate the average power of the input and output signals $f(t)$ and $y(t)$. Hint: You can use either side of Parseval's theorem to calculate the average power of a periodic signal, but depending on the signal, one side often is easier to evaluate than the other.

6.10 Show that the compact trigonometric Fourier series of $f(t)$ shown in Problem 6.9 is

$$f(t) = \frac{1}{2} + \sum_{n=1}^{\infty} \frac{1}{n\pi} \cos(2\pi nt + \frac{\pi}{2}).$$

6.11 Confirm the Fourier series

$$y(t) = \frac{4}{\pi}[\cos(t) - \frac{1}{3}\cos(3t) + \frac{1}{5}\cos(5t) - \frac{1}{7}\cos(7t) + \cdots]$$

for the square wave described and discussed in Example 6.19 of Section 6.3.3.

6.12 The input–output relation for a system with input $f(t)$ is given as

$$y(t) = 6f(t) + f^2(t) - f^3(t).$$

Determine the total harmonic distortion (THD) of the system response to a pure cosine input. Hint:

$$\cos^3 \theta = \frac{1}{4}(3\cos\theta + \cos(3\theta)).$$

Is the system linear or nonlinear? Explain.

6.13 Let $f(t)$ be a real-valued periodic signal with fundamental frequency ω_o. We will approximate $f(t)$ with another function

$$f_N(t) \equiv \frac{\hat{a}_o}{2} + \sum_{n=1}^{N} (\hat{a}_n \cos(n\omega_o t) + \hat{b}_n \sin(n\omega_o t))$$

where (real-valued) coefficients $\hat{a}_m$ and $\hat{b}_m$ are selected to minimize the average power P_e in the *error signal*

$$e(t) \equiv f_N(t) - f(t).$$

That is, to determine the optimal $\hat{a}_m$ and $\hat{b}_m$ we minimize

$$P_e \equiv \frac{1}{T} \int_T e^2(t)\, dt.$$

We will do so by setting $\frac{\partial P_e}{\partial \hat{a}_m}$ and $\frac{\partial P_e}{\partial \hat{b}_m}$ to zero and solving for $\hat{a}_m$ and $\hat{b}_m$.

(a) Confirm that

$$\frac{\partial P_e}{\partial \hat{a}_o} = \frac{1}{T} \int_T e(t) \, dt = \frac{\hat{a}_o}{2} - \frac{1}{T} \int_T f(t) \, dt$$

and, for $1 \le m \le N$,

$$\frac{\partial P_e}{\partial \hat{a}_m} = \frac{2}{T} \int_T e(t) \cos(m\omega_o t) \, dt = \hat{a}_m - \frac{2}{T} \int_T f(t) \cos(m\omega_o t) \, dt$$

and

$$\frac{\partial P_e}{\partial \hat{b}_m} = \frac{2}{T} \int_T e(t) \sin(m\omega_o t) \, dt = \hat{b}_m - \frac{2}{T} \int_T f(t) \sin(m\omega_o t) \, dt.$$

(b) Solve for the optimizing $\hat{a}_m$ and $\hat{b}_m$, and show that the optimal $f_N(t)$ can be rewritten as

$$f_N(t) = \sum_{m=-N}^{N} F_n e^{jn\omega_o t}$$

with

$$F_n = \frac{1}{T} \int_T f(t) e^{-jn\omega_o t} \, dt.$$

(c) Assuming that $f(t)$ satisfies the Dirichlet Conditions, what is the value of P_e in the limit as $N \to \infty$? Explain.

7

Fourier Transform and LTI System Response to Energy Signals

In this chapter we will talk about the Beatles and Beethoven. (See Figure 7.11.) But before that, even before beginning Section 7.1, it is worthwhile to take stock of where we are and where we are going.

We earlier developed the phasor technique for finding the response of dissipative linear time-invariant circuits to co-sinusoidal inputs. We then proceeded to add the concept of Fourier series to our toolbox for the case with periodic inputs. We saw that periodic signals are superpositions of harmonically related co-sinusoids (the frequencies are integer multiples of some fundamental frequency $\omega_0 = 2\pi/T$, where T is the period of the signal). In practice, most of the signals we encounter in the real world are not periodic. We shall call such signals *aperiodic*. For aperiodic signals, we might ask whether there is some form of frequency representation. That is, can we write an aperiodic signal $f(t)$ as a sum of co-sinusoids? The answer is yes, but in this case the frequencies are not harmonically related. In fact, most aperiodic signals consist of a continuum of frequencies, not just a set of discrete frequencies. The derivation that follows leads us to the frequency representation for aperiodic signals.

Frequency-domain description of aperiodic finite-energy signals: Fourier and inverse-Fourier transforms; energy and energy spectrum; low-pass, band-pass, and high-pass signals; bandwidth; linear system response to energy signals

Begin by considering any periodic function $f(t)$ with period $T = \frac{2\pi}{\omega_o}$ and Fourier series

$$f(t) = \sum_{n=-\infty}^{\infty} F_n e^{jn\omega_o t},$$

where

$$F_n = \frac{1}{T} \int_{-T/2}^{T/2} f(t)e^{-jn\omega_o t}\,dt = \frac{\omega_o}{2\pi} \int_{-\pi/\omega_o}^{\pi/\omega_o} f(t)e^{-jn\omega_o t}\,dt.$$

This also can be expressed as

$$f(t) = \frac{1}{2\pi} \sum_{n=-\infty}^{\infty} F(n\omega_o)e^{jn\omega_o t}\omega_o,$$

with

$$F(\omega) \equiv \int_{-\pi/\omega_o}^{\pi/\omega_o} f(t)e^{-j\omega t}\,dt.$$

This expression for $f(t)$ is valid no matter how large or small the fundamental frequency $\omega_o = \frac{2\pi}{T}$ is, so long as the Dirichlet conditions are satisfied by $f(t)$. In the case of a vanishingly small ω_o, however, the period $T = \frac{2\pi}{\omega_o}$ becomes infinite and $f(t)$ loses its periodic character. In that limit—as $\omega_o \to 0$ and $f(t)$ becomes *aperiodic*—the series $f(t)$ and function $F(\omega)$ converge to[1]

$$f(t) = \frac{1}{2\pi} \int_{-\infty}^{\infty} F(\omega)e^{j\omega t}\,d\omega$$

and

$$F(\omega) = \int_{-\infty}^{\infty} f(t)e^{-j\omega t}\,dt,$$

respectively.

The function $F(\omega)$, just introduced, is known as the *Fourier transform* of the function $f(t)$. Similarly, the formula for computing $f(t)$ from $F(\omega)$ is called the *inverse Fourier transform* of $F(\omega)$. Notice that the inverse Fourier transform expresses $f(t)$ as a continuous sum (integral) of co-sinusoids (complex exponentials), where $F(\omega)$ is the weighting applied to frequency ω. The foregoing formulas for the inverse

---

[1]Remember from calculus that, given a function, say, $z(x)$, $\int_{-\infty}^{\infty} z(x)dx \equiv \lim_{\Delta x \to 0} \sum_{n=-\infty}^{\infty} z(n\Delta x)\Delta x$; the definite integral of $z(x)$ is nothing but an infinite sum of infinitely many infinitesimals $z(n\Delta x)\Delta x$, amounting to the area under the curve $z(x)$.

Fourier transform and the (forward) Fourier transform will be applied so frequently throughout the remainder of this text that they should be committed to memory.

We will come to understand the implications of the Fourier transform as we proceed in this chapter. None is more important than the following observation: We have just seen that any aperiodic signal with a Fourier transform $F(\omega)$ is a weighted linear superposition of exponentials

$$e^{j\omega t} = \cos(\omega t) + j\sin(\omega t)$$

of *all* frequencies ω. So, since

$$e^{j\omega t} \longrightarrow \boxed{\text{LTI}} \longrightarrow H(\omega)e^{j\omega t},$$

and since LTI systems allow superposition, it follows that

$$\frac{1}{2\pi}\int_{-\infty}^{\infty} F(\omega)e^{j\omega t}\,d\omega \longrightarrow \boxed{\text{LTI}} \longrightarrow \frac{1}{2\pi}\int_{-\infty}^{\infty} H(\omega)F(\omega)e^{j\omega t}\,d\omega.$$

This indicates that for an LTI system with frequency response $H(\omega)$, and input $f(t)$ with Fourier transform $F(\omega)$, the corresponding system output $y(t)$ is simply the inverse Fourier transform of the product $H(\omega)F(\omega)$. This relationship is illustrated in Figure 7.1.

This chapter explores and expands upon the concepts just introduced, namely the inverse Fourier transform representation of aperiodic signals $f(t)$ and the input–output relation for LTI systems shown in Figure 7.1. In Section 7.1 we will discuss the existence conditions and general properties of the Fourier transform $F(\omega)$ and also compile a table of Fourier transform pairs "$f(t) \leftrightarrow F(\omega)$" for signals $f(t)$ commonly encountered in signal processing applications. In Section 7.2 we will examine the concept of signal energy W and energy spectrum $|F(\omega)|^2$ and discuss a signal classification based on energy spectrum types. Finally, Section 7.3 will address applications of the input–output rule stated in Figure 7.1.

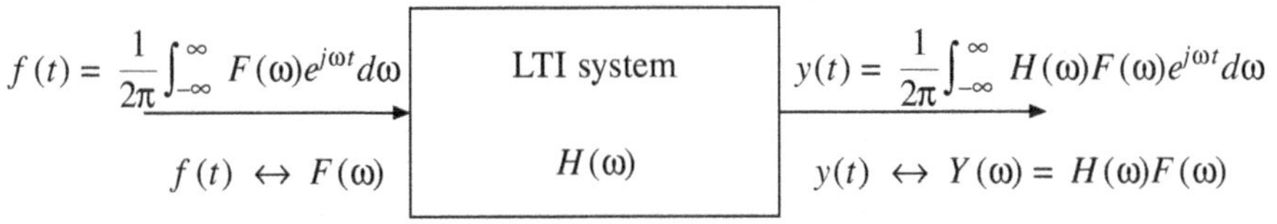

$$f(t) = \frac{1}{2\pi}\int_{-\infty}^{\infty} F(\omega)e^{j\omega t}\,d\omega \qquad \boxed{\begin{array}{c} \text{LTI system} \\[4pt] H(\omega) \end{array}} \qquad y(t) = \frac{1}{2\pi}\int_{-\infty}^{\infty} H(\omega)F(\omega)e^{j\omega t}\,d\omega$$

$$f(t) \leftrightarrow F(\omega) \qquad\qquad\qquad\qquad y(t) \leftrightarrow Y(\omega) = H(\omega)F(\omega)$$

Figure 7.1 Input–output relation for dissipative LTI system $H(\omega)$, with aperiodic input $f(t)$. System $H(\omega)$ converts the Fourier transform $F(\omega)$ of input $f(t)$ to the Fourier transform $Y(\omega)$ of the output $y(t)$, according to the rule $Y(\omega) = H(\omega)F(\omega)$. See Section 7.1 for a description of the "$\leftrightarrow$" notation.

7.1   Fourier Transform Pairs $f(t) \leftrightarrow F(\omega)$ and Their Properties[2]

Signals $f(t)$ and their Fourier transforms $F(\omega)$ satisfying the relations

$$f(t) = \frac{1}{2\pi} \int_{-\infty}^{\infty} F(\omega)e^{j\omega t}\,d\omega \quad \text{and} \quad F(\omega) = \int_{-\infty}^{\infty} f(t)e^{-j\omega t}\,dt$$

Fourier transform pairs

are said to be *Fourier transform pairs*. To indicate a Fourier transform pair, we will use the notation

$$f(t) \quad \leftrightarrow \quad F(\omega).$$

For instance,

$$e^{-|t|} \quad \leftrightarrow \quad \frac{2}{1+\omega^2},$$

as demonstrated later in this section. This pairing is unique, because there exists no time signal $f(t)$ other than $e^{-|t|}$, having the Fourier transform $F(\omega) = \frac{2}{1+\omega^2}$.

Existence of Fourier transform paris

For the Fourier transform of $f(t)$ to exist, it is *sufficient* that $f(t)$ be absolutely integrable,[3] that is, satisfy

$$\int_{-\infty}^{\infty} |f(t)|\,dt < \infty,$$

and for the convergence[4] of the inverse Fourier transform of $F(\omega)$ to $f(t)$ it is sufficient that $f(t)$ satisfy the remaining Dirichlet conditions over any finite interval. However, absolute integrability—which is satisfied by all signals that can be generated in the lab[5]—is not a necessary condition for a Fourier pairing $f(t) \leftrightarrow F(\omega)$ to be true.

---

[2] The term "transform" generally is used in mathematics to describe a reversible conversion. If $F(\omega)$ is a transform of $f(t)$, then there exists an inverse process that converts $F(\omega)$ uniquely back into $f(t)$.

[3] **Proof:** Note that

$$|F(\omega)| = |\int_{-\infty}^{\infty} f(t)e^{-j\omega t}\,dt| \le \int_{-\infty}^{\infty} |f(t)e^{-j\omega t}|\,dt = \int_{-\infty}^{\infty} |f(t)|\,dt.$$

Thus

$$\int_{-\infty}^{\infty} |f(t)|\,dt < \infty$$

is sufficient to ensure that $|F(\omega)| < \infty$, and a bounded $F(\omega)$ exists.

[4] At discontinuity points of $f(t)$, the inverse Fourier transform will converge to the midpoints between the bottoms and tops of the jumps, as in the case of the Fourier series.

[5] In the lab or in a radio station, we can generate only finite-duration signals having finite values. Thus, for a lab signal $f(t)$, the *area under* $|f(t)|$ is finite and the signal is absolutely integrable.

Some signals $f(t)$ that are not absolutely integrable still satisfy the relations

$$f(t) = \frac{1}{2\pi} \int_{-\infty}^{\infty} F(\omega)e^{j\omega t}\, d\omega \quad \text{and} \quad F(\omega) = \int_{-\infty}^{\infty} f(t)e^{-j\omega t}\, dt$$

for a given $F(\omega)$, for instance, when $F(\omega)$ satisfies the Dirichlet conditions. In that case, $f(t)$ and $F(\omega)$ form a Fourier transform pair $f(t) \leftrightarrow F(\omega)$ just the same. Therefore, the input–output relation for LTI systems shown in Figure 7.1 has a very wide range of applicability (which we will explore in Sections 7.2 and 7.3).

Table 7.1 lists some of the general properties of Fourier transform pairs. Many important Fourier transform pairs are listed in Table 7.2. In this section we will prove some of the properties listed in Table 7.1 and verify some of the transform pairs shown in Table 7.2. Detailed discussions of some of the entries in these tables will be delayed until their specific uses are needed. You can ignore the right column of Table 7.2 until it is needed in Chapter 9. Some of the entries in the left column of Table 7.2 contain unfamiliar functions such as $u(t)$, rect(t), sinc(t), and $\triangle(t)$. These functions are defined next.

Unit-step $u(t)$:   The unit-step function $u(t)$ is defined[6] as

$$u(t) \equiv \begin{cases} 1, & \text{for } t > 0, \\ 0, & \text{for } t < 0, \end{cases}$$

and is plotted in Figure 7.2a. Figures 7.2b, 7.2c, and 7.2d show $u(t-2)$ (a shifted unit step), $u(1-t)$ (a reversed and shifted unit step), and the *signum*, or *sign*, function

$$\text{sgn}(t) \equiv 2u(t) - 1.$$

The unit step is not absolutely integrable, but it still has a Fourier transform that can be defined, (which will be identified in Chapter 9).

Unit rectangle rect(t):   The unit rectangle rect(t) is defined as

$$\text{rect}(t) \equiv \begin{cases} 1, & \text{for } |t| < \frac{1}{2}, \\ 0, & \text{for } |t| > \frac{1}{2}, \end{cases}$$

and is plotted in Figure 7.3a. Figures 7.3b, 7.3c and 7.3d show rect$(t - \frac{1}{2})$ (a delayed rect), rect$(\frac{t}{3})$ (a rect stretched by a factor of 3), and rect$(\frac{t+1}{2})$ (a rect stretched by a factor of 2 and shifted by 1 unit to the left). The unit rectangle is absolutely integrable, and we will derive its Fourier transform later in this section.

---

[6]In general, two analog signals, say, $f(t)$ and $g(t)$, can be regarded equal so long as their values differ by a finite amount at no more than a finite number (or perhaps a countably infinite number) of discrete points in time t. In keeping with this definition of equivalence—applicable to Fourier series as well as transforms—there is no need to specify the value of the unit-step $u(t)$ at the single point $t = 0$. This value can be any finite number (e.g., 0.5), and its choice will not affect any of our results. The situation is similar for the points of discontinuity of the unit rectangle rect(t) at $t = \pm 0.5$.

	Name:	Condition:	Property:
1	Amplitude scaling	$f(t) \leftrightarrow F(\omega)$, constant K	$Kf(t) \leftrightarrow KF(\omega)$
2	Addition	$f(t) \leftrightarrow F(\omega), g(t) \leftrightarrow G(\omega), \cdots$	$f(t) + g(t) + \cdots \leftrightarrow$ $F(\omega) + G(\omega) + \cdots$
3	Hermitian	Real $f(t) \leftrightarrow F(\omega)$	$F(-\omega) = F^*(\omega)$
4	Even	Real and even $f(t)$	Real and even $F(\omega)$
5	Odd	Real and odd $f(t)$	Imaginary and odd $F(\omega)$
6	Symmetry	$f(t) \leftrightarrow F(\omega)$	$F(t) \leftrightarrow 2\pi f(-\omega)$
7	Time scaling	$f(t) \leftrightarrow F(\omega)$, real c	$f(ct) \leftrightarrow \frac{1}{|c|}F(\frac{\omega}{c})$
8	Time shift	$f(t) \leftrightarrow F(\omega)$	$f(t - t_o) \leftrightarrow F(\omega)e^{-j\omega t_o}$
9	Frequency shift	$f(t) \leftrightarrow F(\omega)$	$f(t)e^{j\omega_o t} \leftrightarrow F(\omega - \omega_o)$
10	Modulation	$f(t) \leftrightarrow F(\omega)$	$f(t)\cos(\omega_o t) \leftrightarrow$ $\frac{1}{2}F(\omega - \omega_o) + \frac{1}{2}F(\omega + \omega_o)$
11	Time derivative	Differentiable $f(t) \leftrightarrow F(\omega)$	$\frac{df}{dt} \leftrightarrow j\omega F(\omega)$
12	Freq derivative	$f(t) \leftrightarrow F(\omega)$	$-jtf(t) \leftrightarrow \frac{d}{d\omega}F(\omega)$
13	Time convolution	$f(t) \leftrightarrow F(\omega), g(t) \leftrightarrow G(\omega)$	$f(t) * g(t) \leftrightarrow F(\omega)G(\omega)$
14	Freq convolution	$f(t) \leftrightarrow F(\omega), g(t) \leftrightarrow G(\omega)$	$f(t)g(t) \leftrightarrow \frac{1}{2\pi}F(\omega) * G(\omega)$
15	Compact form	Real $f(t)$	$f(t) =$ $\frac{1}{2\pi}\int_0^\infty 2|F(\omega)|\cos(\omega t +$ $\angle F(\omega))d\omega$
16	Parseval, Energy W	$f(t) \leftrightarrow F(\omega)$	$W \equiv \int_{-\infty}^{\infty}|f(t)|^2 dt =$ $\frac{1}{2\pi}\int_{-\infty}^{\infty}|F(\omega)|^2 d\omega$

Table 7.1 Important properties of the Fourier transform.

	$f(t) \leftrightarrow F(\omega)$		
1	$e^{-at}u(t) \leftrightarrow \frac{1}{a+j\omega}, \; a > 0$	14	$\delta(t) \leftrightarrow 1$
2	$e^{at}u(-t) \leftrightarrow \frac{1}{a-j\omega}, \; a > 0$	15	$1 \leftrightarrow 2\pi\delta(\omega)$
3	$e^{-a\|t\|} \leftrightarrow \frac{2a}{a^2+\omega^2}, \; a > 0$	16	$\delta(t - t_o) \leftrightarrow e^{-j\omega t_o}$
4	$\frac{a^2}{a^2+t^2} \leftrightarrow \pi a e^{-a\|\omega\|}, \; a > 0$	17	$e^{j\omega_o t} \leftrightarrow 2\pi\delta(\omega - \omega_o)$
5	$te^{-at}u(t) \leftrightarrow \frac{1}{(a+j\omega)^2}, \; a > 0$	18	$\cos(\omega_o t) \leftrightarrow \pi[\delta(\omega - \omega_o) + \delta(\omega + \omega_o)]$
6	$t^n e^{-at}u(t) \leftrightarrow \frac{n!}{(a+j\omega)^{n+1}}, \; a > 0$	19	$\sin(\omega_o t) \leftrightarrow j\pi[\delta(\omega + \omega_o) - \delta(\omega - \omega_o)]$
7	$\text{rect}(\frac{t}{\tau}) \leftrightarrow \tau\,\text{sinc}(\frac{\omega\tau}{2})$	20	$\cos(\omega_o t)u(t) \leftrightarrow$ $\frac{\pi}{2}[\delta(\omega - \omega_o) + \delta(\omega + \omega_o)] + \frac{j\omega}{\omega_o^2-\omega^2}$
8	$\text{sinc}(Wt) \leftrightarrow \frac{\pi}{W}\text{rect}(\frac{\omega}{2W})$	21	$\sin(\omega_o t)u(t) \leftrightarrow$ $j\frac{\pi}{2}[\delta(\omega + \omega_o) - \delta(\omega - \omega_o)] + \frac{\omega_o}{\omega_o^2-\omega^2}$
9	$\triangle(\frac{t}{\tau}) \leftrightarrow \frac{\tau}{2}\text{sinc}^2(\frac{\omega\tau}{4})$	22	$\text{sgn}(t) \leftrightarrow \frac{2}{j\omega}$
10	$\text{sinc}^2(\frac{Wt}{2}) \leftrightarrow \frac{2\pi}{W}\triangle(\frac{\omega}{2W})$	23	$u(t) \leftrightarrow \pi\delta(\omega) + \frac{1}{j\omega}$
11	$e^{-at}\sin(\omega_o t)u(t) \leftrightarrow$ $\frac{\omega_o}{(a+j\omega)^2+\omega_o^2}, \; a > 0$	24	$\sum_{n=-\infty}^{\infty}\delta(t - nT) \leftrightarrow \frac{2\pi}{T}\sum_{n=-\infty}^{\infty}\delta(\omega - n\frac{2\pi}{T})$
12	$e^{-at}\cos(\omega_o t)u(t) \leftrightarrow$ $\frac{a+j\omega}{(a+j\omega)^2+\omega_o^2}, \; a > 0$	25	$\sum_{n=-\infty}^{\infty}f(t)\delta(t - nT) \leftrightarrow$ $\sum_{n=-\infty}^{\infty}\frac{1}{T}F(\omega - n\frac{2\pi}{T})$
13	$e^{-\frac{t^2}{2\sigma^2}} \leftrightarrow \sigma\sqrt{2\pi}\,e^{-\frac{\sigma^2\omega^2}{2}}$		

Table 7.2 Important Fourier transform pairs. The left-hand column includes only *energy signals* $f(t)$ (see Section 7.2), while the right-hand column includes *power signals* and distributions (covered in Chapter 9).

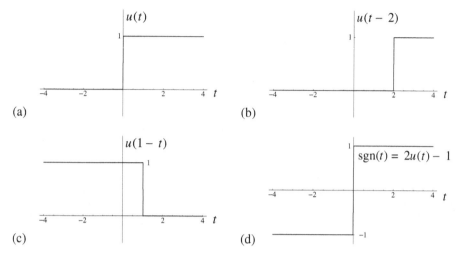

Figure 7.2 (a) The unit step $u(t)$, (b) $u(t - 2)$, (c) $u(1 - t)$, and (d) $\text{sgn}(t) = 2u(t) - 1$.

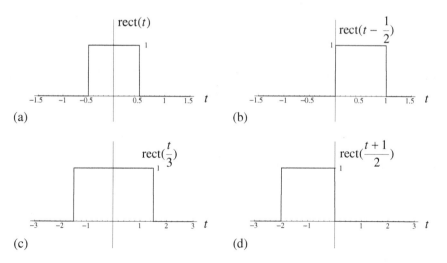

Figure 7.3 (a) The unit rectangle $\text{rect}(t)$, (b) $\text{rect}(t - \frac{1}{2})$, (c) $\text{rect}(\frac{t}{3})$, and (d) $\text{rect}(\frac{t+1}{2})$.

Sinc function sinc(t): The sinc function sinc(t) is defined as

$$\text{sinc}(t) \equiv \frac{\sin(t)}{t}$$

and is plotted in Figure 7.4. Notice that sinc(t) is zero (crosses the horizontal axis) for $t = k\pi$, where k is any integer but 0. The use of l'Hopital's rule[7] indicates that sinc(0) = 1. It can be shown that the sinc function is not absolutely integrable; however, its Fourier transform exists and it will be determined later in this section.

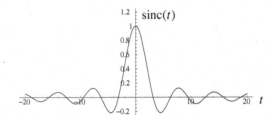

Figure 7.4 sinc(t) versus t. Notice that sinc(t) = 0 at $t = k\pi$ s, where k is any integer but 0.

Unit triangle $\triangle(t)$: The unit triangle function $\triangle(t)$ is defined as

$$\triangle(t) \equiv \begin{cases} 1 - 2|t|, & \text{for}|t| \leq \frac{1}{2}, \\ 0, & \text{otherwise}, \end{cases}$$

and is plotted in Figure 7.5a. A shifted unit-triangle function is shown in Figure 7.5b.

Figure 7.5 (a)$\triangle(t)$, and (b) a shifted unit triangle $\triangle(t - \frac{1}{2})$.

---

[7] $\lim\limits_{t \to 0} \text{sinc}(t) = \dfrac{\frac{d}{dt}\sin(t)_{|t=0}}{\frac{d}{dt}t_{|t=0}} = \dfrac{\cos(0)}{1} = 1.$

7.1.1 Fourier transform examples

Example 7.1

Determine the Fourier transform of

$$f(t) = e^{-at}u(t),$$

where $a > 0$ is a constant. The function is plotted in Figure 7.6a for the case $a = 1$. Notice that the $u(t)$ multiplier of e^{-at} makes $f(t)$ zero for $t < 0$.

Solution   Substituting $e^{-at}u(t)$ for $f(t)$ in the Fourier transform formula gives

$$F(\omega) = \int_{-\infty}^{\infty} e^{-at}u(t)e^{-j\omega t}dt = \int_{0}^{\infty} e^{-(a+j\omega)t}dt = \frac{e^{-(a+j\omega)t}\big|_{0}^{\infty}}{-(a+j\omega)}$$

$$= \frac{0-1}{-(a+j\omega)} = \frac{1}{a+j\omega}.$$

Therefore,

$$e^{-at}u(t) \;\leftrightarrow\; \frac{1}{a+j\omega},$$

which is the first entry in Table 7.2. The magnitude and angle of

$$F(\omega) = \frac{1}{a+j\omega}$$

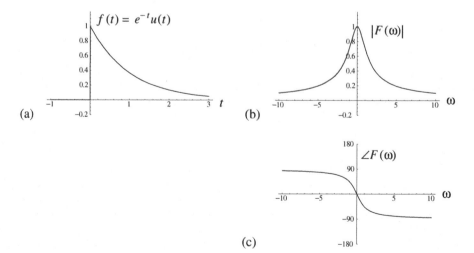

Figure 7.6   (a) $f(t) = e^{-t}u(t)$, (b) $|F(\omega)|$ versus ω, and (c) angle of $F(\omega)$ in degrees versus ω.

are plotted in Figures 7.6b and 7.6c for the case $a = 1$. From the plot of $|F(\omega)|$, we see that $f(t)$ contains all frequencies ω, but that it has more low-frequency content than it does high-frequency content. Notice that for $a < 0$, the preceding integral is not convergent, because in that case $e^{-at} \to \infty$ as $t \to \infty$. Therefore, for $a < 0$, the function $e^{-at}u(t)$ does not have a Fourier transform (and it is not absolutely integrable). For the special case of $a = 0$, $e^{-at}u(t) = u(t)$ is not absolutely integrable, yet it still is possible to define a Fourier transform, as already mentioned. (See Chapter 9.)

Example 7.2

Determine the Fourier transform $G(\omega)$ of the function $g(t) = e^{at}u(-t)$, where $a > 0$ is a constant. The function is plotted in Figure 7.7 for the case $a = 1$.

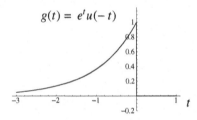

Figure 7.7 $g(t) = e^t u(-t)$.

Solution Notice that $g(t) = f(-t)$, where $f(t) = e^{-at}u(t) \leftrightarrow \frac{1}{a+j\omega}$, $a > 0$. Therefore, we can use the time-scaling property from Table 7.1, with $c = -1$. Thus,

$$G(\omega) = \frac{1}{|-1|}F(\frac{\omega}{-1}) = \frac{1}{a - j\omega}.$$

Hence, we have a new Fourier transform pair,

$$e^{at}u(-t) \leftrightarrow \frac{1}{a - j\omega},$$

which is entry 2 in Table 7.2.

Example 7.3

Determine the Fourier transform $P(\omega)$ of the function $p(t) = e^{-a|t|}$, where $a > 0$ is a constant. The function is plotted in Figure 7.8a for the case $a = 1$.

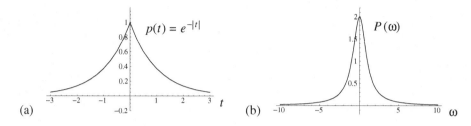

Figure 7.8 (a) $p(t) = e^{-|t|}$, and (b) $P(\omega)$ versus ω.

Solution   Comparing Figures 7.6a, 7.7, and 7.8a, we note that $p(t) = f(t) + g(t)$. Therefore, using the addition property from Table 7.1, we have

$$P(\omega) = F(\omega) + G(\omega) = \frac{1}{a + j\omega} + \frac{1}{a - j\omega}$$

$$= 2\text{Re}[\frac{1}{a + j\omega}] = \frac{2a}{a^2 + \omega^2},$$

which is plotted in Figure 7.8b for the case $a = 1$. Hence, for $a > 0$,

$$e^{-a|t|} \leftrightarrow \frac{2a}{a^2 + \omega^2},$$

which is entry 3 in Table 7.2. This is a case where the Fourier transform turns out to be real, because the time-domain function is real and even (property 4 in Table 7.1).

Example 7.4
Using the symmetry property from Table 7.1 and the result of Example 7.3, determine the Fourier transform $Q(\omega)$ of function

$$q(t) = \frac{a^2}{a^2 + t^2}.$$

Solution   The symmetry property 6 from Table 7.1 states that the Fourier transform of a Fourier transform gives back the original waveform, except reversed and scaled by 2π. Applying this to the result

$$e^{-a|t|} \leftrightarrow \frac{2a}{a^2 + \omega^2}$$

for $a > 0$ of Example 7.3, we can write

$$\frac{2a}{a^2 + t^2} \leftrightarrow 2\pi e^{-a|-\omega|} = 2\pi e^{-a|\omega|}.$$

Therefore,

$$\frac{a^2}{a^2 + t^2} \leftrightarrow \pi a e^{-a|\omega|}$$

for $a > 0$, which is entry 4 in Table 7.2. Hence,

$$Q(\omega) = \pi a e^{-a|\omega|}$$

for $a > 0$. If the sign of a is changed, then $q(t)$ is not altered and no changes can result to the values of $Q(\omega)$. To enforce that, we write

$$Q(\omega) = \pi |a| e^{-|a\omega|},$$

which is valid for all a.

The Fourier transform properties listed in Table 7.1 can be derived from the Fourier transform and inverse Fourier transform definitions, as illustrated in two cases in the next two examples. Notice that all of the Fourier transforms calculated in Examples 7.1 through 7.4 are consistent with property 3,

$$F(-\omega) = F^*(\omega),$$

which is valid for all real-valued $f(t)$. After reading Examples 7.5 and 7.6, try to derive property 3 on your own.

Example 7.5
Confirm the symmetry property listed in Table 7.1.

Solution To confirm this property, it is sufficient to show that, given the inverse Fourier transform integral

$$f(t) = \frac{1}{2\pi} \int_{-\infty}^{\infty} F(\omega) e^{j\omega t} d\omega,$$

it follows that the Fourier transform of function $F(t)$ is

$$\int_{-\infty}^{\infty} F(t) e^{-j\omega t} dt = 2\pi f(-\omega).$$

The first integral given can be rewritten (after ω is changed to the dummy variable x) as

$$f(t) = \frac{1}{2\pi} \int_{-\infty}^{\infty} F(x) e^{jxt} dx.$$

Replacing t with $-\omega$ on both sides and multiplying the result by 2π, we obtain

$$2\pi f(-\omega) = 2\pi \frac{1}{2\pi} \int_{-\infty}^{\infty} F(x)e^{jx(-\omega)}dx = \int_{-\infty}^{\infty} F(x)e^{-j\omega x}dx$$

$$= \int_{-\infty}^{\infty} F(t)e^{-j\omega t}dt.$$

Hence,

$$\int_{-\infty}^{\infty} F(t)e^{-j\omega t}dt = 2\pi f(-\omega),$$

as claimed.

Example 7.6

Confirm the time-derivative property listed in Table 7.1.

Solution   Given the inverse transform

$$f(t) = \frac{1}{2\pi} \int_{-\infty}^{\infty} F(\omega)e^{j\omega t}d\omega,$$

it follows that

$$\frac{df}{dt} = \frac{d}{dt}\frac{1}{2\pi} \int_{-\infty}^{\infty} F(\omega)e^{j\omega t}d\omega = \frac{1}{2\pi} \int_{-\infty}^{\infty} \{j\omega F(\omega)\}e^{j\omega t}d\omega.$$

This expression indicates that $\frac{df}{dt}$ is the inverse Fourier transform of $j\omega F(\omega)$, as stated by the time-derivative property.

Example 7.7

Using the frequency-derivative property from Table 7.1 and entry 1 from Table 7.2, confirm entry 5 in Table 7.2.

Solution   Entry 1 in Table 7.2 is

$$e^{-at}u(t) \leftrightarrow \frac{1}{a + j\omega}, \quad a > 0.$$

The frequency derivative of the right side is

$$\frac{d}{d\omega}\frac{1}{a + j\omega} = \frac{d}{d\omega}(a + j\omega)^{-1} = -(a + j\omega)^{-2}j = \frac{-j}{(a + j\omega)^2},$$

which, according to the frequency-derivative property, is the Fourier transform of

$$-jte^{-at}u(t).$$

Thus,

$$te^{-at}u(t) \leftrightarrow \frac{1}{(a+j\omega)^2}, \quad a > 0,$$

which is entry 5 in Table 7.2.

We next will determine the Fourier transforms of $\text{rect}(\frac{t}{\tau})$ and $\text{sinc}(Wt)$. The results are of fundamental importance in signal processing, as we will see later on.

Fourier transforms of rect and sinc

Example 7.8
Determine the Fourier transform of $f(t) = \text{rect}(\frac{t}{\tau})$.

Solution   Since $\text{rect}(\frac{t}{\tau})$ equals 1 for $-\frac{\tau}{2} < t < \frac{\tau}{2}$ and 0 for $|t| > \frac{\tau}{2}$, it follows that

$$F(\omega) = \int_{-\infty}^{\infty} \text{rect}(\frac{t}{\tau})e^{-j\omega t}dt = \int_{-\tau/2}^{\tau/2} e^{-j\omega t}dt$$

$$= \frac{e^{-j\omega\frac{\tau}{2}} - e^{j\omega\frac{\tau}{2}}}{-j\omega} = \frac{\sin(\frac{\omega\tau}{2})}{\frac{\omega}{2}}.$$

This result usually is written as

$$F(\omega) = \frac{\sin(\frac{\omega\tau}{2})}{\frac{\omega}{2}} = \tau \frac{\sin(\frac{\omega\tau}{2})}{\frac{\omega\tau}{2}} = \tau\,\text{sinc}(\frac{\omega\tau}{2}).$$

Hence,

$$\text{rect}(\frac{t}{\tau}) \leftrightarrow \tau\,\text{sinc}(\frac{\omega\tau}{2}),$$

which is entry 7 in Table 7.2.[8]

Example 7.9
Determine the Fourier transform of $g(t) = \text{sinc}(Wt)$.

Solution   Applying the symmetry property to the result of Example 7.8, we write

$$\tau\,\text{sinc}(\frac{t\tau}{2}) \leftrightarrow 2\pi\,\text{rect}(\frac{-\omega}{\tau}).$$

---

[8]It can be shown that the inverse Fourier transform of $\tau\,\text{sinc}(\frac{\omega\tau}{2})$ gives $\text{rect}(\frac{t}{\tau})$, with values of $\frac{1}{2}$ at the two points of discontinuity. This is a standard feature with Fourier transforms. When either a forward or inverse Fourier transform produces a waveform with discontinuities, the values at the points of discontinuity are always the midpoints between the bottoms and tops of the jumps.

Replacing τ with $2W$ and using the fact that rect is an even function yields

$$\text{sinc}(Wt) \;\leftrightarrow\; \frac{\pi}{W}\text{rect}(\frac{\omega}{2W}),$$

which is entry 8 in Table 7.2.[9]

Four functions, $\text{rect}(\frac{t}{\tau})$ for $\tau = 1$ (solid) and 2 (dashed), and $\text{sinc}(Wt)$ for $W = 2\pi$ (solid) and π (dashed), are plotted in Figure 7.9a. Their Fourier transforms $\tau\text{sinc}(\frac{\omega\tau}{2})$ and $\frac{\pi}{W}\text{rect}(\frac{\omega}{2W})$ are plotted in Figure 7.9b. Notice that the narrower signals (solid) in Figure 7.9a are paired with the broader Fourier transforms (also solid) in Figure 7.9b, while the broader signals (dashed) correspond to the narrower Fourier transforms. This inverse relationship always holds with Fourier transform pairs, as indicated by property 7 in Table 7.1. Narrower signals have broader Fourier transforms because narrower signals must have more high-frequency content in order to exhibit fast transitions in time. Likewise, broader signals are smoother and have more low-frequency content. Try identifying all the Fourier transform pairs in Figure 7.9 and mark them for future reference.

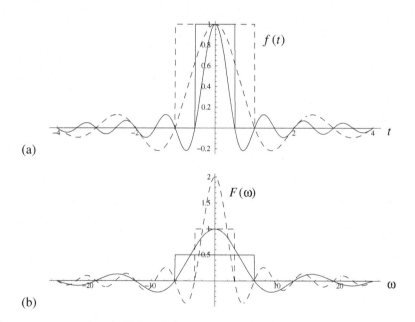

Figure 7.9 Four signals $f(t)$ and their Fourier transforms $F(\omega)$ are shown in (a) versus t (s) and (b) versus ω ($\frac{\text{rad}}{\text{s}}$), respectively.

---

[9] It can be shown that $\text{sinc}(Wt)$ is not absolutely integrable but yet we have found its Fourier transform using the symmetry property. An alternative way of confirming the Fourier transform of $\text{sinc}(Wt)$ is to show that the inverse Fourier transform of $\frac{\pi}{W}\text{rect}(\frac{W}{2W})$ is $\text{sinc}(Wt)$—see Exercise Problem 7.4.

Example 7.10

Consider a function $f(t)$ defined as the time derivative of $\triangle(\frac{t}{\tau})$, which also can be expressed as

$$f(t) \equiv \frac{d}{dt}\triangle(\frac{t}{\tau}) = \frac{2}{\tau}\text{rect}(\frac{t+\tau/4}{\tau/2}) - \frac{2}{\tau}\text{rect}(\frac{t-\tau/4}{\tau/2})$$

in terms of shifted and scaled rect functions. (See Exercise 7.5.) Using this relationship, verify[10] the Fourier transform of $\triangle(\frac{t}{\tau})$ specified in Table 7.2.

Solution   Using the time-shift and time-scaling properties of the Fourier transform and

$$\text{rect}(\frac{t}{\tau}) \leftrightarrow \tau\text{sinc}(\frac{\omega\tau}{2}),$$

we first note that

$$\text{rect}(\frac{t\pm\tau/4}{\tau/2}) = \text{rect}(2\frac{(t\pm\tau/4)}{\tau}) \leftrightarrow \frac{1}{2}\tau\text{sinc}(\frac{\omega\tau}{4})e^{\pm j\omega\tau/4}.$$

Thus, the Fourier transform of

$$f(t) = \frac{d}{dt}\triangle(\frac{t}{\tau})$$

is

$$F(\omega) = \text{sinc}(\frac{\omega\tau}{4})(e^{j\omega\tau/4} - e^{-j\omega\tau/4}) = j2\sin(\frac{\omega\tau}{4})\text{sinc}(\frac{\omega\tau}{4}).$$

Subsequently, using the time-derivative property of the Fourier transform, we find

$$\triangle(\frac{t}{\tau}) \leftrightarrow \frac{F(\omega)}{j\omega} = \frac{j2}{j\omega}\sin(\frac{\omega\tau}{4})\text{sinc}(\frac{\omega\tau}{4}) = \frac{\tau}{2}\text{sinc}^2(\frac{\omega\tau}{4}),$$

Fourier transform of a triangle

in agreement with item 9 in Table 7.2.

Item 10 in Table 7.2, showing the Fourier transform of

$$\text{sinc}^2(\frac{Wt}{2}),$$

can be obtained from item 9 using the symmetry property of the Fourier transform.

---

[10]Another approach to the same problem (less tricky, but more tedious) is illustrated in Example 7.11.

Example 7.11

In Example 7.10 the Fourier transform of $\triangle(\frac{t}{\tau})$ was obtained by a sequence of clever tricks related to some of the properties of the Fourier transform. Alternatively, we can get the same result in a straightforward way by performing the Fourier transform integral

$$\int_{-\infty}^{\infty} \triangle(\frac{t}{\tau}) e^{-j\omega t} dt = 2 \int_{0}^{\infty} \triangle(\frac{t}{\tau}) \cos(\omega t) dt = \frac{2}{\omega} \int_{0}^{\tau/2} \triangle(\frac{t}{\tau}) \frac{d}{dt}[\sin(\omega t)] \, dt,$$

where we make use of the fact that $\triangle(\frac{t}{\tau})$ is an even function that vanishes for $t > \frac{\tau}{2}$. Integrating by parts, we have

$$\int_{0}^{\tau/2} \triangle(\frac{t}{\tau}) \frac{d}{dt}[\sin(\omega t)] \, dt = 0 - \int_{0}^{\tau/2} \frac{-1}{\tau/2} \sin(\omega t) dt = \frac{2}{\tau} \int_{0}^{\tau/2} \sin(\omega t) dt$$

$$= -\frac{2}{\tau\omega} \cos(\omega t) \Big|_{0}^{\tau/2} = \frac{2}{\tau\omega}(1 - \cos(\frac{\omega\tau}{2})) = \frac{4}{\tau\omega} \sin^2(\frac{\omega\tau}{4}).$$

Hence,

$$\int_{-\infty}^{\infty} \triangle(\frac{t}{\tau}) e^{-j\omega t} dt = \frac{2}{\omega} \int_{0}^{\tau/2} \triangle(\frac{t}{\tau}) d[\sin(\omega t)]$$

$$= \frac{2}{\omega} \frac{4}{\tau\omega} \sin^2(\frac{\omega\tau}{4}) = \frac{\tau}{2} \text{sinc}^2(\frac{\omega\tau}{4}),$$

in agreement with item 9 in Table 7.2.

7.2   Frequency-Domain Description of Signals

7.2.1   Signal energy and Parseval's theorem

Signal energy

The concept of average power is not applicable to aperiodic signals, because aperiodic signals lack a standard time measure, like a period T, for normalizing the signal energy in a meaningful way. Instead, the energetics (and cost) of aperiodic signals is described by the total signal energy

$$W \equiv \int_{-\infty}^{\infty} |f(t)|^2 dt.$$

Similar to the case with periodic signals and the Fourier series, it is possible to calculate the integral of the squared magnitude of an aperiodic signal in terms of its Fourier representation. The difference in the aperiodic case is that the range of integration, extends from $-\infty$ to $+\infty$, as shown, not across just a single period, and we have a Fourier transform rather than a set of Fourier series coefficients.

The corresponding Parseval's theorem—also known as Rayleigh's theorem—for the aperiodic case states that

$$W \equiv \int_{-\infty}^{\infty} |f(t)|^2 dt = \frac{1}{2\pi} \int_{-\infty}^{\infty} |F(\omega)|^2 d\omega.$$

Parseval's theorem

This identity[11] is the last entry in Table 7.1, and it provides an alternative means for calculating the energy W in a signal. Parseval's theorem holds for all signals with finite W (i.e., $W < \infty$), which are referred to as *energy signals*. The entire left column of Table 7.2 is composed of such signals $f(t)$ with finite energy W.

Energy signals

All aperiodic signals that can be generated in a lab or at a radio station are necessarily energy signals, because generating a signal with infinite W is not physically possible (since the cost for the radio station would be unbounded). Parseval's theorem indicates that the energy of a signal is spread across the frequency space in a manner described by $|F(\omega)|^2$. Hence, in analogy with the power spectrum $|F_n|^2$ (for periodic signals), we refer to $|F(\omega)|^2$ as the *energy spectrum*. The notion of energy spectrum provides a useful basis for signal classification and bandwidth definitions, which we will discuss next.

Energy spectrum $|F(\omega)|^2$

7.2.2 Signal classification based on energy spectrum $|F(\omega)|^2$

Figures 7.10a and 7.10b show two different types of energy spectra $|F_1(\omega)|^2$ and $|F_2(\omega)|^2$ associated with possible aperiodic signals. The signal spectrum $|F_1(\omega)|^2$ shown in Figure 7.10a is large for small values of ω and then vanishes as ω increases, which indicates that the energy in signal $f_1(t) \leftrightarrow F_1(\omega)$ is concentrated at low frequencies. Thus, $f_1(t)$ has the spectral characteristics of a signal expected at the output of a low-pass filter and is said to be a *low-pass signal*. The spectrum $|F_2(\omega)|^2$ of signal $f_2(t)$ vanishes at both low and high frequencies, and therefore the energy in $f_2(t) \leftrightarrow F_2(\omega)$ is concentrated in an intermediate frequency band where $|F_2(\omega)|^2$ is relatively large. Thus, $f_2(t)$ is called a *band-pass signal*.

Low-pass and band-pass signals

Classification of signals according to their spectral shapes has an important and useful purpose. When we design a radio transmitter circuit, for instance, we have no knowledge of the exact signal the circuit will be transmitting during actual use; yet we know that the microphone or CD player output will be a low-pass signal in the audio frequency range that we somehow must convert into a higher-frequency band-pass signal before sending it to the transmitting antenna. (See Chapter 8.) Thus, we design and build circuits and systems to process signals having certain types of spectra

---

[11]**Proof of Parseval's theorem:** Assuming that $F(\omega)$ exists,

$$\int_{-\infty}^{\infty} |f(t)|^2 dt = \int_{-\infty}^{\infty} f(t) f^*(t) dt = \int_{t=-\infty}^{\infty} f(t) \left\{ \frac{1}{2\pi} \int_{-\infty}^{\infty} F(\omega) e^{j\omega t} d\omega \right\}^* dt$$

$$= \frac{1}{2\pi} \int_{\omega=-\infty}^{\infty} F^*(\omega) \int_{-\infty}^{\infty} f(t) e^{-j\omega t} dt \, d\omega = \frac{1}{2\pi} \int_{\omega=-\infty}^{\infty} F^*(\omega) F(\omega) d\omega = \frac{1}{2\pi} \int_{-\infty}^{\infty} |F(\omega)|^2 d\omega.$$

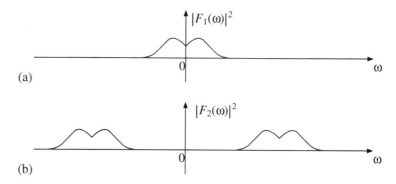

(a)

(b)

Figure 7.10 Energy spectra of possible (a) low-pass and (b) band-pass signals.

and spectral widths (see the next section), without requiring detailed knowledge of the signal. For instance, a transmitter that we build should work equally well for Beethoven's 5th and the Beatles' "A Hard Day's Night," even though the two signals have little in common other than their energy spectra shown in Figures 7.11a and 7.11b. Most audio signal spectra have the same general shape (low-pass, up to about 20 kHz or less) as the example audio spectra shown in Figure 7.11.

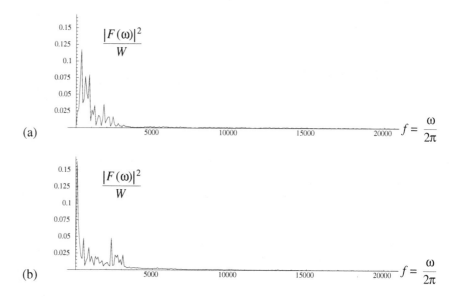

(a)

(b)

Figure 7.11 The normalized energy spectrum $\frac{|F(\omega)|^2}{W}$ of approximately the first 6 s of (a) *Beethoven's* "Symphony No. 5 in C minor" ("da-da-da-daaa ... ") and (b) the *Beatles'* "A Hard Day's Night" ("BWRRAAANNG! It's been a ha-rd da-y's night ...), plotted against frequency $f = \frac{\omega}{2\pi}$ in the range $f \in [0, 22050]$ Hz. Note that the energy spectrum (i.e., signal energy per unit frequency) is negligible in each case for $f > 15$ kHz (1 Hz=$2\pi \frac{\text{rad}}{\text{s}}$). The procedure for calculating these normalized energy spectra is described in Section 9.4.

7.2.3 Signal bandwidth

Signal *bandwidth* describes the width of the energy spectrum of low-pass and band-pass signals. We next will discuss how signal bandwidth can be defined and determined.

Low-pass signal bandwidth: By convention, the bandwidth of a low-pass signal $f(t)$ corresponds to a positive frequency

$$\omega = \Omega = 2\pi B$$

beyond which the energy spectrum $|F(\omega)|^2$ is very small. For example, for the audio signals with the energy spectra shown in Figure 7.11, we might define the bandwidth to be $B \approx 15\,\text{kHz}$, since for $f = \frac{\omega}{2\pi} > 15\,\text{kHz}$ the energy spectra $|F(\omega)|^2$ are negligible.

There are several standardized bandwidth definitions that are appropriate for quantitative work. One of them, the *3-dB bandwidth*, requires that

$$\frac{|F(\Omega)|^2}{|F(0)|^2} = \frac{1}{2},$$

3-dB bandwidth

or, equivalently,

$$10\log(\frac{|F(\Omega)|^2}{|F(0)|^2}) = -3\,\text{dB},$$

meaning that the bandwidth $\Omega = 2\pi B$ is the frequency where the energy spectrum $|F(\omega)|^2$ falls to one-half the spectral value $|F(0)|^2$ at DC. Another definition for $\Omega = 2\pi B$ requires that

$$\frac{1}{2\pi} \int_{-\Omega}^{\Omega} |F(\omega)|^2 d\omega = rW,$$

where r is some number such as 0.95 or 0.99 that is close to 1. Using this criterion for Ω with $r = 0.95$, for example, we refer to $\Omega = 2\pi B$ as the 95% bandwidth of a low-pass signal $f(t)$, because with $r = 0.95$, the frequency band $|\omega| \le \Omega\,\frac{\text{rad}}{\text{s}}$, or $|f| \le B$ Hz, contains 95% of the total signal energy W.

95% bandwidth

The bandwidths $\Omega = 2\pi B$, just defined, characterize the half-width of the signal energy spectrum curve $|F(\omega)|^2$ (i.e., the full width over only positive ω). This convention of associating the signal bandwidth with the width of the spectrum over just positive frequencies ω is sensible, because, for real-valued $f(t)$ (with $F(-\omega) = F^*(\omega)$), the inverse Fourier transform can be expressed as (see Exercise Problem 7.9)

$$f(t) = \frac{1}{2\pi} \int_0^\infty 2|F(\omega)| \cos(\omega t + \angle F(\omega))d\omega.$$

So, clearly, a real-valued $f(t)$ can be written as a superposition of co-sinusoids $2|F(\omega)|\cos(\omega t + \angle F(\omega))$ with frequencies $\omega \geq 0$, and therefore, only nonnegative frequencies are pertinent.[12]

Example 7.12
The Fourier transform of signal

$$f(t) = \text{rect}(\frac{t}{\tau})$$

is

$$F(\omega) = \tau \text{sinc}(\frac{\omega\tau}{2}).$$

Hence, the corresponding energy spectrum is

$$|F(\omega)|^2 = \tau^2 \text{sinc}^2(\frac{\omega\tau}{2}),$$

which is shown in Figure 7.12 for $\tau = 1$ s. Since $|F(\omega)|^2 = \tau^2 \text{sinc}^2(\frac{\omega\tau}{2})$ has relatively small values for $\omega > \frac{2\pi}{\tau}$, where $\omega = \frac{2\pi}{\tau}$ is the first zero-crossing of $\text{sinc}(\frac{\omega\tau}{2})$, we might decide to define

$$\Omega = \frac{2\pi}{\tau}$$

(or $B = \frac{1}{\tau}$) as the signal bandwidth. Show that this choice for $\Omega = 2\pi B$ corresponds to approximately the 90% bandwidth of $\text{rect}(\frac{t}{\tau})$.

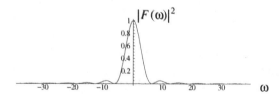

Figure 7.12   Energy spectrum $|F(\omega)|^2$ of the low-pass signal $f(t) = \text{rect}(t)$ versus ω in $\frac{\text{rad}}{\text{s}}$.

---

[12]There are real systems, however, containing signals that are modeled as being complex. For example, certain communications transmitters and receivers have two channels, and it is mathematically convenient to represent the signals in the two channels as the real and imaginary parts of a single complex signal. In this case the Fourier transform does not have the usual symmetry, and it is preferred to define a two-sided bandwidth covering both negative and positive frequencies.

Solution   Given that $f(t) = \text{rect}(\frac{t}{\tau})$, the signal energy is

$$W = \int_{-\infty}^{\infty} |\text{rect}(\frac{t}{\tau})|^2 dt = \int_{t=-\frac{\tau}{2}}^{\frac{\tau}{2}} dt = \tau.$$

Hence,

$$\Omega = \frac{2\pi}{\tau}$$

is the $r\%$ bandwidth of the signal, where r satisfies the condition

$$\frac{1}{2\pi} \int_{-\frac{2\pi}{\tau}}^{\frac{2\pi}{\tau}} \tau^2 \text{sinc}^2(\frac{\omega\tau}{2}) d\omega = r\tau.$$

Thus, after a change of variables with

$$x \equiv \frac{\omega\tau}{2},$$

and using the fact that the integrand is even, we find

$$r = \frac{2}{\pi} \int_{x=0}^{\pi} \text{sinc}^2(x) dx \approx 0.903,$$

where the numerical value on the right is obtained by numerical integration. According to this result, $\Omega = \frac{2\pi}{\tau}$, corresponding to the frequency of the first null in the energy spectrum of the signal $\text{rect}(\frac{t}{\tau})$, is the 90.3% bandwidth of the signal.

The result of Example 7.12 is well worth remembering: the 90% bandwidth of a pulse of duration τ s is $B = \frac{1}{\tau}$ Hz. Hence, a 1 μs pulse has a 1 MHz (i.e., 10^6 Hz) bandwidth. To view a 1 μs pulse on an oscilloscope, the scope bandwidth needs to be larger than 1 MHz.

Band-pass signal bandwidth: Recall that band-pass signals have energy spectra with shapes similar to those in Figure 7.10b. Because, for real-valued $f(t)$, we can write

$$f(t) = \frac{1}{2\pi} \int_0^{\infty} 2|F(\omega)| \cos(\omega t + \angle F(\omega)) d\omega,$$

the bandwidth $\Omega = 2\pi B$ of such signals is once again defined to characterize the width of the energy spectrum $|F(\omega)|^2$ over only positive frequencies ω.

For instance, the 95% bandwidth of a band-pass signal is defined to be

$$\Omega = \omega_u - \omega_l,$$

with $\omega_u > \omega_l > 0$, such that

$$\frac{1}{2\pi} \int_{\omega_l}^{\omega_u} |F(\omega)|^2 d\omega = 0.95\frac{W}{2}.$$

The right-hand side of this constraint is 95% of just half the signal energy W, because the integral on the left is computed over only positive ω.

Example 7.13

Determine the 95% bandwidths of the signals $f(t)$ and $g(t)$ with the energy spectra shown in Figures 7.13a and 7.13c.

Solution   According to Parseval's theorem, the energy W of signal $f(t)$ is simply the area under the $|F(\omega)|^2$ curve, scaled by $\frac{1}{2\pi}$. Using the formula for the area of a triangle, we have

$$W = \frac{1}{2\pi} \frac{2\pi \times 1}{2} = \frac{1}{2}.$$

To determine the 95% bandwidth, it is simplest to compute the total signal energy *outside* the signal bandwidth (i.e., $|\omega| > \Omega$) and to then set this equal to 5% of W, or $0.05W = 0.025$.

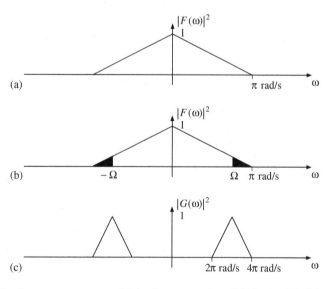

Figure 7.13   Energy spectrum of (a) a low-pass signal $f(t) \leftrightarrow F(\omega)$, (b) portion of the energy spectrum in (a) lying outside the bandwidth Ω, and (c) energy spectrum of a band-pass signal $g(t) \leftrightarrow G(\omega)$.

The energy outside $|\omega| > \Omega$ equals $\frac{1}{2\pi}$ times the combined areas of the right and left tips of the $|F(\omega)|^2$ triangle shown in Figure 7.13b, which is

$$\frac{1}{2\pi}(\pi - \Omega)\frac{(\pi - \Omega)}{\pi}.$$

Setting this quantity equal to 0.025 yields

$$(\pi - \Omega)^2 = 0.05\pi^2$$

so that

$$\Omega = \pi(1 - \sqrt{.05}) \approx 0.7764\pi \ \frac{\text{rad}}{\text{s}}.$$

Now, a comparison of $|F(\omega)|^2$ in Figure 7.13a and $|G(\omega)|^2$ in Figure 7.13c shows that $|G(\omega)|^2$, for positive ω, is a shifted replica of $|F(\omega)|^2$. Hence, for positive ω, $|G(\omega)|^2$ is twice as wide as $|F(\omega)|^2$, and therefore, the 95% bandwidth of the band-pass signal $g(t)$ is twice that of $f(t)$, i.e.

$$\Omega \approx 1.5528\pi \ \frac{\text{rad}}{\text{s}}.$$

Example 7.14
What is the 100% bandwidth of the signal $g(t)$ defined in Example 7.13?

Solution   For 100% bandwidth, we observe that $\omega_u = 4\pi$ and $\omega_l = 2\pi$ rad/s. Hence, the 100% bandwidth is simply

$$\Omega = \omega_u - \omega_l = 2\pi \ \frac{\text{rad}}{\text{s}},$$

or $B = 1$ Hz.

7.3   LTI Circuit and System Response to Energy Signals

In the opening section of this chapter we obtained the input–output relation

$$f(t) = \frac{1}{2\pi}\int_{-\infty}^{\infty}F(\omega)e^{j\omega t}d\omega \longrightarrow \boxed{\text{LTI}} \longrightarrow y(t) = \frac{1}{2\pi}\int_{-\infty}^{\infty}H(\omega)F(\omega)e^{j\omega t}d\omega.$$

(See Figure 7.1.) This also may be expressed as

$$y(t) \leftrightarrow Y(\omega) = H(\omega)F(\omega).$$

In either case, the system output is the inverse Fourier transform of the product of the system frequency response and the Fourier transform of the input.

This relation was derived as a straightforward application of

$$e^{j\omega t} \longrightarrow \boxed{\text{LTI}} \longrightarrow H(\omega)e^{j\omega t}.$$

Because this latter rule describes a steady-state response (see Section 5.2), we may think that the relation $y(t) \leftrightarrow Y(\omega) = H(\omega)F(\omega)$ describes just the steady-state response $y(t)$ of a dissipative LTI system to an input $f(t) \leftrightarrow F(\omega)$. However, the relation is more powerful than that: Because in dissipative systems transient components of the zero-state response to inputs $\cos(\omega t)$ and $\sin(\omega t)$ applied at $t = -\infty$ have vanished at all finite times t, the rule

$$e^{j\omega t} \longrightarrow \boxed{\text{LTI}} \longrightarrow H(\omega)e^{j\omega t}$$

Zero-state response

$y(t) \leftrightarrow H(\omega)F(\omega)$

actually describes the zero-state response *for all finite times t*. Consequently, the inverse Fourier transform of $Y(\omega) = H(\omega)F(\omega)$ represents, for all finite t, the *entire* zero-state response of system $H(\omega)$—not just the steady-state part of it—to input $f(t) \leftrightarrow F(\omega)$.

In summary, Figure 7.1 illustrates how the *zero-state response* can be calculated for dissipative LTI systems, for all finite t. We shall make use of this relation in the next several examples.

Example 7.15

The input of an LTI system

$$H(\omega) = \frac{1}{1 + j\omega}$$

is

$$f(t) = e^{-t}u(t),$$

shown in Figure 7.14a. Determine the system zero-state response $y(t)$, using the input–output rule shown in Figure 7.1.

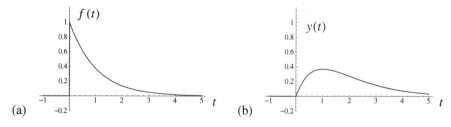

(a)   (b)

Figure 7.14   (a) Input and (b) output signals of the system examined in Example 7.15.

Solution   Since

$$f(t) = e^{-t}u(t) \leftrightarrow F(\omega) = \frac{1}{1 + j\omega}$$

(entry 1 in Table 7.2), the Fourier transform of output $y(t)$ is

$$Y(\omega) = H(\omega)F(\omega) = \frac{1}{1 + j\omega}\frac{1}{1 + j\omega} = \frac{1}{(1 + j\omega)^2}.$$

According to entry 5 in Table 7.2,

$$te^{-t}u(t) \leftrightarrow \frac{1}{(1 + j\omega)^2}.$$

Therefore, the system zero-state response is

$$y(t) = te^{-t}u(t),$$

as plotted in Figure 7.14b.

Example 7.16

Figure 7.15a shows the input

$$f(t) = \text{sinc}(t) \leftrightarrow F(\omega) = \pi\,\text{rect}(\frac{\omega}{2}).$$

Suppose this input is applied to an LTI system having the frequency response

$$H(\omega) = \text{rect}(\omega).$$

Determine the system zero-state response $y(t)$.

Solution   Given that $F(\omega) = \pi\,\text{rect}(\frac{\omega}{2})$ and $H(\omega) = \text{rect}(\omega)$, we have

$$Y(\omega) = H(\omega)F(\omega) = \text{rect}(\omega)\pi\,\text{rect}(\frac{\omega}{2}) = \pi\,\text{rect}(\omega)$$

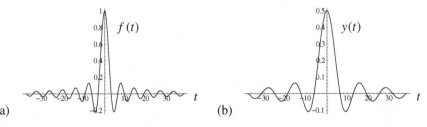

(a)   (b)

Figure 7.15   (a) Input and (b) output signals of the system examined in Example 7.16.

as the Fourier transform of the system response $y(t)$. Taking the inverse Fourier transform of $Y(\omega)$, we find that

$$y(t) = \frac{1}{2\pi} \int_{-\infty}^{\infty} Y(\omega)e^{j\omega t} d\omega = \frac{1}{2\pi} \int_{-\infty}^{\infty} \pi\,\text{rect}(\omega)e^{j\omega t} d\omega$$

$$= \frac{1}{2} \int_{-1/2}^{1/2} e^{j\omega t} d\omega = \frac{e^{jt/2} - e^{-jt/2}}{2jt}.$$

This result simplifies to

$$y(t) = \frac{1}{2}\text{sinc}(\frac{t}{2}),$$

which is plotted in Figure 7.15b. Notice that the system broadens the input $f(t)$ by a factor of 2 by halving its bandwidth.

Example 7.17
Signal

$$f(t) = g_1(t) + g_2(t),$$

where $G_1(\omega)$ and $G_2(\omega)$ are depicted in Figures 7.16a and 7.16b, and is passed through a band-pass system with the frequency response $H(\omega)$ shown in Figure 7.16c. Determine the system zero-state output $y(t)$ in terms of $g_1(t)$ and $g_2(t)$.

Solution Since $f(t) = g_1(t) + g_2(t)$,

$$f(t) \leftrightarrow F(\omega) = G_1(\omega) + G_2(\omega).$$

Therefore, the Fourier transform of the output $y(t)$ is

$$Y(\omega) = H(\omega)F(\omega) = H(\omega)G_1(\omega) + H(\omega)G_2(\omega).$$

From Figure 7.16 we can determine that

$$H(\omega)G_1(\omega) = 0$$

and

$$H(\omega)G_2(\omega) = \frac{1}{2}G_2(\omega).$$

Hence,

$$Y(\omega) = \frac{1}{2}G_2(\omega)$$

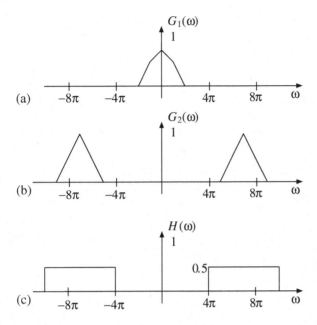

Figure 7.16 (a) Fourier transforms of signals $g_1(t)$ and $g_2(t)$, and (b) the frequency response $H(\omega)$ of the system examined in Example 7.16 with an input $f(t) = g_1(t) + g_2(t)$.

so that

$$y(t) = \frac{1}{2}g_2(t).$$

Thus, the system filters out the component $g_1(t)$ from the input $f(t)$ and delivers a scaled-down replica of $g_2(t)$ as the output.

Example 7.18

An input $f(t)$ is passed through a system having frequency response

$$H(\omega) = e^{-j\omega t_o}.$$

Determine the system zero-state output $y(t)$.

Solution

$$Y(\omega) = H(\omega)F(\omega) = e^{-j\omega t_o}F(\omega).$$

Therefore, using the time-shift property from Table 7.1, we find that the output is

$$y(t) = f(t - t_o),$$

which is a delayed copy of the input $f(t)$.

Example 7.19
The input of an LTI system

$$H(\omega) = \frac{1}{1 + j\omega}$$

is

$$f(t) = e^t u(-t)$$

shown in Figure 7.17a. What is the system zero-state output $y(t)$?

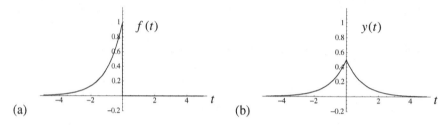

Figure 7.17   (a) Input and (b) output signals of the system examined in Example 7.19.

Solution   Since

$$f(t) = e^t u(-t) \leftrightarrow F(\omega) = \frac{1}{1 - j\omega}$$

(entry 2 in Table 7.2), the Fourier transform of the output is

$$Y(\omega) = H(\omega)F(\omega) = \frac{1}{1 + j\omega}\frac{1}{1 - j\omega} = \frac{1}{1 + \omega^2}.$$

According to entry 3 in Table 7.2,

$$e^{-|t|} \leftrightarrow \frac{2}{1 + \omega^2}.$$

Therefore, the system output is

$$y(t) = \frac{1}{2}e^{-|t|},$$

which is plotted in Figure 7.17b.

Example 7.20

The input of an LTI system

$$H(\omega) = \frac{1}{1+j\omega}$$

is

$$f(t) = \text{rect}(t).$$

What is the system output $y(t)$?

Solution   Since

$$f(t) = \text{rect}(t) \leftrightarrow F(\omega) = \text{sinc}(\frac{\omega}{2}),$$

the Fourier transform of the output is

$$Y(\omega) = H(\omega)F(\omega) = \frac{1}{1+j\omega}\text{sinc}(\frac{\omega}{2}).$$

We cannot find an appropriate entry in Table 7.2 to determine $y(t)$. Thus, we must work out the inverse Fourier transform

$$y(t) = \frac{1}{2\pi}\int_{-\infty}^{\infty}\frac{1}{1+j\omega}\text{sinc}(\frac{\omega}{2})e^{j\omega t}d\omega$$

by hand. But, the integral looks too complicated to carry out. In Chapter 9, we will learn a simpler way of finding $y(t)$ for this problem. So, let us leave the answer in integral form. (It is the right answer, but it is not in a form suitable for visualizing the output.)

We will close the chapter with a small surprise (plus one more example), which will illustrate the power and generality of the Fourier transform method, even though, as we found out in Example 7.20, the application of the method may sometimes be difficult.

Here is our surprise: Figure 7.18 shows an inductor L with voltage $v(t)$, current $i(t)$, and v–i relation

$$v(t) = L\frac{di}{dt}.$$

Suppose that this inductor is a component of some dissipative LTI circuit in the laboratory, and therefore its voltage $v(t)$ and current $i(t)$ are lab signals (i.e., absolutely integrable, finite energy, etc.) Thus, $v(t)$ and $i(t)$ have Fourier transforms $V(\omega)$ and $I(\omega)$, respectively. Since $i(t) \leftrightarrow I(\omega)$, the time-derivative property from Table 7.1

$$v(t) = L\frac{di}{dt}$$

$$+ \quad L \underset{i(t)}{\text{———}} \quad - \qquad\qquad \leftrightarrow \qquad\qquad V(\omega) = j\omega L I(\omega)$$

$$+ \quad L \underset{I(\omega)}{\text{———}} \quad -$$

Figure 7.18 v–i and $V(\omega)$–$I(\omega)$ relations for an inductor L.

indicates that $\frac{di}{dt} \leftrightarrow j\omega I(\omega)$. Thus, the amplitude-scaling property (with $K = L$) implies that

$$V(\omega) = j\omega L I(\omega),$$

or

$$V(\omega) = Z I(\omega),$$

with

$$Z = j\omega L.$$

Fourier
$V(\omega)$–$I(\omega)$
relations

Notice that we effectively have obtained a Fourier $V(\omega)$–$I(\omega)$ relation (see Figure 7.18 again) for the inductor, which has the same form as the phasor V–I relation for the same element. The $V(\omega)$–$I(\omega)$ relations for a capacitor C and resistor R are similar to the relation for the inductor, but with impedances $Z = \frac{1}{j\omega C}$ and $Z = R$, respectively, just as in the phasor case.

So, it is no wonder that the frequency-response function $H(\omega)$ derived by the phasor method also describes the circuit response to arbitrary inputs. The next example will show how the Fourier transform method can be applied to directly analyze a circuit.

Example 7.21
Determine the response $y(t) \leftrightarrow Y(\omega)$ of the circuit shown in Figure 7.19a to an arbitrary input $f(t) \leftrightarrow F(\omega)$.

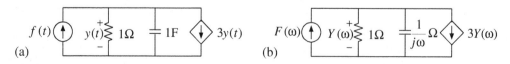

(a)

(b)

Figure 7.19   (a) An LTI circuit with an arbitrary input $f(t)$, and (b) the Fourier equivalent of the same circuit.

Solution   Building on what we just learned, in Figure 7.19b we construct the *equivalent Fourier transform circuit* and then proceed with the node-voltage method, as usual. The KCL equation for the top node is

$$F(\omega) = \frac{Y(\omega)}{1} + \frac{Y(\omega)}{\frac{1}{j\omega}} + 3Y(\omega) = (4 + j\omega)Y(\omega).$$

Thus,

$$Y(\omega) = \frac{1}{4 + j\omega}F(\omega),$$

which is the Fourier transform of the system zero-state response $y(t)$. Hence, the system frequency response is

$$H(\omega) = \frac{1}{4 + j\omega},$$

and, using the inverse Fourier transform,

$$y(t) = \frac{1}{2\pi} \int_{-\infty}^{\infty} \frac{1}{4 + j\omega}F(\omega)e^{j\omega t}d\omega.$$

This result is valid for any laboratory input $f(t) \leftrightarrow F(\omega)$.

EXERCISES

7.1   **(a)**   Given that $f(t) = e^{-a(t-t_o)}u(t - t_o)$, where $a > 0$, determine the Fourier transform $F(\omega)$ of $f(t)$.

(b)   Given that

$$g(t) = \frac{1}{a + jt},$$

where $a > 0$, determine the Fourier transform $G(\omega)$ of $g(t)$ by using the symmetry property and the result of part (a).

(c)   Confirm the result of part (b) by calculating $g(t)$ from $G(\omega)$, using the inverse Fourier transform integral.

7.2   Let

$$f(t) = \text{rect}(\frac{t}{2}).$$

(a)   Plot $g(t) = f(t - 1)$.

(b)   Determine the Fourier transform $G(\omega)$ by using the time-shift property and the fact that

$$\text{rect}(\frac{t}{T}) \quad \leftrightarrow \quad T\text{sinc}(\frac{\omega T}{2}).$$

(c) Determine $G(\omega)$ by direct Fourier transformation (integration) of $g(t)$, and confirm that (b) is correct.

(d) Taking advantage of Parseval's theorem (Table 7.1, entry 16), determine the signal energy

$$W = \frac{1}{2\pi} \int_{-\infty}^{\infty} |G(\omega)|^2 d\omega.$$

7.3   Determine the Fourier transform $F(\omega)$ of the following signal $f(t)$:

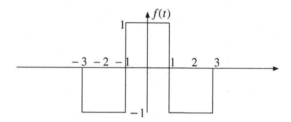

7.4   Determine the inverse Fourier transform of

$$F(\omega) = \frac{\pi}{W} \text{rect}\left(\frac{\omega}{2W}\right)$$

by direct integration. Is $F(\omega)$ absolutely integrable? Does it satisfy the Dirichlet conditions?

7.5   Plot the time derivatives of the unit triangle $\triangle(\frac{t}{\tau})$ and the function

$$f(t) = \frac{2}{\tau} \text{rect}(\frac{t + \tau/4}{\tau/2}) - \frac{2}{\tau} \text{rect}(\frac{t - \tau/4}{\tau/2})$$

to show that they are equivalent. In plotting $f(t)$, superpose the plots of $\frac{2}{\tau} \text{rect}(\frac{t \pm \tau/4}{\tau/2})$, which you obtain by shifting and scaling the graph of $\text{rect}(\frac{t}{\tau})$.

7.6   Given that $f(t) = 5\triangle^2(\frac{t}{5})$, evaluate the Fourier transform $F(\omega)$ at $\omega = 0$.

7.7   **(a)** Show that for real-valued signals $f(t)$, the Fourier transform $F(\omega)$ satisfies the property

$$F(-\omega) = F^*(\omega).$$

(b) Using this result, show that for real-valued $f(t)$, we have $|F(-\omega)| = |F(\omega)|$ and $\angle F(-\omega) = -\angle F(\omega)$ (i.e. that the magnitude of the Fourier transform is even and the phase is odd).

7.8   On an exam, you are asked to calculate $F(0)$ for some real-valued signal $f(t)$. You obtain the answer $F(0) = 4 - j2$. Explain why, for sure, you have made a mistake in your calculation.

7.9 Show that, given a real-valued signal $f(t)$, the inverse Fourier transform integral can be expressed as

$$f(t) = \frac{1}{2\pi} \int_0^\infty 2|F(\omega)| \cos(\omega t + \angle F(\omega)) d\omega.$$

7.10 The bandwidth Ω of a low-pass signal $f(t) \leftrightarrow F(\omega)$ is defined by the constraint

$$\frac{1}{2\pi} \int_{-\Omega}^{\Omega} |F(\omega)|^2 d\omega = 0.8 W_f,$$

where W_f denotes the energy of signal $f(t)$.

(a) What fraction of the signal energy W_f is contained in the frequency band $0 < \omega < \Omega$? Explain.

(b) The signal $f(t)$ is filtered by a linear system with a frequency response $H(\omega)$ satisfying $H(\omega) = 0$ for $|\omega| < \Omega$ and $|H(\omega)| = 1$ for $|\omega| \geq \Omega$. What is the total energy of the system output $y(t)$ in terms of the energy W_f of the input $f(t)$?

7.11 Determine the 3-dB bandwidth and the 95% bandwidth of signals $f(t)$ and $g(t)$ with the following energy spectra:

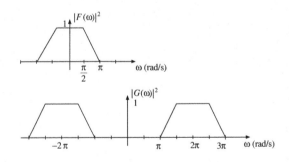

7.12 (a) Let $f(t) = f_1(t) + f_2(t)$ such that $f_1(t) \leftrightarrow F_1(\omega)$ and $f_2(t) \leftrightarrow F_2(\omega)$. Show that

$$f(t) \leftrightarrow F_1(\omega) + F_2(\omega).$$

(b) The input signal of an LTI system with a frequency response $H(\omega) = |H(\omega)|e^{j\chi(\omega)}$ is $f_1(t) + f_2(t)$. Functions $F_1(\omega)$, $F_2(\omega)$, $H(\omega)$, and $\chi(\omega)$ are given graphically as follows:

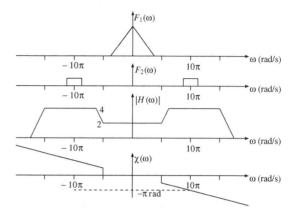

Express the output $y(t)$ of the system as a superposition of scaled and/or shifted versions of $f_1(t)$ and $f_2(t)$. (Hint: $y(t) = y_1(t) + y_2(t)$, with $Y_1(\omega) = H(\omega)F_1(\omega)$ and $Y_2(\omega) = H(\omega)F_2(\omega)$.)

7.13   Determine the response $y(t)$ of the following circuit with an arbitrary input $f(t)$ in the form of an inverse Fourier transform and then evaluate $y(t)$ for the case $f(t) = e^{-\frac{t}{6}}u(t)$ V:

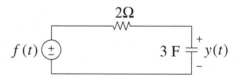

8

Modulation and AM Radio

AM radio communication is based on relatively simple signal processing concepts. The output signal $f(t)$ of a microphone or an audio playback device (see Figure 8.1a) is used to vary, or *modulate*, the amplitude of a cosine signal $\cos(\omega_c t)$. The resulting AM (short for *amplitude modulation*) signal

$$f(t)\cos(\omega_c t)$$

(see Figure 8.1b) is subsequently used to excite a transmitting antenna that converts the signal (a voltage or current) into a propagating radiowave (traveling electric and magnetic field variations). We refer to ω_c as the *carrier frequency* of the AM signal, or the *operating frequency* of the radio station where the signal is generated. In typical cases the carrier frequency $\omega_c \gg \Omega$, where Ω denotes the bandwidth of the audio signal $f(t)$. The task of an AM radio receiver is to pick up the radiowave signal arriving from the broadcast station and convert it into an output $y(t)$ that is proportional to $f(t - t_o)$, where t_o is some small propagation time delay, typically a millisecond or less in practice. Subsequently, $y(t)$ (again, a voltage or current) is converted into sound by a suitable transducer,[1] which ordinarily is a loudspeaker or a set of headphones.

In this chapter we will study some of the details of AM radio communication, using the Fourier series and transform tools introduced in Chapters 6 and 7. In Section 8.1 we will begin by introducing the Fourier transform shift and modulation properties and discussing their relevance to AM radio. Sections 8.2 and 8.3 will describe two different ways of demodulating an AM radio signal acquired by a receiving antenna. Finally, Section 8.4 will examine the components and overall performance details of practical superheterodyne AM receivers.

Voice and music signals modulate radio carriers; modulated carriers are filtered and the outputs are demodulated and detected; how commercial AM transmitters and receivers work

---

[1] A *transducer* converts energy from one form to another—for example, electrical to acoustical.

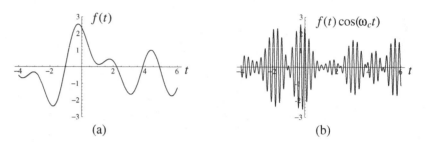

Figure 8.1 (a) A signal $f(t)$, and (b) a corresponding AM signal $f(t)\cos(\omega_c t)$ with a carrier frequency ω_c.

8.1 Fourier Transform Shift and Modulation Properties

Example 8.1

Prove the Fourier time-shift property of Table 7.1.

Solution   The time-shift property states that

$$f(t - t_o) \leftrightarrow F(\omega)e^{-j\omega t_o},$$

where $F(\omega)$ denotes the Fourier transform of $f(t)$. To verify this property, we write the Fourier transform of $f(t - t_o)$ as

$$\int_{-\infty}^{\infty} f(t - t_o)e^{-j\omega t}dt = \int_{-\infty}^{\infty} f(x)e^{-j\omega(x+t_o)}dx = e^{-j\omega t_o}\int_{-\infty}^{\infty} f(x)e^{-j\omega x}dx,$$

where the second line uses the change of variable

$$x = t - t_o \implies t = x + t_o$$

and where $dt = dx$. The last integral is $F(\omega)$, and so we have

$$f(t - t_o) \leftrightarrow F(\omega)e^{-j\omega t_o},$$

as claimed. Note that we already applied this time-shift property to several problems in Chapter 7.

Example 8.2

Prove the Fourier frequency-shift property of Table 7.1.

Solution   The frequency-shift property states that

$$f(t)e^{j\omega_o t} \leftrightarrow F(\omega - \omega_o).$$

This is the dual of the time-shift property. It can be proven as in Example 8.1, except by the use of the inverse Fourier transform. Alternatively, a more direct proof yields

$$\int_{-\infty}^{\infty} f(t)e^{j\omega_o t} e^{-j\omega t} dt = \int_{-\infty}^{\infty} f(t)e^{-j(\omega-\omega_o)t} dt = F(\omega - \omega_o),$$

as claimed.

The frequency-shift property will play a major role in this chapter. In particular, it forms the basis for the *modulation property* in Table 7.1. The modulation property is derived as follows. Evaluating the frequency-shift property for $\omega_o = \pm\omega_c$ gives

$$f(t)e^{j\omega_c t} \leftrightarrow F(\omega - \omega_c)$$

and

$$f(t)e^{-j\omega_c t} \leftrightarrow F(\omega + \omega_c).$$

Summing these results, we obtain

$$f(t)(e^{j\omega_c t} + e^{-j\omega_c t}) = 2f(t)\cos(\omega_c t) \leftrightarrow F(\omega - \omega_c) + F(\omega + \omega_c).$$

Hence,

$$f(t)\cos(\omega_c t) \leftrightarrow \frac{1}{2}F(\omega - \omega_c) + \frac{1}{2}F(\omega + \omega_c),$$

which is the modulation property.

The implication of the modulation property is illustrated in Figure 8.2. The top figure shows the Fourier transform $F(\omega)$ of some signal $f(t)$. The bottom figure then shows the Fourier transform of $f(t)\cos(\omega_c t)$, where the replicas of $F(\omega)$ have half the height of the original (as depicted by the dashed curve in the top figure) and are shifted to the right and left by ω_c. The modulation process cannot be described in terms of a frequency response function $H(\omega)$, because, unlike the LTI filtering operations discussed in Section 7.3, modulation is a time-varying operation that shifts the energy content of signal $f(t)$ from its *baseband* to an entirely new location in the frequency domain. Figure 8.2, illustrates how the modulation process generates a bandpass signal $f(t)\cos(\omega_c t)$ from a low-pass signal $f(t)$. We will refer to $f(t)\cos(\omega_c t)$ as an AM signal with carrier ω_c.

Modulation property

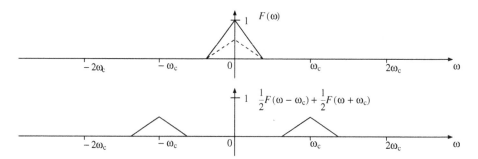

Figure 8.2 An example illustrating the modulation property: Fourier transform of some low-pass signal $f(t)$ (top) and Fourier transform of the AM signal $f(t)\cos(\omega_c t)$ (bottom).

In another example, shown in Figure 8.3, modulation generates a multi-band signal $g(t)\cos(\omega_c t)$ from a band-pass signal $g(t)$. Notice that, as in the earlier example, we obtain the Fourier transform of $g(t)\cos(\omega_c t)$ by shifting half (in amplitude) of the Fourier transform of $g(t)$ to the right by an amount ω_c, shifting half to the left by the same amount, and then summing.

The system symbol for the AM modulation process is shown in Figure 8.4. The multiplying unit in the diagram is known as a *mixer*. As already mentioned, in AM radio communication an audio signal $f(t)$ is mixed with a carrier signal $\cos(\omega_c t)$ and transmitted as a band-pass radiowave (a time-varying electric field traveling through space). The purpose of modulation prior to transmission is twofold:

Mixer

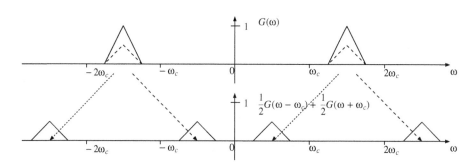

Figure 8.3 Another example illustrating the modulation property: Fourier transform of some band-pass signal $g(t)$ (top) and Fourier transform of modulated multi-band signal $g(t)\cos(\omega_c t)$ (bottom). The short- and long-dashed arrows from top to bottom suggest how the right and left shifts of half of $G(\omega)$ result in the Fourier transform of the modulated signal $g(t)\cos(\omega_c t)$.

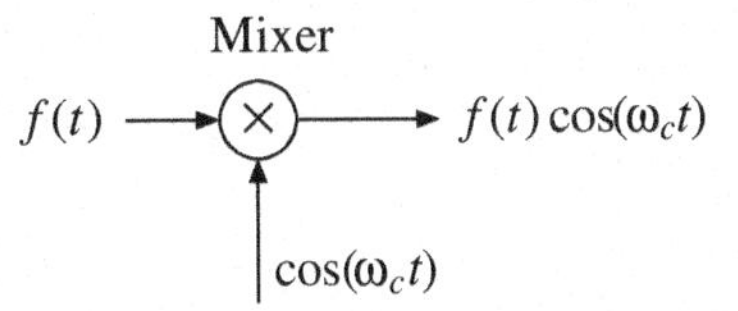

Mixer

$$f(t) \longrightarrow \boxed{\times} \longrightarrow f(t)\cos(\omega_c t)$$

$$\cos(\omega_c t)$$

Figure 8.4 System symbol for AM modulation process. The multiplier unit is known as a *mixer*. Shifting the frequency content of a signal to a new location in ω-space, by the use of a mixer, also is known as *heterodyning*.

(1) Radio antennas, which convert voltage and current signals into radiowaves (and vice versa), essentially are high-pass systems[2] with negligible amplitude response $|H_{\text{ant}}(\omega)|$ for $|\omega| < \frac{\pi c}{2L}$, where $c = 3 \times 10^8$ m/s is the speed of light and L is the physical size of the antenna.[3] For $L = 75$ m, for instance, the antenna performs poorly unless

$$\omega \geq \frac{\pi \, 3 \times 10^8}{150} = 2\pi \, 10^6 \, \frac{\text{rad}}{\text{s}},$$

or, equivalently, frequencies are above 1 MHz. No practical antenna with a reasonable physical size exists to transmit an audio signal $f(t)$ directly. By contrast, a modulated AM signal, $f(t)\cos(\omega_c t)$, can be efficiently radiated with an antenna having a 75-m length, when the carrier frequency is in the 1-MHz frequency range.

(2) Even if efficient antennas were available in the audio frequency range, radio communication within the audio band would be problematic. The reason for this is that all signal transmissions then would occupy the same frequency band. Thus, it would be impossible to extract the signal for a single station of interest from the superposed signals of all other stations operating in the same geographical region. In other words, the signals would *interfere* with one another. To overcome this difficulty, modulation is essential.

In the United States, different AM stations operating in the same vicinity are assigned different individual carrier frequencies—chosen from a set spaced 10 kHz or $2\pi \times 10^4$ rad/s apart, within the frequency range 540 to 1700 kHz. With a 10-kHz operation bandwidth, each AM radio station can broadcast a voice signal of up to 5-kHz bandwidth. (Can you explain why?)

Example 8.3
A mixer is used to multiply

$$1 + \sum_{n=1}^{\infty} \frac{1}{n} \cos(n\omega_0 t)$$

---

[2]Examined and modeled in courses on electromagnetics.
[3]For a monopole antenna over a conducting ground plane.

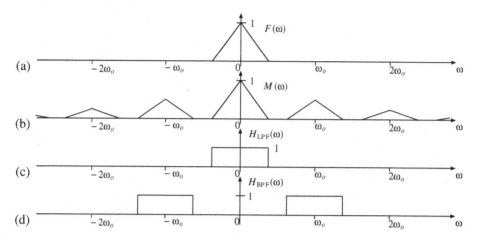

Figure 8.5   (a) Fourier transform of a low-pass signal $f(t)$, (b) $M(\omega)$ computed in Example 8.3, and (c) a low-pass filter $H_{\mathrm{LPF}}(\omega)$, and (d) a band-pass filter $H_{\mathrm{BPF}}(\omega)$.

with a low-pass signal $f(t)$. The Fourier transform of $f(t)$ is plotted in Figure 8.5a. Plot the Fourier transform $M(\omega)$ of the mixer output

$$m(t) = f(t)\{1 + \sum_{n=1}^{\infty} \frac{1}{n} \cos(n\omega_o t)\}.$$

Solution   We first expand $m(t)$ as

$$m(t) = f(t) + \sum_{n=1}^{\infty} \frac{1}{n} f(t) \cos(n\omega_o t).$$

Next, we apply the addition and modulation properties of the Fourier transform to this expression to obtain

$$M(\omega) = F(\omega) + \sum_{n=1}^{\infty} \frac{1}{2n}\{F(\omega - n\omega_o) + F(\omega + n\omega_o)\}.$$

This result is plotted in Figure 8.5b. Notice that $f(t)$ can be extracted from $m(t)$ with the low-pass filter $H_{\mathrm{LPF}}(\omega)$ shown in Figure 8.5c. On the other hand, filtering $m(t)$ with the band-pass filter $H_{\mathrm{BPF}}(\omega)$, shown in Figure 8.5d, would generate a band-pass AM signal $f(t)\cos(\omega_o t)$ having carrier ω_o.

The antenna for an AM radio receiver typically captures the signals from tens of radio stations simultaneously. The receiver uses a *front-end* band-pass filter with an adjustable passband to respond to just the single AM signal $f(t)\cos(\omega_c t)$ to be "tuned in." The band-pass filter output then is converted into an audio signal that is proportional to $f(t - t_o)$. This conversion process, known as *demodulation*, is discussed in the next two sections.

8.2 Coherent Demodulation of AM Signals

The block diagram in Figure 8.6a depicts a possible AM communication system. The mixer on the left idealizes an AM radio transmitter. The mixer on the right and the low-pass filter $H_{\mathrm{LPF}}(\omega)$ constitute a *coherent-demodulation*, or *coherent-detection*, AM receiver. Because the transmitter and receiver are located at different sites, the transmitted AM signal

$$f(t)\cos(\omega_c t)$$

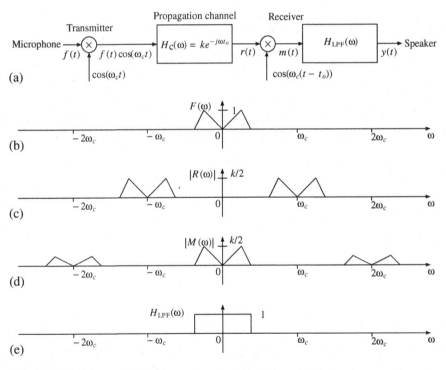

Figure 8.6 (a) A possible AM communication system using a coherent-demodulation receiver, (b) the Fourier transform of the audio signal $f(t)$, (c) the Fourier transform magnitude of the receiver input signal $r(t)$, (d) the Fourier transform magnitude of the mixer output $m(t)$ in the receiver, and (e) the frequency response of the low-pass filter in the receiver.

arrives at the receiver after traveling the intervening distance through some channel (usually, through air in the form of a radiowave). In the block diagram the center box

$$H_c(\omega) = ke^{-j\omega t_o}$$

represents an ideal propagation channel from the transmitter to the receiver, which only delays the AM signal by an amount t_o and scales it in amplitude by a constant factor k. Hence, the receiver input labeled as $r(t)$ is

$$r(t) = kf(t - t_o)\cos(\omega_c(t - t_o)).$$

Figures 8.6b and 8.6c show plots of a possible $F(\omega)$ and $|R(\omega)|$, where

$$f(t) \leftrightarrow F(\omega) \ \text{ and } \ r(t) \leftrightarrow R(\omega).$$

Let us now examine how demodulation takes place as in Figure 8.6a. The mixer output in the receiver is

$$\begin{aligned}
m(t) &= r(t)\cos(\omega_c(t - t_o)) \\
&= kf(t - t_o)\cos^2(\omega_c(t - t_o)) \\
&= f(t - t_o)\frac{k}{2}\{1 + \cos(2\omega_c(t - t_o))\}.
\end{aligned}$$

Therefore, using a combination of Fourier addition, modulation, and time-shift properties, we determine its Fourier transform $M(\omega)$ as

$$M(\omega) = \frac{k}{2}F(\omega)e^{-j\omega t_o} + \frac{k}{4}\{F(\omega - 2\omega_c) + F(\omega + 2\omega_c)\}e^{-j\omega t_o},$$

where the first term is the Fourier transform of

$$\frac{k}{2}f(t - t_o)$$

and the second term is the transform of

$$\frac{k}{2}f(t - t_o)\cos(2\omega_c(t - t_o)).$$

A plot of $|M(\omega)|$ is shown in Figure 8.6d.

Clearly, the first term of $m(t)$,

$$\frac{k}{2}f(t - t_o),$$

is the delayed audio signal that we hope to recover. To extract this signal, we pass $m(t)$ through the low-pass filter $H_{\text{LPF}}(\omega)$ shown in Figure 8.6e. The result is

$$Y(\omega) = H_{\text{LPF}}(\omega)M(\omega) = \frac{k}{2}F(\omega)e^{-j\omega t_o},$$

implying that

$$y(t) = \frac{k}{2}f(t - t_o),$$

the desired audio signal at the loudspeaker input.[4]

The crucial and most difficult step in coherent demodulation is the mixing of the incoming signal $r(t)$ with $\cos(\omega_c(t - t_o))$. The difficulty lies in obtaining $\cos(\omega_c(t - t_o))$ for mixing purposes. A locally generated $\cos(\omega_c t + \theta)$ in the receiver with the right frequency ω_c, but an arbitrary phase shift $\theta \neq -\omega_c t_o$, will not work, because even small fluctuations of the propagation delay t_o will translate to large phase-shift variations of the carrier and will cause $y(t)$ to fluctuate. Coherent detection receivers are thus required to extract $\cos(\omega_c(t - t_o))$ from the incoming signal $r(t)$ before the mixing can take place. This requirement increases the complexity and, therefore, the cost of coherent demodulation receivers. The term *coherent* demodulation or detection refers to the requirement that the phase shift θ of the mixing signal $\cos(\omega_c t + \theta)$ be coherent (same as or different by a constant amount) with the phase shift $-\omega_c t_o$ of the incoming carrier.

The receiver complexity can be reduced if the incoming signal is of the form

$$r(t) = k(f(t - t_o) + \alpha)\cos(\omega_c(t - t_o)),$$

that is, if it contains a constant-amplitude cosine component $k\alpha \cos(\omega_c(t - t_o))$ in addition to the primary term $kf(t - t_o)\cos(\omega_c(t - t_o))$ carrying the voice signal $f(t)$. In commercial AM radio, we achieve this by adding a constant offset $\alpha > 0$ to the voice signal $f(t)$ before modulating the carrier $\cos(\omega_c t)$. (See Figure 8.7a.) For sufficiently large offsets α satisfying

$$f(t) + \alpha > 0 \text{ for all } t,$$

a simple envelope detection procedure, described in the next section, works well, obviating the need to generate $\cos(\omega_c(t - t_o))$ within the receiver.

8.3   Envelope Detection of AM Signals

Consider the modified AM transmitter system shown in Figure 8.7a. In the modified transmitter, a DC offset α is added to the voice signal $f(t)$, and the sum is used

---

[4]In practice, $y(t)$ is amplified by an audio amplifier (which is not shown) prior to being applied to the loudspeaker.

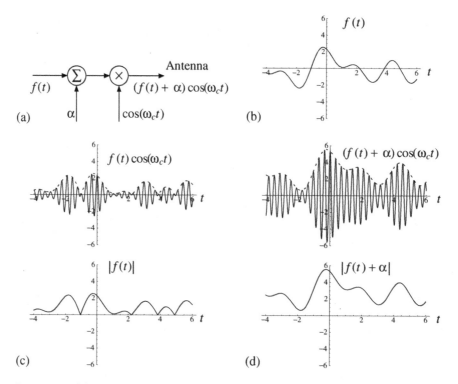

Figure 8.7 (a) An idealized AM transmitter with offset (α) insertion, (b) a possible voice signal $f(t)$, (c) AM signal $f(t)\cos(\omega_c t)$ and its envelope $|f(t)|$, and (d) AM signal $(f(t) + \alpha)\cos(\omega_c t)$ with a DC offset α and its envelope $|f(t) + \alpha| = f(t) + \alpha$. Notice that the envelope function in (d) is an offset version of $f(t)$ plotted in (b); the envelope in (c), however, does not resemble $f(t)$.

to modulate the amplitude of a carrier $\cos(\omega_c t)$. A possible waveform for $f(t)$ is shown in Figure 8.7b. The incoming signal to an AM receiver, from the transmitter in Figure 8.7a, will be

$$r(t) = (f(t) + \alpha)\cos(\omega_c t),$$

assuming, for simplicity, a propagation channel with $k = 1$ and zero time delay t_o. This signal is plotted in Figures 8.7c and 8.7d for the cases $\alpha = 0$ and $\alpha > \max |f(t)|$. The bottoms of Figures 8.7c and 8.7d show the envelopes of $r(t)$ for the same two cases, where the envelope of $r(t)$ is defined as

$$|f(t) + \alpha|$$

(also indicated by the dashed lines superposed upon the $r(t)$ curves in Figures 8.7c and 8.7d).

Except for a DC offset α, the envelope signal shown in Figure 8.7d is the same **Envelope** as the desired signal $f(t)$. By contrast, the envelope $|f(t)|$ shown in Figure 8.7c, **of AM** corresponding to the case $\alpha = 0$, does not resemble $f(t)$, because of the "rectification **signal** effect" of the absolute-value operation. We next will describe an envelope detector system that can extract from the AM signal shown at the top of Figure 8.7d the desired envelope

$$|f(t) + \alpha| = f(t) + \alpha.$$

The detector is designed to work when

$$\alpha > \max |f(t)|.$$

Figure 8.8a shows an ideal envelope detector that consists of a *full-wave rectifier* **Envelope** followed by a low-pass filter $H_{\mathrm{LPF}}(\omega)$. The figure also includes plots of a possible AM **detection** input signal $r(t)$ (same as in Figure 8.7d), the rectifier output $p(t) = |r(t)|$, and the output $q(t)$ that follows the peaks of $p(t)$ and therefore is equal to the envelope of the input $r(t)$.

Assuming that $\alpha > \max |f(t)|$ so that $f(t) + \alpha > 0$ for all t, we have $|f(t) + \alpha| = f(t) + \alpha$, and the rectifier output is

$$p(t) = |r(t)| = |(f(t) + \alpha)\cos(\omega_c t)| = |f(t) + \alpha||\cos(\omega_c t)|$$

$$= (f(t) + \alpha)|\cos(\omega_c t)|,$$

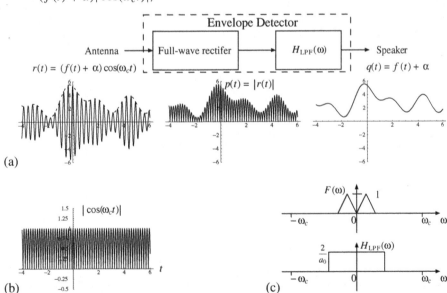

Figure 8.8 (a) Block diagram of an ideal envelope detector, as well as example plots of an input signal $r(t)$, its rectified version $p(t) = |r(t)|$, and the detector output $q(t)$, (b) plot of a full-wave rectified AM carrier, and (c) example $F(\omega)$ and frequency response of an ideal low-pass filter included in the envelope detector system.

as shown in the figure. Clearly, after rectification the desired AM signal envelope $f(t) + \alpha$ appears as the amplitude of the rectified cosine $|\cos(\omega_c t)|$ shown in Figure 8.8b.

We next will see that we can extract $f(t) + \alpha$ from $p(t)$ by passing $p(t)$ through a low-pass filter $H_{LPF}(\omega)$ having the shape shown in Figure 8.8c. The rectified cosine $|\cos(\omega_c t)|$ of Figure 8.8b is a periodic function with period

$$T = \frac{T_c}{2} = \frac{2\pi/\omega_c}{2} = \frac{\pi}{\omega_c}$$

and fundamental frequency

$$\omega_o = \frac{2\pi}{T} = \frac{2\pi}{\pi/\omega_c} = 2\omega_c.$$

Thus, it can be expanded in a Fourier series as

$$|\cos(\omega_c t)| = \frac{a_0}{2} + \sum_{n=1}^{\infty} a_n \cos(n2\omega_c t)$$

with an appropriate set of Fourier coefficients a_n. (See Example 6.7 in Section 6.2.4, where it was found that $a_0 = \frac{4}{\pi}$.) Hence, the rectifier output, or the input to filter $H_{LPF}(\omega)$, can be expressed as

$$p(t) = (f(t) + \alpha)|\cos(\omega_c t)| \equiv p_1(t) + p_2(t),$$

where

$$p_1(t) \equiv \frac{a_0}{2} f(t) + \sum_{n=1}^{\infty} a_n f(t) \cos(n2\omega_c t)$$

and

$$p_2(t) \equiv \frac{a_0}{2} \alpha + \sum_{n=1}^{\infty} a_n \alpha \cos(n2\omega_c t).$$

Now, the response of the filter $H_{LPF}(\omega)$ to the input $p_2(t)$ is

$$q_2(t) = \alpha,$$

since (see Figure 8.8c)

$$H_{LPF}(0) = \frac{2}{a_o}$$

and

$$H_{LPF}(n2\omega_c) = 0 \text{ for } n \geq 1.$$

To determine the filter response $q_1(t)$ to input $p_1(t)$, we first observe (using the Fourier addition and modulation properties) that

$$P_1(\omega) = \frac{a_0}{2}F(\omega) + \sum_{n=1}^{\infty} \frac{a_n}{2}\{F(\omega - n2\omega_c) + F(\omega + n2\omega_c)\}.$$

Because only the first term is within the passband of $H_{\text{LPF}}(\omega)$, it follows that

$$Q_1(\omega) = H_{\text{LPF}}(\omega)P_1(\omega) = F(\omega),$$

implying that

$$q_1(t) = f(t).$$

Therefore, using superposition, we find the filter output due to input $p(t) = p_1(t) + p_2(t)$ to be

$$q(t) = q_1(t) + q_2(t) = f(t) + \alpha,$$

which is the desired envelope of the envelope detector input

$$r(t) = (f(t) + \alpha)\cos(\omega_c t).$$

In summary, an envelope detector performs exactly as intended: It detects the envelope of an AM signal corresponding to an audio signal that is offset by some suitable DC level.

Example 8.4

Is envelope detection a linear or nonlinear process? Explain.

Solution   Envelope detection is a nonlinear process, because, in general, the envelope of

$$(f_1(t) + f_2(t))\cos(\omega_c t)$$

will be different from the sum of envelopes $|f_1(t)|$ and $|f_2(t)|$ of signals

$$f_1(t)\cos(\omega_c t)$$

and

$$f_2(t)\cos(\omega_c t),$$

respectively. The envelope of $(f_1(t) + f_2(t))\cos(\omega_c t)$ is $|f_1(t) + f_2(t)|$, but

$$|f_1(t) + f_2(t)| \neq |f_1(t)| + |f_2(t)|,$$

unless, for each value of t, $f_1(t)$ and $f_2(t)$ have the same algebraic sign.

Example 8.5

Suppose that the input signal of an envelope detector is

$$r(t) = (f(t) + \alpha)\cos(\omega_c(t - t_o)),$$

where $t_o > 0$. What will be the detector output, assuming that $f(t) + \alpha > 0$?

Solution   The detector output still will be

$$f(t) + \alpha,$$

because the amplitude coefficients of the Fourier series of $|\cos(\omega_c(t - t_o))|$ and $|\cos(\omega_c t)|$ are identical. Hence, envelope detection is not sensitive to a phase shift of the carrier signal.

We have just described the operation of an ideal envelope detector. Figure 8.9 shows a practical envelope detector circuit that offers an approximation to the ideal system in Figure 8.8a (which consists of both a full-wave rectifier and an ideal low-pass filter). The practical and very simple detector consists of a diode in series with a parallel RC network. The capacitor voltage $q(t)$ in the circuit will closely approximate the envelope of an AM input $r(t)$ if the RC time constant of the circuit is appropriately selected (long compared with the carrier period $\frac{2\pi}{\omega_c}$ and short compared with the inverse of the bandwidth of the envelope of $r(t)$). Whenever the capacitor voltage $q(t)$ is larger than the instantaneous value of the AM input $r(t)$, the diode behaves as an open circuit and the capacitor discharges through resistor R with a time constant RC. When the decaying capacitor voltage $q(t)$ drops below $r(t)$, the diode starts conducting, the capacitor begins recharging, and $q(t)$ is pulled up to follow $r(t)$. The decay and growth cycle repeats after $r(t)$ dips below $q(t)$ (once every period of $\cos(\omega_c t)$), and the overall result is an output that closely approximates the envelope of $r(t)$. The output will contain a small ripple component (a deviation from the true envelope) with an energy content concentrated near the frequency ω_c and above. This component cannot be converted to sound by audio loudspeakers and headphones; consequently, its presence can be ignored for all practical purposes.

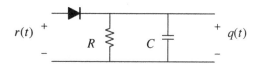

Figure 8.9 A practical envelope detector. When $r(t) > q(t)$, the diode conducts and the capacitor C charges up to a voltage $q(t)$ that remains close to the envelope of $r(t)$.

8.4   Superheterodyne AM Receivers with Envelope Detection

Figure 8.10 depicts a portion of the energy spectrum of a possible AM receiver input versus frequency $f = \frac{\omega}{2\pi} \geq 0$. The spectrum consists of features 10 kHz apart, where each feature is the energy spectrum of a different AM signal broadcast from a different radio station. A practical AM receiver, employing an envelope detector, needs to select one of these AM signals by using a band-pass filter placed at the receiver front end. One option is to use a band-pass filter with a variable, or *tunable*, center frequency. However, such filters are difficult to build when the *bandwidth-to-center-frequency ratio* $\frac{B}{f}$ is of the order 1% or less. With $B = 10$ kHz and $f = 1$ MHz (a typical frequency in the AM broadcast band), this ratio[5] is exactly 1%.

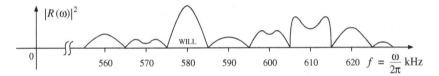

Figure 8.10   A sketch of the energy spectrum of a portion of the AM broadcast band. The University of Illinois AM station WILL operates at a carrier frequency of 580 kHz, near the low end of the AM broadcast band that extends from 540 to 1700 kHz. Each AM broadcast occupies a bandwidth of 10 kHz. In the United States, frequency allocations are regulated by the FCC (Federal Communications Commission).

A practical way to circumvent this difficulty is to use a band-pass filter with a fixed center frequency that is below the AM band, for example, at

$$f = f_{\text{IF}} = \frac{\omega_{\text{IF}}}{2\pi} = 455 \text{ kHz},$$

and to *shift* or *heterodyne* the frequency band of the desired AM signal into the pass-band of this filter, called an *IF filter*, which is short for *intermediate frequency* filter.

Intermediate frequency (IF)

As shown in Figure 8.11, we can do this by mixing the receiver input signal $r(t)$ with a *local oscillator* signal $\cos(\omega_{\text{LO}}t)$ of an appropriate *frequency* ω_{LO} prior to entry of $r(t)$ into the IF filter. To understand how this procedure works, let us first assume that in Figure 8.11

Local oscillator (LO)

$$r(t) = f(t)\cos(\omega_c t),$$

---

[5]This is also the inverse of the *quality factor Q* introduced and discussed in Chapter 12.

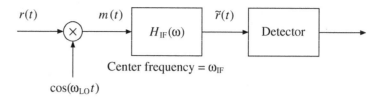

Figure 8.11 Heterodyning an incoming signal $r(t)$ into the IF band by the use of a local-oscillator signal $\cos(\omega_{LO}t)$.

representing a single AM signal. The mixer output then can be represented as[6]

$$m(t) = f(t)\cos(\omega_c t)\cos(\omega_{LO}t)$$

$$= \frac{1}{2}f(t)\cos((\omega_{LO} - \omega_c)t) + \frac{1}{2}f(t)\cos((\omega_{LO} + \omega_c)t).$$

Choosing

$$\omega_{LO} = \omega_c + \omega_{IF}$$

so that

$$\omega_{LO} - \omega_c = \omega_{IF},$$

we find that the expression for $m(t)$ reduces to

$$m(t) = \frac{1}{2}f(t)\cos(\omega_{IF}t) + \frac{1}{2}f(t)\cos((2\omega_c + \omega_{IF})t),$$

which is equivalent to the sum of two AM station outputs where the audio signal $f(t)$ is transmitted by two carriers, ω_{IF} and $2\omega_c + \omega_{IF}$.

Now, the first term of $m(t)$ lies within the passband of an IF filter constructed about a center frequency ω_{IF}, whereas the second term lies in the filter stopband. Therefore, in Figure 8.11 the IF filter output $\tilde{r}(t)$ is just the first term of $m(t)$ and the detector output will be proportional to the desired $f(t)$ (or $f(t) + \alpha$, if a DC offset is included in the incoming AM signal).

---

[6]We use here the *sum and difference* trigonometric identity

$$\cos(a)\cos(b) = \frac{1}{2}\cos(b-a) + \frac{1}{2}\cos(b+a);$$

since cosine is an even function, the order of the difference (i.e., $b - a$ versus $a - b$) does not matter. To verify the identity, note that

$$\cos(a)\cos(b) = \frac{e^{ja} + e^{-ja}}{2}\frac{e^{jb} + e^{-jb}}{2} = \frac{e^{j(a+b)} + e^{-j(a+b)}}{4} + \frac{e^{j(b-a)} + e^{-j(b-a)}}{4},$$

which obviously equals one-half times the sum of two cosines with sum and difference arguments.

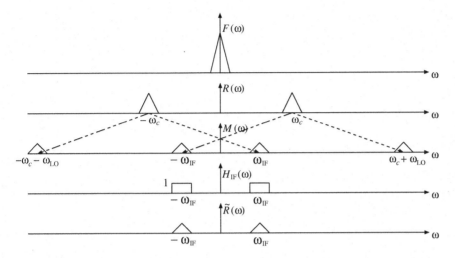

Figure 8.12 Fourier-domain view of the mixing and filtering operations shown in Figure 8.11. $R(\omega)$ represents the Fourier transform of the input to the system.

Figure 8.12 helps illustrate, in the Fourier domain, the mixing and filtering operations shown in Figure 8.11. Figure 8.12 depicts possible Fourier transforms of signals $f(t)$, $r(t)$, and $m(t)$, as well as an ideal IF-filter frequency response $H_{\mathrm{IF}}(\omega)$. Notice that $R(\omega)$ and $M(\omega)$ can be obtained from $F(\omega)$ via successive uses of the Fourier modulation property with frequency shifts $\pm\omega_c$ and $\pm\omega_{\mathrm{LO}} = \pm(\omega_c + \omega_{\mathrm{IF}})$. Comparison of $\tilde{R}(\omega)$ and $R(\omega)$ indicates that, at the IF filter output, the signal $\tilde{r}(t)$ is just a carrier-shifted version of the signal $r(t)$.

In commercial AM receivers,

$$f_{\mathrm{IF}} = \frac{\omega_{\mathrm{IF}}}{2\pi} = 455\,\mathrm{kHz}$$

is used as the standard intermediate frequency. So, to listen to WILL, which broadcasts from the University of Illinois with the carrier frequency

$$f_c = \frac{\omega_c}{2\pi} = 580\,\mathrm{kHz},$$

we must tune the local oscillator to

$$f_{\mathrm{LO}} = \frac{\omega_{\mathrm{LO}}}{2\pi} = 580 + 455 = 1035\,\mathrm{kHz}.$$

To listen to WTKA in Ann Arbor, with $f_c = 1050\,\mathrm{kHz}$, we need

$$f_{\mathrm{LO}} = 1050 + 455 = 1505\,\mathrm{kHz}.$$

In commercial AM receivers, f_{LO} is controlled by a tuning knob that changes the capacitance C in an LC-based oscillator circuit. The knob adjusts C so that

$$\frac{1}{\sqrt{LC}} = 2\pi f_{LO}.$$

In the presence of multiple AM signals (with different carriers), the receiver just described will not perform as desired because of an *image station* problem. To understand the image station problem, examine Figure 8.13. In this figure, the M-shaped component of $R(\omega)$ depicts the Fourier transform of an AM signal being broadcast by a second radio station with a carrier $\omega_c + 2\omega_{IF}$ that accompanies the AM signal from the first radio station with carrier ω_c. When the signal with carrier $\omega_c + 2\omega_{IF}$ is mixed with the LO signal

$$\cos(\omega_{LO}t) = \cos((\omega_c + \omega_{IF})t)$$

tuned to carrier ω_c, its Fourier transform (the M-shaped component) is also shifted into the IF band, as illustrated in Figure 8.13 (a straightforward consequence of the Fourier modulation property applied to $R(\omega)$). Thus, the signal from the second station, called the image station, interferes with the ω_c carrier signal, unless $r(t)$ is band-pass filtered prior to the LO mixer to eliminate the image station signal. This filtering can be accomplished with a *preselector* band-pass filter placed in front of the LO mixer. The preselector filter must have a variable center frequency ω_c, but it need not be a high-quality filter. The preselector is permitted to have a wide bandwidth that varies with ω_c, so long as the bandwidth remains smaller than about $2\omega_{IF}$, since ω_c and the image carrier $\omega_c + 2\omega_{IF}$ are $2\omega_{IF}$ apart (and thus a narrower preselector bandwidth is unnecessary). Since the bandwidth-to-center-frequency ratio

Image station, interference

$$\frac{2\omega_{IF}}{\omega_c} = \frac{2f_{IF}}{f_c}$$

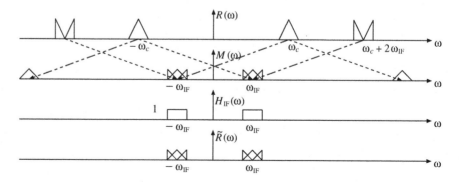

Figure 8.13 Similar to Figure 8.12 but illustrating the image station problem.

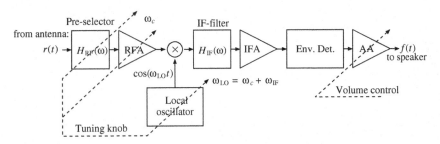

Figure 8.14 Block diagram of a superheterodyne AM receiver. RFA, IFA, and AA represent RF amplifier, IF amplifier, and audio amplifier, respectively.

of the preselector is close to 100% in the AM band, its construction is relatively simple and inexpensive, despite the variable center–frequency requirement. The tuning knob in commercial AM receivers simultaneously controls both the LO frequency $\omega_c + \omega_{IF}$ and the preselector center frequency ω_c.

Figure 8.14 shows the block diagram of a *superheterodyne*, or *superhet*, AM receiver that incorporates the features discussed. In addition to the preselector (to eliminate the image station), the LO mixer (to heterodyne the station of interest into the IF band), the IF filter (to eliminate adjacent channel stations), and the envelope detector, the diagram includes three blocks identified as RFA, IFA, and AA, which are RF, IF, and audio amplifiers, respectively. The audio amplifier is controlled by an external volume knob to adjust the sound output level of the receiver. The amplifier also blocks the DC component α from the envelope detector. In commercial AM receivers the same DC component is used to control the gains of the RF and IF amplifiers to maintain a constant output level under slowly time-varying propagation or reception conditions (such as the decrease in the incoming signal level that you experience when you are traveling in your car away from a broadcast station).

We note that in superhet receivers,

$$\omega_{LO} = \omega_c + \omega_{IF}$$

is not the only possible LO choice. A local oscillator signal $\cos(\omega_{LO}t)$ with frequency

$$\omega_{LO} = \omega_c - \omega_{IF}$$

also works, and ω_{IF} can be specified below, within, or even above the broadcast band. Some of these possibilities are explored in Exercise Problem 8.8.

However, the standard IF frequency of 455 kHz for commercial AM receivers and the *high-LO* standard $\omega_{LO} = \omega_c + \omega_{IF}$ have advantages: First, the low-IF (i.e., IF below the AM broadcast band) leads to a reasonably large $\frac{B}{f_{IF}}$ ratio that lessens the cost of the IF filter circuit. Second, the high-LO choice (i.e., $\omega_{LO} = \omega_c + \omega_{IF}$) has the advantage over *low-LO* in that it requires the generation of LO frequencies only in the range from 995 to 2155 kHz, a reasonable 2-to-1 tuning ratio as opposed to a

Superhet

High-LO versus low-LO

more demanding 15-to-1 ratio (85 to 1245 kHz) that would be required by a low-LO system based on $\omega_{LO} = \omega_c - \omega_{IF}$.

EXERCISES

8.1   Verify the frequency-shift property

$$f(t)e^{j\omega_o t} \leftrightarrow F(\omega - \omega_o)$$

by taking the inverse Fourier transform of $F(\omega - \omega_o)$.

8.2   Given that

$$f(t)e^{\pm j\omega_o t} \leftrightarrow F(\omega \mp \omega_o),$$

determine the Fourier transform of

$$f(t)\sin(\omega_o t).$$

8.3   Given that

$$\cos(\omega_o t + \theta) = \cos(\omega_o t)\cos(\theta) - \sin(\omega_o t)\sin(\theta),$$

determine the Fourier transform of

$$f(t)\cos(\omega_o t + \theta).$$

Hint: Use the multiplication, addition, and modulation Fourier properties, as well as the frequency-shift property of Problem 8.1.

8.4   Given that

$$f(t)\cos(\omega_o t) \leftrightarrow \frac{1}{2}F(\omega - \omega_o) + \frac{1}{2}F(\omega + \omega_o),$$

determine the Fourier transform of

$$f(t + \frac{\theta}{\omega_o})\cos(\omega_o t + \theta).$$

Hint: Use the time-shift property.

8.5   A linear system with frequency response $H(\omega)$ is excited with an input

$$f(t) \leftrightarrow F(\omega).$$

$H(\omega)$ and $F(\omega)$ are plotted below:

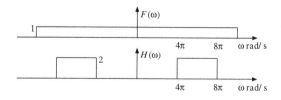

(a) Sketch the Fourier transform $Y(\omega)$ of the system output $y(t)$ and calculate the energy W_y of $y(t)$.

(b) It is observed that output $q(t)$ of the following system equals $y(t)$ determined in part (a).

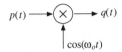

Sketch $P(\omega)$ and determine ω_o.

8.6 It was indicated that the coherent demodulation scheme depicted in Figure 8.6a does not perform properly when the phase of the demodulating carrier signal is mismatched to the phase of the modulating carrier. In Figure 8.6a, suppose $k = 1$ and $t_o = 0$, but that the demodulating carrier is $\cos(\omega_o t + \theta)$, where θ is the phase mismatch.

(a) Find an expression for $y(t)$ in terms of $f(t)$ and θ.

(b) For what values of θ is the amplitude of $y(t)$ the largest and smallest?

(c) Explain how a time-varying θ would affect the sound that you hear coming from the loudspeaker. This should help you understand why a coherent-demodulation receiver requires precise tracking of the carrier phase.

8.7 Consider the system

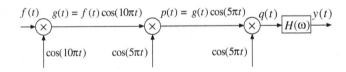

where $F(\omega)$ and $H(\omega)$ are as follows:

(a) Express $q(t)$ in terms of $p(t)$.

(b) Sketch the Fourier transforms $G(\omega)$, $P(\omega)$, $Q(\omega)$, and $Y(\omega)$.

(c) Express $y(t)$ in terms of $f(t)$.

8.8   We wish to heterodyne an AM signal

$$f(t)\cos(\omega_c t)$$

into an IF band, with a center frequency ω_{IF}, by mixing it with a signal

$$\cos(\omega_{LO} t).$$

Assume that an IF filter is present with twice the bandwidth of the low-pass signal $f(t)$, as in our discussion of AM receiver systems in Section 8.4. Determine all of the usable values of ω_{LO} if

(a) $\omega_c = 2\pi 10^6$ and $\omega_{IF} = 5\pi 10^6$ rad/s.

(b) $\omega_c = 4\pi 10^6$ and $\omega_{IF} = \pi 10^6$ rad/s.

8.9   For each possible choice of ω_{LO} in Problem 8.8(a), determine the carrier frequency of the corresponding image station.

8.10   What would be a disadvantage of using a very low IF, say, $f_{IF} = 20$ kHz, in AM reception? Also, under what circumstances could such a low IF be tolerated? Hint: Think of image station interference issues.

9

Convolution, Impulse, Sampling, and Reconstruction

Figure 9.1 is a replica of Figure 5.9a from Chapter 5. In Chapters 5 through 7 we developed a *frequency-domain* description of how dissipative LTI systems, such as $H(\omega) = \frac{1}{1+j\omega}$, convert their inputs $f(t)$ to outputs $y(t)$. The description relies on the fact that all signals that can be generated in the lab can be expressed as sums of co-sinusoids (a discrete or continuous sum (integral), depending on whether the signal is periodic or nonperiodic). LTI systems scale the amplitudes of co-sinusoidal inputs with the amplitude response $|H(\omega)|$ and shift their phases by the phase response $\angle H(\omega)$. When all co-sinusoidal components of an input $f(t)$ are modified according to this simple rule, their sum is the system response $y(t)$. That, in essence, is the frequency-domain description of how LTI systems work.

There is an alternative *time-domain* description of the same process, which is the main topic of this chapter. Notice how response $y(t)$ in Figure 9.1 appears as an "ironed-out" version of input $f(t)$; it is as if $y(t)$ is some sort of "running average" of the input. In this chapter we will learn that is exactly the case for a low-pass filter. More generally, the output $y(t)$ of any LTI system is a weighted linear superposition of the present and past values of the input $f(t)$, where the specific type of running average of $f(t)$ is controlled by a function $h(t)$ related to the frequency response $H(\omega)$. For reasons that will become clear, we will refer to $h(t)$ as the system *impulse response,* and we will use the term *convolution* to describe the mathematical operation between $f(t)$ and $h(t)$ that yields $y(t)$.

A time-domain perspective on signal processing: convolution, impulse $\delta(t)$, impulse response $h(t)$; Fourier transform of power signals; sampling and signal reconstruction.

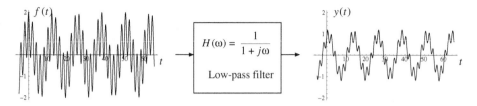

Figure 9.1   An example of a LTI system response to an input $f(t)$.

In Section 9.1 we will examine the convolution properties of the Fourier transform (properties 13 and 14 in Table 7.1) and learn about the convolution operation and its properties. In Section 9.2 we will introduce a new signal $\delta(t)$, known as the *impulse,* and learn how to use it in convolution and Fourier transform calculations. Section 9.3 extends the idea of Fourier transforms to signals with infinite energy, but finite power—signals such as sines, cosines, and the unit step. These so-called *power signals* are shown to have Fourier transforms that can be expressed in terms of impulses in the frequency domain. The chapter concludes with further applications of convolution and the impulse, including, in Section 9.4, discussions of sampling (analog-to-digital conversion) and signal reconstruction.

9.1   Convolution

Given a pair of signals, say, $h(t)$ and $f(t)$, we call a new signal $y(t)$, defined as

$$y(t) \equiv \int_{-\infty}^{\infty} h(\tau)f(t - \tau)d\tau,$$

Convolution   the *convolution* of $h(t)$ and $f(t)$ and denote it symbolically as

$$y(t) = h(t) * f(t).$$

As we later shall see, we can perform many LTI system calculations by *convolving* the system input $f(t)$ with the inverse Fourier transform $h(t)$ of the system frequency response $H(\omega)$. That possibility motivates our study of convolution in this section.

Example 9.1
Let $y(t) \equiv h(t) * u(t)$, the convolution of some $h(t)$ with the unit step $u(t)$.
Express $y(t)$ in terms of $h(t)$ only.

Solution   Since

$$y(t) = h(t) * u(t) = \int_{-\infty}^{\infty} h(\tau)u(t - \tau)d\tau$$

and

$$u(t - \tau) = \begin{cases} 1, & \tau < t, \\ 0, & \tau > t, \end{cases}$$

by replacing $u(t - \tau)$ in the integrand by 1 and changing the upper integration limit from ∞ to t, we obtain the answer:

$$y(t) = \int_{-\infty}^{t} h(\tau)d\tau.$$

Example 9.2
Determine the function

$$y(t) = u(t) * u(t).$$

Solution   Using the result of Example 9.1 with $h(t) = u(t)$, we get

$$y(t) = \int_{-\infty}^{t} u(\tau)d\tau = \begin{cases} 0, & t < 0, \\ \int_{0}^{t} d\tau = t, & t > 0, \end{cases} = tu(t).$$

So, the convolution of a unit step with another unit step produces a ramp signal $tu(t)$.

9.1.1   Fourier convolution properties

We are just about to show that if $h(t) \leftrightarrow H(\omega)$, $f(t) \leftrightarrow F(\omega)$, and

$$Y(\omega) = H(\omega)F(\omega),$$

then the inverse Fourier transform $y(t)$ of $Y(\omega)$ is the convolution

$$y(t) = h(t) * f(t).$$

As a consequence, convolution provides an alternative means of finding the zero-state response $y(t)$ of a system having frequency response $H(\omega)$. This alternative approach to finding $y(t)$ is an application of the *time-convolution* property of the Fourier transform (see item 13 in Table 7.1), which effectively states that

$$h(t) * f(t) \leftrightarrow H(\omega)F(\omega).$$

Convolution in time domain implies multiplication in frequency domain

Verification of time-convolution property: We first Fourier transform $h(t) * f(t)$ as

$$\int_{-\infty}^{\infty} \{h(t) * f(t)\}e^{-j\omega t}dt = \int_{t=-\infty}^{\infty} \left\{ \int_{\tau=-\infty}^{\infty} h(\tau)f(t-\tau)d\tau \right\} e^{-j\omega t}dt,$$

and, subsequently, exchange the order of t and τ integrations to obtain

$$\int_{-\infty}^{\infty} \{h(t) * f(t)\}e^{-j\omega t}dt = \int_{\tau=-\infty}^{\infty} h(\tau) \left\{ \int_{t=-\infty}^{\infty} f(t-\tau)e^{-j\omega t}dt \right\} d\tau.$$

By the Fourier time-shift property, the expression in curly brackets on the right side is recognized as $F(\omega)e^{-j\omega\tau}$, where $F(\omega)$ is the Fourier transform of $f(t)$. Hence,

$$\int_{-\infty}^{\infty} \{h(t) * f(t)\}e^{-j\omega t}dt = \int_{\tau=-\infty}^{\infty} h(\tau)F(\omega)e^{-j\omega\tau}d\tau$$

$$= F(\omega) \int_{\tau=-\infty}^{\infty} h(\tau)e^{-j\omega\tau}d\tau = F(\omega)H(\omega).$$

So, as claimed,

$$h(t) * f(t) \leftrightarrow H(\omega)F(\omega).$$

It similarly is true that the Fourier transform of

$$f(t) * h(t) \equiv \int_{-\infty}^{\infty} f(\tau)h(t-\tau)d\tau$$

is

$$F(\omega)H(\omega) = H(\omega)F(\omega).$$

This indicates that

$$h(t) * f(t) = f(t) * h(t),$$

Convolution is commutative

which means that the convolution operation is *commutative*. Other properties of convolution will be discussed in the next section.

We can verify the Fourier *frequency-convolution* property (item 14 in Table 7.1),

$$f(t)g(t) \leftrightarrow \frac{1}{2\pi}F(\omega) * G(\omega),$$

Multiplication in time-domain implies convolution in frequency-domain

in a similar maner, except by starting with the inverse Fourier transform integral of $F(\omega) * G(\omega)$. In this property, the convolution

$$F(\omega) * G(\omega) \equiv \int_{-\infty}^{\infty} F(\Omega)G(\omega - \Omega)d\Omega,$$

can be replaced with

$$F(\omega) * G(\omega) = G(\omega) * F(\omega) \equiv \int_{-\infty}^{\infty} G(\Omega)F(\omega - \Omega)d\Omega,$$

because convolution is a commutative operation.

9.1.2 The meaning and properties of convolution

To appreciate the meaning of the convolution operation

$$h(t) * f(t) = \int_{-\infty}^{\infty} h(\tau)f(t - \tau)d\tau,$$

let us first examine

$$y(t) \equiv h(t) * f(t)$$

at some specific instant in time, say, at $t = 2$. Evaluating $y(t)$ at $t = 2$, we obtain

$$y(2) = [h(t) * f(t)]_{|t=2} = \int_{-\infty}^{\infty} h(\tau)f(2 - \tau)d\tau.$$

This equation indicates that $y(2)$ is a weighted linear superposition of $f(2)$ (corresponding to $\tau = 0$ in $f(2 - \tau)$) and all other values of $f(t)$ before and after $t = 2$ (corresponding to positive and negative τ in $f(2 - \tau)$, respectively) with different weights $h(\tau)$; $f(2)$ is weighted by $h(0)$, $f(1)$ by $h(1)$, $f(0)$ by $h(2)$, $f(-1)$ by $h(3)$, and so forth. The same interpretation, of course, holds for $y(t)$ at every value of t; $y(t) = h(t) * f(t)$ is a weighted linear superposition of present ($\tau = 0$), past ($\tau > 0$), and future ($\tau < 0$) values $f(t - \tau)$ of the signal $f(t)$ with $h(\tau)$ weightings.

Table 9.1 lists some of the important properties of convolution. The first three properties indicate that convolution is commutative (as we already have seen), distributive, and associative. Convolution is distributive because   **Distributive**

$$f(t) * (g(t) + h(t)) = \int_{-\infty}^{\infty} f(\tau)(g(t - \tau) + h(t - \tau))d\tau$$

$$= \int_{-\infty}^{\infty} f(\tau)g(t - \tau)d\tau + \int_{-\infty}^{\infty} f(\tau)h(t - \tau)d\tau$$

$$= f(t) * g(t) + f(t) * h(t).$$

To show that convolution is associative, we can use the Fourier time-convolution   **Associative**
property to note that

$$f(t) * (g(t) * h(t)) \leftrightarrow F(\omega)(G(\omega)H(\omega)) = (F(\omega)G(\omega))H(\omega)$$

Commutative	$h(t) * f(t) = f(t) * h(t)$
Distributive	$f(t) * (g(t) + h(t)) = f(t) * g(t) + f(t) * h(t)$
Associative	$f(t) * (g(t) * h(t)) = (f(t) * g(t)) * h(t)$
Shift	$h(t) * f(t) = y(t) \ \Rightarrow \ h(t - t_o) * f(t) = h(t) * f(t - t_o) = y(t - t_o)$
Derivative	$h(t) * f(t) = y(t) \ \Rightarrow \ \frac{d}{dt}h(t) * f(t) = h(t) * \frac{d}{dt}f(t) = \frac{d}{dt}y(t)$
Reversal	$h(t) * f(t) = y(t) \ \Rightarrow \ h(-t) * f(-t) = y(-t)$
Start-point	If $h(t) = 0$ for $t < t_{sh}$ and $f(t) = 0$ for $t < t_{sf}$ then $y(t) = h(t) * f(t) = 0$ for $t < t_{sy} = t_{sh} + t_{sf}$.
End-point	If $h(t) = 0$ for $t > t_{eh}$ and $f(t) = 0$ for $t > t_{ef}$ then $y(t) = h(t) * f(t) = 0$ for $t > t_{ey} = t_{eh} + t_{ef}$.
Width	$h(t) * f(t) = y(t) \ \Rightarrow \ T_y = T_h + T_f$ where T_h, T_f and T_y denote the widths of $h(t)$, $f(t)$, and $y(t)$.

Table 9.1 Convolution and its properties.

as well as

$$(f(t) * g(t)) * h(t) \leftrightarrow (F(\omega)G(\omega))H(\omega).$$

Thus, by the uniqueness of the Fourier transform, it follows that

$$f(t) * (g(t) * h(t)) = (f(t) * g(t)) * h(t).$$

Shift property

The convolution shift property also can be verified using the properties of the Fourier transform: Since

$$f(t - t_o) \leftrightarrow F(\omega)e^{-j\omega t_o},$$

it follows that

$$h(t) * f(t - t_o) \leftrightarrow H(\omega)F(\omega)e^{-j\omega t_o} \equiv Y(\omega)e^{-j\omega t_o},$$

where $Y(\omega) = H(\omega)F(\omega)$ has inverse Fourier transform $y(t) = h(t) * f(t)$. But the inverse Fourier transform of $Y(\omega)e^{-j\omega t_o}$ is $y(t - t_o)$, so

$$h(t) * f(t - t_o) = y(t - t_o),$$

as claimed. Also, using the commutative property of convolution, we can see that

$$h(t - t_o) * f(t) = y(t - t_o)$$

must be true.

To verify the derivative property, note that

$$\frac{d}{dt} y(t) = \frac{d}{dt}[h(t) * f(t)] \leftrightarrow j\omega[H(\omega)F(\omega)] = H(\omega)[j\omega F(\omega)],$$

which implies that

$$h(t) * \frac{df}{dt} = \frac{dy}{dt}.$$

Likewise,

$$\frac{dh}{dt} * f(t) = \frac{dy}{dt}$$

is true, since convolution is commutative.

Example 9.3
Given that

$$f(t) * g(t) = p(t),$$

where $p(t) = \triangle(\frac{t}{2})$ (as shown in Figure 9.2a), determine and plot

$$c(t) = f(t) * (g(t) - g(t - 2)).$$

Solution   Using the distributive property, we find that

$$c(t) = f(t) * (g(t) - g(t - 2)) = f(t) * g(t) - f(t) * g(t - 2).$$

Since $f(t) * g(t) = p(t)$, the shift property indicates that

$$f(t) * g(t - 2) = p(t - 2).$$

Therefore,

$$c(t) = p(t) - p(t - 2) = \triangle(\frac{t}{2}) - \triangle(\frac{t-2}{2}),$$

which is plotted in Figure 9.2b.

We will not prove the start-point, end-point, and width properties, but after every example that follows you should check that the convolution results are consistent with these properties. The width property is relevant only when both $h(t)$ and $f(t)$ have finite widths over which they have nonzero values. As an example, $h(t) = \text{rect}(t)$ has unit width (width=1), because outside the interval $-\frac{1}{2} < t < \frac{1}{2}$ the value of $\text{rect}(t)$ is

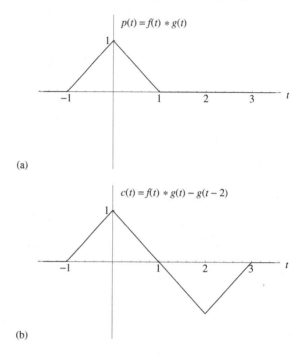

(a)

(b)

Figure 9.2   (a) $p(t) = f(t) * g(t) = \triangle(\frac{t}{2})$, and (b) $c(t) = f(t) * (g(t) - g(t-2))$.

zero. Likewise, $f(t) = \triangle(\frac{t}{2})$ has a width of two units. Hence, the width property tells us that if $h(t) = \text{rect}(t)$ and $f(t) = \triangle(\frac{t}{2})$ were convolved, the result $y(t)$ would be $1 + 2 = 3$ units wide. Also, functions $h(t) = \text{rect}(t)$ and $f(t) = \triangle(\frac{t}{2})$ start at times $t = -\frac{1}{2}$ and $t = -1$, respectively, which means that the convolution $\text{rect}(t) * \triangle(\frac{t}{2})$ will start at time $t = (-\frac{1}{2}) + (-1) = -\frac{3}{2}$ (according to the start-point property).

Example 9.4
Given that

$$c(t) = \text{rect}(\frac{t-5}{2}) * \triangle(\frac{t-8}{4}),$$

determine the width and start-time of $c(t)$.

Solution   Since the widths of $\text{rect}(\frac{t-5}{2})$ and $\triangle(\frac{t-8}{4})$ are 2 and 4, respectively, the width of $c(t)$ must be $2 + 4 = 6$. Since the start times of $\text{rect}(\frac{t-5}{2})$ and $\triangle(\frac{t-8}{4})$ are 4 and 6, respectively, the start time of $c(t)$ must be $4 + 6 = 10$. You should try to find the convolution $c(t) = \text{rect}(\frac{t-5}{2}) * \triangle(\frac{t-8}{4})$ after reading the next section (see Exercise Problem 9.5) and then verify the width and start time.

1	$h(t) * u(t) = \int_{-\infty}^{t} h(\tau)d\tau$ for any $h(t)$	4	$h(t) * \delta(t) = h(t)$ for any $h(t)$
2	$\text{rect}(\frac{t}{T}) * \text{rect}(\frac{t}{T}) = T\triangle(\frac{t}{2T})$	5	$h(t) * \delta(t - t_o) = h(t - t_o)$ for any $h(t)$
3	$u(t) * u(t) = tu(t)$		

Table 9.2 A short list of frequently encountered convolutions. Signals $\delta(t)$ and $\delta(t - t_o)$ will be discussed in Section 9.2.

9.1.3 Convolution examples and graphical convolution

Example 9.5
Find the convolution

$$y(t) = u(t) * e^t.$$

Solution Because convolution is commutative, $y(t) = u(t) * e^t = e^t * u(t)$. Therefore, using the general result

$$h(t) * u(t) = \int_{-\infty}^{t} h(\tau)d\tau$$

established in Example 9.1 (the same result is also included in Table 9.2 that lists some commonly encountered convolutions), we find that

$$y(t) = e^t * u(t) = \int_{-\infty}^{t} e^\tau d\tau = e^t - e^{-\infty} = e^t.$$

Example 9.6
Given that

$$h(t) = e^{-t}u(t),$$
$$f(t) = e^{-2t}u(t),$$

and $y(t) = h(t) * f(t)$, determine $y(1)$.

Solution Since we can calculate $y(t)$ with the formula

$$y(t) = \int_{-\infty}^{\infty} h(\tau)f(t - \tau)d\tau = \int_{-\infty}^{\infty} [e^{-\tau}u(\tau)][e^{-2(t-\tau)}u(t - \tau)]d\tau,$$

for any value of t, it follows that

$$y(1) = \int_{-\infty}^{\infty} [e^{-\tau}u(\tau)][e^{-2(1-\tau)}u(1-\tau)]d\tau = \int_{0}^{1} e^{-\tau}e^{-2(1-\tau)}d\tau.$$

To obtain the last step we used the fact that

$$u(\tau)u(1-\tau) = \begin{cases} 1, & 0 < \tau < 1, \\ 0, & \tau < 0 \text{ or } \tau > 1, \end{cases}$$

and replaced $u(\tau)u(1 - \tau)$ by 1 after changing the integration limits to 0 and 1 as shown. Continuing from where we left off, we get

$$y(1) = \int_{0}^{1} e^{-\tau}e^{-2(1-\tau)}d\tau = e^{-2}\int_{0}^{1} e^{\tau}d\tau$$

$$= e^{-2}(e^1 - 1) = e^{-1} - e^{-2} \approx 0.233.$$

Example 9.7

Repeat Example 9.6 to determine $y(t) = h(t) * f(t)$ for all values of t. The signals $h(t)$ and $f(t)$ are plotted in Figures 9.3a and b.

Solution   Once again,

$$y(t) = \int_{-\infty}^{\infty} h(\tau)f(t-\tau)d\tau = \int_{-\infty}^{\infty} e^{-\tau}u(\tau)e^{-2(t-\tau)}u(t-\tau)d\tau$$

$$= e^{-2t}\int_{-\infty}^{\infty} e^{\tau}u(\tau)u(t-\tau)d\tau.$$

Now, generalizing from Example 9.6, we obtain

$$u(\tau)u(t-\tau) = \begin{cases} 1, & 0 < \tau < t, \\ 0, & \tau < 0 \text{ or } \tau > t. \end{cases}$$

This follows because $u(t - \tau) = 0$ for $\tau > t$, as shown in Figure 9.3c. Therefore, assuming that $t > 0$, we get

$$y(t) = e^{-2t}\int_{-\infty}^{\infty} e^{\tau}u(\tau)u(t-\tau)d\tau = e^{-2t}\int_{0}^{t} e^{\tau}d\tau$$

$$= e^{-2t}(e^t - 1) = e^{-t} - e^{-2t}.$$

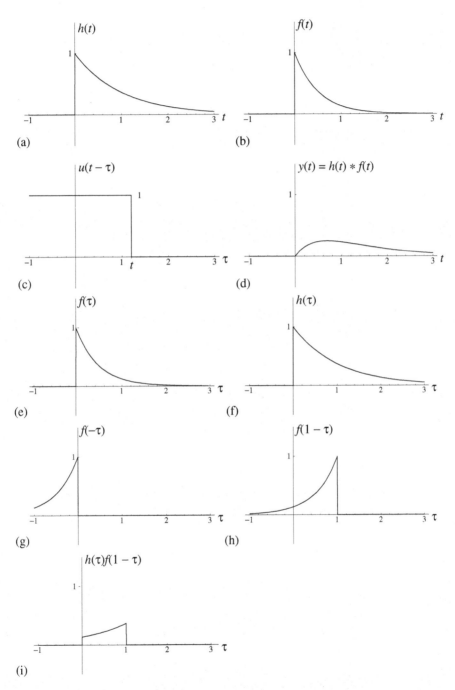

Figure 9.3   (a) $h(t) = e^{-t}u(t)$, (b) $f(t) = e^{-2t}u(t)$, (c) $u(t - \tau)$ vs τ, and (d) $y(t) = h(t) * f(t) = (e^{-t} - e^{-2t})u(t)$. (e)–(i) are self explanatory.

For $t < 0$, $u(\tau)u(t - \tau)$ is identically zero, so that $y(t) = 0$. Therefore, a formula for $y(t)$ that holds for all t is

$$y(t) = (e^{-t} - e^{-2t})u(t),$$

which is plotted in Figure 9.3d.

Graphical convolution

Figures 9.3e through 9.3i will provide some further insight about the convolution results of Examples 9.6 and 9.7: Figures 9.3e and 9.3f show plots of $f(\tau) = e^{-2\tau}u(\tau)$ and $h(\tau) = e^{-\tau}u(\tau)$ versus τ. Furthermore, Figures 9.3g and 9.3h show plots of $f(-\tau)$, which is a flipped (reversed) version of Figure 9.3e, and $f(1 - \tau)$, which is the same as Figure 9.3g except shifted to the right by 1 unit in τ. The "area under" the product of the curves shown in Figures 9.3f and 9.3h—that is, the area under the $h(\tau)f(1 - \tau)$ curve shown in Figure 9.3i—is the result of the convolution calculation $\int_{-\infty}^{\infty} h(\tau)f(1 - \tau)d\tau$ performed in Example 9.6 and also the result of Example 9.7 evaluated at $t = 1$. For other values of t, the convolution result is still the area under the product curve $h(\tau)f(t - \tau)$.

As you can see, visualizing the waveforms at the various steps in the convolution process can be challenging. When contemplating evaluation of the convolution integral

$$y(t) = \int_{-\infty}^{\infty} h(\tau)f(t - \tau)d\tau,$$

we *strongly* recommend that, prior to performing the calculation, you make the following series of plots:

(1) Plot $f(\tau)$ versus τ and $h(\tau)$ versus τ (looking ahead to Figures 9.4a and 9.4b, for instance), where τ is the variable of integration.
(2) Plot $f(-\tau)$ versus τ by flipping the plot of $f(\tau)$ about the vertical axis (as in Figure 9.4c).
(3) Plot $f(t - \tau)$ versus τ by shifting the graph of step 2 to the right by some amount t, as shown in Figure 9.4d. (Note that the $\tau = -1$ mark in Figure 9.4c becomes the $\tau = t - 1$ mark in Figure 9.4d.)

The next example illustrates how these plots are helpful in performing convolution, in this case a convolution of the waveforms $2\text{rect}(t - 5.5)$ and $u(t - 1)$.

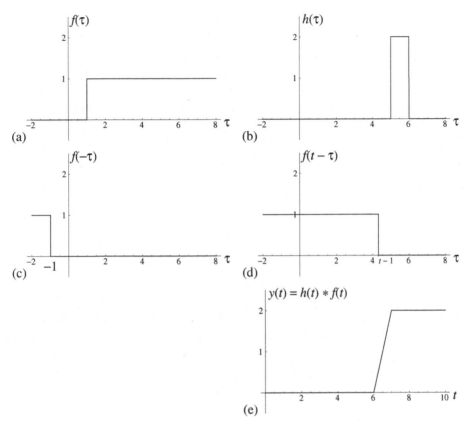

Figure 9.4   (a) $f(\tau)$ vs τ, (b) $h(\tau)$ vs τ, (c) $f(-\tau)$ vs τ, (d) $f(t-\tau)$ vs τ, and (e) $y(t) = h(t) * f(t)$. See Example 9.8.

Example 9.8

Given that $h(t) = 2\text{rect}(t - 5.5)$ and $f(t) = u(t - 1)$, determine $y(t) = h(t) * f(t)$.

Solution   The plots of $h(\tau)$ and $f(t - \tau)$ of the integrand

$$h(\tau)f(t - \tau)$$

of the convolution integral

$$\int_{-\infty}^{\infty} h(\tau)f(t - \tau)d\tau$$

are shown in Figures 9.4b and 9.4d, respectively. (We obtained these plots by following steps 1 through 3 outlined previously). The convolution integral is simply the area under the product of these two curves shown in

Figures 9.4b and 9.4d. As explained next, and as seen from the figures, the area under the product curve depends on the value of t:

- First, for $t - 1 < 5$, the $h(\tau)$ and $f(t - \tau)$ curves do not "overlap," their product is zero, and therefore, $h(t) * f(t) = 0$ for $t < 6$.
- Next, for $5 < t - 1 < 6$, the same two curves overlap between $\tau = 5$ and $\tau = t - 1 > 5$. In this interval only, the product of the two curves is nonzero and equals 2. The area under the product of the curves is $2 \times ((t - 1) - 5) = 2(t - 6)$ for $6 < t < 7$, and hence

$$h(t) * f(t) = 2(t - 6)$$

for the same time interval.
- Finally, for $t - 1 > 6$, or $t > 7$, the two curves fully overlap, and the area under their product is 2. Hence, for $t > 7$,

$$h(t) * f(t) = 2.$$

Therefore,

$$y(t) = 2\mathrm{rect}(t - 5.5) * u(t - 1) = \begin{cases} 0, & t < 6, \\ 2(t - 6), & 6 < t < 7, \\ 2, & t > 7, \end{cases}$$

as shown in Figure 9.4e.

Example 9.9
Figures 9.5a and 9.5b show plots of two waveforms $f(\tau)$ and $h(\tau)$, versus τ. Determine the convolution $y(t) = h(t) * f(t)$.

Solution First, flipping Figure 9.5a, we obtain $f(-\tau)$ shown in Figure 9.5c. Next, we shift Figure 9.5c to the right by an amount t to obtain the graph of $f(t - \tau)$, shown in Figure 9.5d. The convolution $y(t) = h(t) * f(t)$ is the area under the product of the curves shown in Figures 9.5b and 9.5d, which, of course, depends on the value of t as in Example 9.8.

- For $t < 0$, the curves do not overlap, and hence, $y(t) = h(t) * f(t) = 0$.
- For $0 < t < 1$, the curves partially overlap, between $\tau = 0$ and t, and hence,

$$y(t) = \int_0^t 2\tau d\tau = t^2.$$

Note that in obtaining the result, we used the fact that $f(t - \tau) = 1$ and $h(\tau) = 2\tau$, which are valid for $0 < \tau < t < 1$.

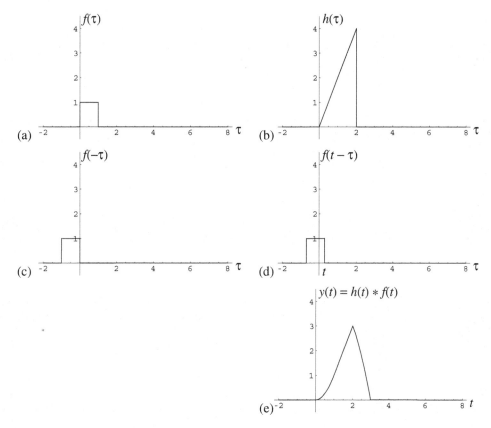

Figure 9.5 (a) $f(\tau)$ vs τ, (b) $h(\tau)$ vs τ, (c) $f(-\tau)$ vs τ, (d) $f(t-\tau)$ vs τ, and (e) $y(t) = h(t) * f(t)$. See Example 9.9.

- For $1 < t < 2$, the curves overlap between $\tau = t - 1$ and $\tau = t$, and hence,

$$y(t) = \int_{t-1}^{t} 2\tau d\tau = t^2 - (t - 1)^2.$$

- For $2 < t < 3$, the curves partially overlap, only between $\tau = t - 1$ and $t = 2$, and hence,

$$y(t) = \int_{t-1}^{2} 2\tau d\tau = 4 - (t - 1)^2.$$

- Finally, for $t > 3$, there is no overlap and $y(t) = 0$.

Thus, overall,

$$y(t) = h(t) * f(t) = \begin{cases} 0, & t < 0, \\ t^2, & 0 < t < 1, \\ t^2 - (t-1)^2 = 2t - 1, & 1 < t < 2, \\ 4 - (t-1)^2, & 2 < t < 3, \\ 0, & t > 3, \end{cases}$$

as plotted in Figure 9.5e.

Example 9.10
Determine $y(t) = h(t) * h(t)$ where $h(t) = \text{rect}(t)$.

Solution   Figures 9.6a and 9.6b display $h(\tau) = \text{rect}(\tau)$ and $h(t - \tau) = \text{rect}(t - \tau)$ versus τ, respectively.

- For $t + 0.5 < -0.5$, or $t < -1$, the two curves do not overlap and $h(t) * h(t) = 0$.
- For $-0.5 < t + 0.5 < 0.5$, or $-1 < t < 0$, the curves overlap between $\tau = -0.5$ and $t + 0.5$. Hence, for $-1 < t < 0$,

$$h(t) * h(t) = \int_{-0.5}^{t+0.5} 1 \cdot 1 d\tau = t + 1.$$

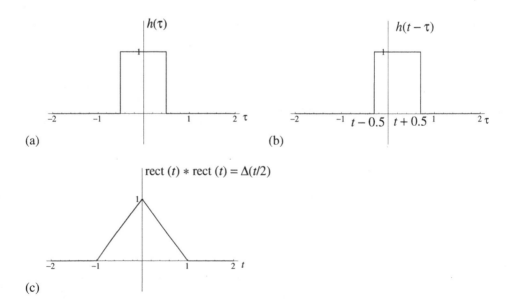

(a)

(b)

(c)

Figure 9.6   (a) $h(\tau) = \text{rect}(\tau)$ vs τ, (b) $h(t - \tau) = \text{rect}(t - \tau)$ vs τ, and (c) $\text{rect}(t) * \text{rect}(t) = \Delta(\frac{t}{2})$.

- Next, for $-0.5 < t - 0.5 < 0.5$, or $0 < t < 1$, the overlap is between $\tau = t - 0.5$ and 0.5. Hence, for $0 < t < 1$,

$$h(t) * h(t) = \int_{t-0.5}^{0.5} 1 \cdot 1 d\tau = 1 - t.$$

- Finally, for $t - 0.5 > 0.5$, or $t > 1$, there is no overlap and thus $h(t) * h(t) = 0$.

  Overall,

$$h(t) * h(t) = \text{rect}(t) * \text{rect}(t) = \begin{cases} 0, & t < -1, \\ t + 1, & -1 < t < 0, \\ 1 - t, & 0 < t < 1, \\ 0, & t > 1, \end{cases} = \triangle(\frac{t}{2}),$$

  as shown in Figure 9.6c.

The result of Example 9.10 is a special case of entry 2 in Table 9.2. According to the entry, the *self-convolution* of a rectangle of width T and unit height is a *triangle* of width $2T$ and apex height T; that is,

$$\text{rect}(\frac{t}{T}) * \text{rect}(\frac{t}{T}) = T\triangle(\frac{t}{2T}).$$

Let us apply the Fourier time-convolution property to this convolution identity. Since from Table 7.2 we know

$$\text{rect}(\frac{t}{T}) \leftrightarrow T\text{sinc}(\frac{\omega T}{2}),$$

the time-convolution property implies that

$$\text{rect}(\frac{t}{T}) * \text{rect}(\frac{t}{T}) \leftrightarrow T^2\text{sinc}^2(\frac{\omega T}{2}).$$

But, we just saw that

$$\text{rect}(\frac{t}{T}) * \text{rect}(\frac{t}{T}) = T\triangle(\frac{t}{2T});$$

therefore,

$$T\triangle(\frac{t}{2T}) \leftrightarrow T^2\text{sinc}^2(\frac{\omega T}{2}).$$

Letting $\tau = 2T$, this relation reduces to

$$\triangle(\frac{t}{\tau}) \leftrightarrow \frac{\tau}{2}\text{sinc}^2(\frac{\omega\tau}{4}),$$

which verifies entry 9 in Table 7.2. Furthermore, entry 10 in the same table can be obtained from this result by using the symmetry property of the Fourier transform.

Example 9.11

Given that $h(t) = \text{rect}(t)$ and $f(t) = \text{rect}(\frac{t}{2})$, determine $y(t) = h(t) * f(t)$.

Solution   This problem is easy if we notice that

$$\text{rect}(\frac{t}{2}) = \text{rect}(t + \frac{1}{2}) + \text{rect}(t - \frac{1}{2});$$

that is, a rect 2 units wide is the same as a pair of side-by-side shifted rects with unit widths, shifted by $\pm\frac{1}{2}$ units. Thus,

$$y(t) = \text{rect}(t) * \text{rect}(\frac{t}{2})$$

$$= \text{rect}(t) * (\text{rect}(t + \frac{1}{2}) + \text{rect}(t - \frac{1}{2}))$$

$$= \triangle(\frac{t + \frac{1}{2}}{2}) + \triangle(\frac{t - \frac{1}{2}}{2}),$$

by the distributive and time-shift properties of convolution and the result of Example 9.10. Figures 9.7a through 9.7c show the plots of $\triangle(\frac{t \pm \frac{1}{2}}{2})$ and

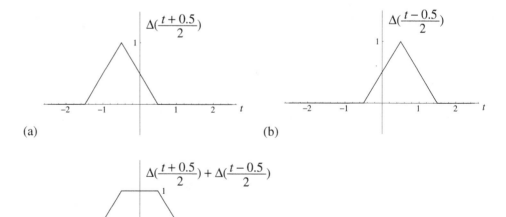

(a)

(b)

(c)

Figure 9.7   (a) $\triangle(\frac{t+0.5}{2})$, (b) $\triangle(\frac{t-0.5}{2})$, and (c) $\text{rect}(t) * \text{rect}(\frac{t}{2}) = \triangle(\frac{t+0.5}{2}) + \triangle(\frac{t-0.5}{2})$.

$y(t)$. You should try finding the same convolution directly, without first decomposing rect($\frac{t}{2}$) into a sum of two rects.

Example 9.12
Convolve the two functions shown in Figures 9.8a and 9.8b.

Solution   Clearly, according to Figure 9.8a, $f(t) = \text{rect}(\frac{t-1}{2})$. Also, according to Figure 9.8b, $h(t) = f(t) - f(t-2)$. Thus, the required convolution is

$$h(t) * f(t) = f(t) * h(t) = f(t) * (f(t) - f(t-2))$$
$$= f(t) * f(t) - f(t) * f(t-2).$$

Next, we observe that since

$$\text{rect}(\frac{t}{2}) * \text{rect}(\frac{t}{2}) = 2\triangle(\frac{t}{4}),$$

we can apply the time-shift property of convolution to write

$$f(t) * f(t) = \text{rect}(\frac{t-1}{2}) * \text{rect}(\frac{t-1}{2}) = 2\triangle(\frac{t-2}{4}).$$

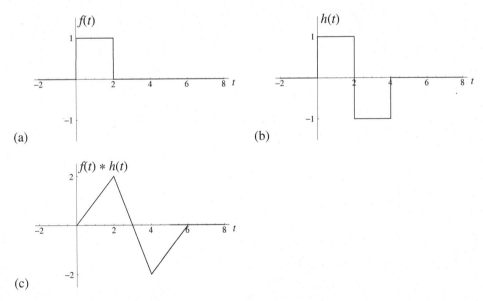

(a)

(b)

(c)

Figure 9.8   (a) $f(t)$, (b) $h(t)$, and (c) their convolution $f(t) * h(t) = h(t) * f(t)$.

Using this result and applying the time-shift property once again, we see that

$$h(t) * f(t) = f(t) * f(t) - f(t) * f(t - 2)$$

$$= 2\Delta(\frac{t-2}{4}) - 2\Delta(\frac{t-4}{4}),$$

as shown in Figure 9.8c.

Example 9.13

Suppose that the input of an LTI system having frequency response

$$H(\omega) = \frac{1}{1 + j\omega}$$

is $f(t) = \text{rect}(t)$. Determine the zero-state response $y(t)$ by using the convolution formula $y(t) = h(t) * f(t)$, where $h(t)$ is the inverse Fourier transform of $H(\omega)$.

Solution   From Table 7.2 we note that

$$e^{-t}u(t) \leftrightarrow \frac{1}{1 + j\omega}.$$

Thus, $H(\omega) = \frac{1}{1+j\omega}$ implies that $h(t) = e^{-t}u(t)$. We now can proceed by computing the convolution directly (making the required plots first). Instead, begin by noting that

$$f(t) = \text{rect}(t) = u(t + \frac{1}{2}) - u(t - \frac{1}{2}).$$

Thus, the system zero-state response can be found as

$$y(t) = h(t) * f(t) = e^{-t}u(t) * \text{rect}(t) = e^{-t}u(t) * [u(t + \frac{1}{2}) - u(t - \frac{1}{2})]$$

$$= q(t + \frac{1}{2}) - q(t - \frac{1}{2}),$$

where

$$q(t) \equiv e^{-t}u(t) * u(t) = \int_{-\infty}^{t} e^{-\tau}u(\tau)d\tau = u(t)(1 - e^{-t}).$$

Thus,

$$y(t) = u(t + \frac{1}{2})(1 - e^{-(t+\frac{1}{2})}) - u(t - \frac{1}{2})(1 - e^{-(t-\frac{1}{2})}),$$

where we have made use of the time-shift property of convolution. The resulting output $y(t)$ is shown in Figure 9.9. Notice that the low-pass filter "smooths" the rectangular input signal into the shape shown in the figure.

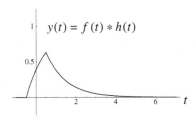

Figure 9.9 The response of low-pass filter $H(\omega) = \frac{1}{1+j\omega}$ to input $f(t) = \text{rect}(t)$.

9.2 Impulse $\delta(t)$

9.2.1 Definition and properties of the impulse

Convolution is a mathematical operation, just as is multiplication. In the case of multiplication, there is a special number called *one*. Multiplying any number by *one* just reproduces the original number. That is, multiplication by *one* is the identity operation; it does not change the value of a number. In the case of convolution, we might ask whether there is any signal that plays a role analogous to that of *one*. More precisely, is there any signal that, when convolved with an arbitrary signal $f(t)$, always reproduces $f(t)$? In mathematical terms, we are seeking a signal $p(t)$ satisfying

$$p(t) * f(t) = f(t).$$

It turns out that, in general, there is no waveform that exactly satisfies this identity for an arbitrary $f(t)$. However, there are waveforms that will produce an approximation that is as fine as we like. In particular, consider the rectangular pulse signal defined by

$$p_\epsilon(t) \equiv \frac{1}{\epsilon}\text{rect}(\frac{t}{\epsilon}).$$

(See Figure 9.10.) This pulse has unit area and, for small enough ϵ, the pulse is very tall and narrow. In the limit, as ϵ approaches zero, it can be shown that[1]

$$\lim_{\epsilon \to 0}\{p_\epsilon(t) * f(t)\} = f(t),$$

---

[1] **Verification:**

$$p_\epsilon(t) * f(t) \equiv \frac{1}{\epsilon}\text{rect}(\frac{t}{\epsilon}) * f(t) = \int_{-\infty}^{\infty} \frac{1}{\epsilon}\text{rect}(\frac{\tau}{\epsilon})f(t-\tau)d\tau = \frac{\int_{-\epsilon/2}^{\epsilon/2} f(t-\tau)d\tau}{\epsilon}.$$

For $\epsilon = 0$, $p_\epsilon(t) * f(t)$ is indeterminate because both the numerator and denominator of the last expression reduce to zero. However, $\int_{-\epsilon/2}^{\epsilon/2} f(t-\tau)d\tau \approx \epsilon f(t)$ for small ϵ, and therefore, as claimed,

$$\lim_{\epsilon \to 0}\{p_\epsilon(t) * f(t)\} = \lim_{\epsilon \to 0}\frac{\int_{-\epsilon/2}^{\epsilon/2} f(t-\tau)d\tau}{\epsilon} = \lim_{\epsilon \to 0}\frac{\epsilon f(t)}{\epsilon} = f(t),$$

which also can be verified more rigorously using l'Hopital's rule.

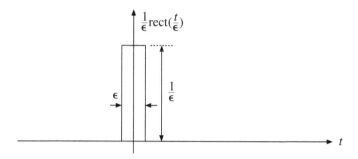

Figure 9.10 Pulse signal $p_\epsilon(t) \equiv \frac{1}{\epsilon}\mathrm{rect}(\frac{t}{\epsilon})$. Note that the area under $p_\epsilon(t)$ is 1 for all values of the pulse width ϵ, and as ϵ decreases $p_\epsilon(t)$ gets "thinner" and "taller." Function $p_\epsilon(t)$ has the property that, given an arbitrary function $f(t)$, $\lim_{\epsilon \to 0}\{p_\epsilon(t) * f(t)\} = f(t)$.

which indicates that $p_\epsilon(t)$, for very small ϵ, is essentially the *identity pulse* we are seeking.

It is convenient (and useful, as we shall see) to denote the left-hand side of this identity, $\lim_{\epsilon \to 0}\{p_\epsilon(t) * f(t)\}$, as $\delta(t) * f(t)$, and think of $\delta(t)$ as a special signal—essentially, $p_\epsilon(t)$, for exceedingly small ϵ—having the property

$$\delta(t) * f(t) = f(t).$$

(See entry 4 in Table 9.2.) Other properties of $\delta(t)$, some of which are listed in Table 9.3, include

$$\delta(t) \leftrightarrow 1;$$

this Fourier transform property of $\delta(t)$ is a consequence of applying the Fourier time-convolution property to the identity $\delta(t) * f(t) = f(t)$. Furthermore, given that $\delta(t) \leftrightarrow 1$, the inverse Fourier transform of 1 must be

$$\delta(t) = \frac{1}{2\pi} \int_{-\infty}^{\infty} e^{j\omega t} d\omega.$$

Impulse $\delta(t)$

We will refer to the signal $\delta(t)$ with the properties shown in Table 9.3 as an *impulse*. The properties of the impulse are a set of instructions specifying how various calculations involving the impulse and other signals can be performed. These instructions are necessary for computational purposes, since a numerical interpretation of $\delta(t) = \frac{1}{2\pi} \int_{-\infty}^{\infty} e^{j\omega t} d\omega$ is not possible because the integral $\int_{-\infty}^{\infty} e^{j\omega t} d\omega$ does not converge. Signals such as $\delta(t)$ that lack a numerical interpretation—because, in

Distributions

essence, they are non-realizable—are known as *distributions*[2] (instead of functions)

---

[2]The impulse $\delta(t)$ and its properties were first used as shortcuts by Paul Dirac in the 1920s in his quantum mechanics calculations. Soon after, a new branch of mathematics, known as *distribution theory*, was developed to provide a firm foundation for the impulse and its applications. The impulse distribution $\delta(t)$ also is known as the *Dirac delta*, or simply, *delta*.

Name	Impulse properties:	Shifted-impulse properties:
Convolution	$\delta(t) * f(t) = f(t)$	$\delta(t - t_o) * f(t) = f(t - t_o)$
Sifting	$\int_{-\infty}^{\infty} \delta(t) f(t) dt = f(0)$ and $\int_a^b \delta(t) f(t) dt$ $= \begin{cases} f(0) & \text{if } a < 0 < b \\ 0, & \text{otherwise} \end{cases}$	$\int_{-\infty}^{\infty} \delta(t - t_o) f(t) dt = f(t_o)$ and $\int_a^b \delta(t - t_o) f(t) dt$ $= \begin{cases} f(t_o) & \text{if } a < t_o < b \\ 0, & \text{otherwise} \end{cases}$
Sampling	$f(t)\delta(t) = f(0)\delta(t)$	$f(t)\delta(t - t_o) = f(t_o)\delta(t - t_o)$
Symmetry	$\delta(-t) = \delta(t)$	$\delta(t_o - t) = \delta(t - t_o)$
Scaling	$\delta(at) = \frac{1}{|a|}\delta(t),\ a \neq 0$	$\delta(a(t - t_o)) = \frac{1}{|a|}\delta(t - t_o),\ a \neq 0$
Area	$\int_{-\infty}^{\infty} \delta(t) dt = 1$ and $\int_a^b \delta(t) dt = \begin{cases} 1 & \text{if } a < 0 < b \\ 0, & \text{otherwise} \end{cases}$	$\int_{-\infty}^{\infty} \delta(t - t_o) dt = 1$ and $\int_a^b \delta(t - t_o) dt = \begin{cases} 1 & \text{if } a < t_o < b \\ 0, & \text{otherwise} \end{cases}$
Definite integral	$\int_{-\infty}^{t} \delta(\tau) d\tau = u(t)$	$\int_{-\infty}^{t} \delta(\tau - t_o) d\tau = u(t - t_o)$
Unit-step derivative	$\frac{du}{dt} = \delta(t)$	$\frac{d}{dt} u(t - t_o) = \delta(t - t_o)$
Derivative	$\frac{d}{dt}\delta(t) * f(t) = \frac{d}{dt} f(t)$	$\frac{d}{dt}\delta(t - t_o) * f(t) = \frac{d}{dt} f(t - t_o)$
Fourier transform	$\delta(t) \leftrightarrow 1$	$\delta(t - t_o) \leftrightarrow e^{-j\omega t_o}$
Graphical symbol		

Table 9.3 The properties and graphical symbols of the impulse and shifted impulse. The derivative of the impulse also is known as the *doublet*. Further properties of the doublet $\delta'(t) \equiv \frac{d}{dt}\delta(t)$, e.g., $\delta'(-t) = -\delta'(t)$, etc., can be enumerated starting with the derivative property for the impulse given in the table.

and are described in terms of their properties such as those listed in Table 9.3 (instead of with plots or tabulated values).

Convolution property

The properties of the impulse distribution $\delta(t)$ can be viewed to be a consequence of its *convolution property*

$$\delta(t) * f(t) = f(t)$$

and should be interpreted as "shorthands" for various expressions involving[3] $p_\epsilon(t)$, just as $\delta(t) * f(t) = f(t)$ is itself such a shorthand. For instance, by applying the shift property of convolution to the statement $\delta(t) * f(t) = f(t)$, we obtain

$$\delta(t - t_o) * f(t) = f(t - t_o),$$

which is a shorthand for

$$\lim_{\epsilon \to 0} \{p_\epsilon(t - t_o) * f(t)\} = f(t - t_o).$$

Shifted impulse

It is convenient to think of $\delta(t - t_o)$ as a signal, a new *distribution* known as a *shifted impulse*. This property of the shifted impulse is listed in the upper right corner of Table 9.3.

Writing out the same property as an explicit convolution integral, we obtain

$$\int_{-\infty}^{\infty} \delta(\tau - t_o) f(t - \tau) d\tau = f(t - t_o).$$

For the special case $f(t) = 1$, this expression reduces to

$$\int_{-\infty}^{\infty} \delta(\tau - t_o) d\tau = 1.$$

Area property

This result is known as the *area property* of the shifted impulse, and it appears in Table 9.3 in the form $\int_{-\infty}^{\infty} \delta(t - t_o) dt = 1$. For the case $t_o = 0$, the same expression yields the area property of the impulse $\delta(t)$.

We also can evaluate the foregoing convolution integral for arbitrary $f(t)$ at time $t = 0$ to obtain

$$\int_{-\infty}^{\infty} \delta(\tau - t_o) f(-\tau) d\tau = f(-t_o),$$

and hence

$$\int_{-\infty}^{\infty} \delta(\tau - t_o) g(\tau) d\tau = g(t_o),$$

---

[3]Although $p_\epsilon(t)$ was defined here as $\frac{1}{\epsilon}\text{rect}(\frac{t}{\epsilon})$, it also can be replaced by other functions such as $\frac{2}{\epsilon}\triangle(\frac{t}{\epsilon})$, $\frac{e^{-\frac{t^2}{2\epsilon^2}}}{\sqrt{2\pi}\epsilon}$, and $\frac{1}{\epsilon}\text{sinc}(\frac{\pi t}{\epsilon})$ peaking at $t = 0$ and satisfying the constraint $\int_{-\infty}^{\infty} p_\epsilon(t) dt = 1$.

with $g(t) \equiv f(-t)$. This last expression is known as the *sifting property* of the shifted   **Sifting**
impulse. Its special case for $t_o = 0$ gives

$$\int_{-\infty}^{\infty} \delta(\tau)g(\tau)d\tau = g(0),$$

the sifting property of the impulse. These last two expressions show that the impulse
$\delta(t)$ and its shifted version $\delta(t - t_o)$ act like sieves and sift out specific values $f(0)$ and
$f(t_o)$ of any function $f(t)$, which they multiply under an integral sign, for instance,
as in

$$\int_{-\infty}^{\infty} \delta(t - t_o)f(t)dt = f(t_o)$$

(so long as $f(t)$ is continuous at $t = t_o$ so that $f(t_o)$ is specified). From the previous
discussion, we know that the sifting property of $\delta(t - t_o)$ is a shorthand for

$$\lim_{\epsilon \to 0} \left\{ \int_{-\infty}^{\infty} p_\epsilon(t - t_o)f(t)dt \right\} = f(t_o).$$

One consequence of sifting is the *sampling property*, namely,   **Sampling**

$$\delta(t - t_o)f(t) = \delta(t - t_o)f(t_o),$$

because if we were to replace $\delta(t - t_o)f(t)$ in the second integral above with
$\delta(t - t_o)f(t_o)$, we would obtain

$$\int_{-\infty}^{\infty} \delta(t - t_o)f(t_o)dt = f(t_o) \int_{-\infty}^{\infty} \delta(t - t_o)dt = f(t_o) \times 1 = f(t_o),$$

using the area property of the shifted impulse. Thus, the sampling property is consistent with the sifting property and is valid. For the special case $t_o = 0$, we obtain

$$\delta(t)f(t) = \delta(t)f(0),$$

which is the sampling property for the impulse. This property of the impulse also
can be viewed as shorthand[4] for the approximation $p_\epsilon(t)f(t) \approx p_\epsilon(t)f(0)$, which
we easily can see is valid for sufficiently small ϵ (provided that $f(t)$ is continuous at
$t = 0$).

---

[4]So far we have emphasized the fact that Table 9.3 comprises shorthand statements about the special
pulse function $p_\epsilon(t)$ in the limit as $\epsilon \to 0$. Each statement is, of course, also a *property* of the distribution
$\delta(t)$, expressed in terms of familiar mathematical symbols such as integral, convolution, and equality signs.

A nuance that needs to be appreciated here is that the statements in the table also provide the special
meaning that these mathematical symbols take on when used with distributions instead of regular functions.
For instance, the $=$ sign in the statement $\delta(t)f(t) = \delta(t)f(0)$ indicates that the distributions on each side of
$=$ have the same effect on any regular function, say, $g(t)$, via any operation such as convolution or integration
defined in Table 9.3 in the very same spirit. The "equality in distribution" that we have just described is
distinct from, say, numerical equality between regular functions, say, $\cos(\omega t) = \sin(\omega t + \pi/2)$.

As a consequence of the sampling property, for example,

$$\cos(t)\delta(t) = \delta(t)$$

because $\cos(0) = 1$. Also,

$$\sin(t)\delta(t) = 0$$

since $\sin(0) = 0$,

$$(1 + t^2)\delta(t - 1) = 2\delta(t - 1)$$

since $(1 + (1)^2) = 2$, and

$$(1 + t^3)\delta(t + 1) = 0$$

since $(1 + (-1)^3) = 0$.

Symmetry and scaling

We will ask you to verify the symmetry and scaling properties in homework problems. These are very useful properties that give you the freedom to replace $\delta(-t)$ with $\delta(t)$, $\delta(5 - t)$ with $\delta(t - 5)$, $\delta(-2t)$ with $\frac{1}{2}\delta(t)$, etc. For example,

$$t^2\delta(2 - t) = t^2\delta(t - 2) = (2)^2\delta(t - 2) = 4\delta(t - 2).$$

Example 9.14

Using the Fourier transform property of the impulse, $\delta(t) \leftrightarrow 1$, determine

$$c(t) = a(t) * b(t)$$

if

$$a(t) = u(t)$$

and

$$B(\omega) = 1 - \frac{1}{1 + j\omega}.$$

Solution   Using the fact that

$$\delta(t) \leftrightarrow 1$$

and

$$e^{-t}u(t) \leftrightarrow \frac{1}{1 + j\omega},$$

we have

$$b(t) = \delta(t) - e^{-t}u(t).$$

Thus,

$$c(t) = u(t) * b(t) = u(t) * \delta(t) - u(t) * e^{-t}u(t)$$
$$= u(t) - e^{-t}u(t) * u(t) = u(t) - (1 - e^{-t})u(t).$$

In the last equality we obtained the first term by using the convolution property of the impulse, whereas the second term represents $\int_{-\infty}^{t} e^{-\tau}u(\tau)d\tau$. Note that with further simplification,

$$c(t) = e^{-t}u(t).$$

Example 9.15
Verify the Fourier transform property of a shifted impulse.

Solution   For the shifted impulse $\delta(t - t_o)$, the corresponding Fourier transform is

$$\int_{-\infty}^{\infty} \delta(t - t_o)e^{-j\omega t}dt = e^{-j\omega t_o},$$

as a consequence of the sifting property.

Example 9.16
Given that $f(t) = \delta(t - t_o)$, determine the energy W_f of signal $f(t)$.

Solution   Since $\delta(t - t_o) \leftrightarrow e^{-j\omega t_o}$, $F(\omega) = e^{-j\omega t_o}$, and therefore the energy spectrum for the signal is $|F(\omega)|^2 = 1$. Hence, using Rayleigh's theorem, we find that

$$W_f = \frac{1}{2\pi} \int_{-\infty}^{\infty} |F(\omega)|^2 d\omega = \frac{1}{2\pi} \int_{-\infty}^{\infty} 1 \cdot d\omega = \infty.$$

Thus, $\delta(t - t_o)$ contains infinite energy and therefore is not an energy signal. (Consequently, it cannot be generated in the lab.)

The versions of the area and sifting properties discussed earlier include integration limits $-\infty$ and ∞. However, alternative forms also included in Table 9.3 indicate that these properties still are valid when the integration limits are finite a and b, so long as $a < t_o < b$. Thus, for instance,

$$\int_{-5}^{8} \delta(t - 2) = 1,$$

$$\int_{4}^{6} \delta(t - 5)2 \cos(\pi t)dt = 2\cos(\pi 5) = -2,$$

and

$$\int_{0^-}^{\infty} \delta(t) e^{-st} dt = e^{-s \cdot 0} = 1.$$

To verify the alternative forms of the sampling property, consider first a function $f(t)w(t)$ where

$$w(t) = \begin{cases} 1, & a < t < b, \\ 0, & t < a \text{ and } t > b, \end{cases}$$

is a unit-amplitude *window function* shown in Figure 9.11a. Starting with the sifting property

$$\int_{-\infty}^{\infty} \delta(t - t_o) f(t) w(t) dt = f(t_o) w(t_o),$$

we can write

$$\int_{a}^{b} \delta(t - t_o) f(t) dt = f(t_o) w(t_o),$$

since $w(t) = 0$ outside $a < t < b$, and 1 within. Now, if $a < t_o < b$, as shown in Figure 9.11a, then $w(t_o) = 1$ and

$$\int_{a}^{b} \delta(t - t_o) f(t) dt = 1 \cdot f(t_o) = f(t_o);$$

otherwise—that is, if $t_o < a$ or $t_o > b$ (see Figure 9.11b, as an example)—then $w(t_o) = 0$ and

$$\int_{a}^{b} \delta(t - t_o) f(t) dt = 0 \cdot f(t_o) = 0.$$

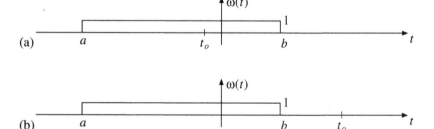

Figure 9.11 A unit-amplitude window function with $t = t_0$ (a) included within the window, and (b) excluded from the window.

Overall,

$$\int_a^b \delta(t - t_o) f(t) dt = \begin{cases} f(t_o), & a < t_o < b, \\ 0, & t_o < a \text{ or } t_o > b, \end{cases}$$

as in Table 9.3.

Use of the modified form of the sifting property with $f(t) = 1$ leads to the modified area properties given in Table 9.3. Furthermore, the modified form of the area property for the shifted impulse allows us to write

$$\int_{-\infty}^t \delta(\tau - t_o) d\tau = \begin{cases} 1, & t > t_o, \\ 0, & t < t_o, \end{cases} = u(t - t_o)$$

and

$$\int_{-\infty}^t \delta(\tau) d\tau = u(t).$$

The last two properties, listed in Table 9.3 as *definite-integral* properties, also imply that the derivatives of the unit step $u(t)$ and the shifted unit step $u(t - t_o)$ can be defined as $\delta(t)$ and $\delta(t - t_o)$, respectively. Of course, $u(t)$ and $u(t - t_o)$ are **Unit-step** discontinuous functions, and therefore they are not differentiable within the ordinary **and its** function space over all t; however, their derivatives can be regarded as the distributions **derivative**

$$\frac{d}{dt} u(t) = \delta(t) \text{ and } \frac{d}{dt} u(t - t_o) = \delta(t - t_o).$$

Example 9.17
Find the derivative of function

$$y(t) = t^2 u(t).$$

Solution $\frac{dy}{dt} = 0$ for $t < 0$, and $\frac{dy}{dt} = 2t$ for $t > 0$. Patching together these two results, we can write

$$\frac{dy}{dt} = 2tu(t)$$

as the answer. Alternatively, using the *product rule of differentation* and properties of the impulse, we obtain

$$\frac{dy}{dt} = \frac{d}{dt}(t^2 u(t)) = 2tu(t) + t^2 \frac{du}{dt} = 2tu(t) + t^2 \delta(t)$$

$$= 2tu(t) + 0^2 \delta(t) = 2tu(t).$$

This second version of the solution is not any faster or better than the first one, but it works and gives the correct result. Moreover, this second method is the only safe approach in certain cases, as illustrated by the next example.

Example 9.18

Find the derivative of

$$z(t) = e^{2t}u(t).$$

Solution $\frac{dz}{dt} = 0$ for $t < 0$ and $\frac{dz}{dt} = 2e^{2t}$ for $t > 0$. If we were to patch these together to write

$$\frac{dz}{dt} = 2e^{2t}u(t),$$

we would be wrong. The reason is that $z(t)$ is discontinuous at $t = 0$, where its derivative is undefined; so, as a consequence, $\frac{dz}{dt}$ is not the function $2e^{2t}u(t)$. The integral of $2e^{2\tau}u(\tau)$ from $-\infty$ to t, over τ, does not lead to $z(t) = e^{2t}u(t)$ as it should if $2e^{2t}u(t)$ were the correct derivative. However, using the product rule and properties of the impulse, we obtain

$$\frac{dz}{dt} = \frac{d}{dt}(e^{2t}u(t)) = 2e^{2t}u(t) + e^{2t}\frac{du}{dt} = 2e^{2t}u(t) + e^{2t}\delta(t)$$

$$= 2e^{2t}u(t) + e^{0}\delta(t) = 2e^{2t}u(t) + \delta(t),$$

which is the right answer (and can be confirmed by integrating $2e^{2\tau}u(\tau) + \delta(\tau)$ from $-\infty$ to t, over τ, to recover $z(t) = e^{2t}u(t)$).

What about the derivative of the impulse $\delta(t)$ itself? Since the impulse is a distribution, its derivative

$$\frac{d}{dt}\delta(t) \equiv \delta'(t)$$

must also be a distribution described by some set of properties. By applying the time-derivative property of convolution to $\delta(t) * f(t) = f(t)$, we find

$$\delta'(t) * f(t) = \frac{df}{dt},$$

Doublet $\delta'(t)$

which is both the time-derivative property of the impulse listed in Table 9.3 and the convolution property of a new distribution $\delta'(t) = \frac{d}{dt}\delta(t)$, known as the *doublet*. The doublet will play a relatively minor role in Chapters 10 and 11, and all we need to remember about it until then is its convolution property just given.

9.2.2 Graphical representation of $\delta(t)$ and $\delta(t - t_o)$

Since $\delta(t)$ and $\delta(t - t_o)$ are not functions, they cannot be plotted in the usual way functions are plotted. Instead, we use *defined* graphical representations for $\delta(t)$ and $\delta(t - t_o)$, which are shown in Table 9.3. By convention, $\delta(t - t_o)$ is depicted as an up-pointing arrow of unit height placed at $t = t_o$ along the t-axis. The length of the arrow is a reminder of the area property, while the placement of the arrow at $t = t_o$ owes to the sampling and sifting properties, as well as to the fact that a graphical symbol for $\delta(t - t_o)$ is also a symbol for $\frac{d}{dt}u(t - t_o)$ (which is numerically 0 everywhere except at $t = t_o$, where it is undefined).

Following this convention, $2\delta(t - 3)$, for instance, can be pictured as an arrow 2 units tall, placed at $t = 3$, while $-3\delta(t + 1)$ can be pictured as a down-pointing arrow of length 3 units, placed at $t = -1$. (See Figures 9.12a and 9.12b.) The graphical representation of

$$f(t) = 2\delta(t) - 3\delta(t + 2) + \delta(t - 4) + \text{rect}(t)$$

is shown in Figure 9.12c. Notice that $f(t)$ is a superposition of a number of impulses and an ordinary function, $\text{rect}(t)$. Even though we cannot assign numerical values to the distribution $f(t)$, we still represent it graphically as a collection of impulse arrows plus the regular graph of $\text{rect}(t)$.

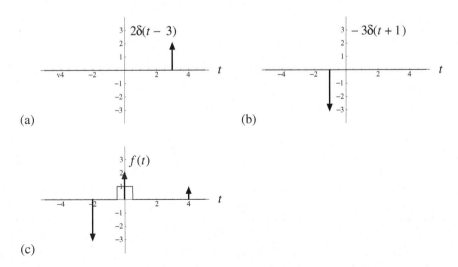

Figure 9.12 Examples of sketches, including the impulse and the shifted impulse.

Example 9.19

Determine the Fourier transform of rect(t), using the Fourier time-derivative property and the fact that

$$\delta(t - t_o) \leftrightarrow e^{-j\omega t_o}.$$

Solution   Since rect(t) $= u(t + \tfrac{1}{2}) - u(t - \tfrac{1}{2})$,

$$\frac{d}{dt}\text{rect}(t) = \frac{d}{dt}[u(t + \frac{1}{2}) - u(t - \frac{1}{2})]$$

$$= \delta(t + \frac{1}{2}) - \delta(t - \frac{1}{2}) \leftrightarrow e^{j\frac{\omega}{2}} - e^{-j\frac{\omega}{2}}.$$

Plots of rect(t) and its derivative are sketched in Figure 9.13.

Now, since by the Fourier time-derivative property the Fourier transform of $\frac{d}{dt}$rect(t) is $j\omega$ times the Fourier transform of rect(t), it follows that

$$\text{rect}(t) \leftrightarrow \frac{1}{j\omega}(e^{j\frac{\omega}{2}} - e^{-j\frac{\omega}{2}}) = \frac{e^{j\frac{\omega}{2}} - e^{-j\frac{\omega}{2}}}{j2 \cdot \frac{\omega}{2}} = \frac{\sin(\omega/2)}{\omega/2} = \text{sinc}(\frac{\omega}{2}),$$

which is the same result that we had established earlier through more conventional means.

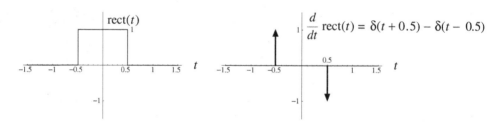

Figure 9.13   (a) rect(t), and (b) $\frac{d}{dt}$rect(t) $= \delta(t + 0.5) - \delta(t - 0.5)$.

9.2.3 Impulse response of an LTI system

It is important to ask how an LTI system would respond to an impulse input $f(t) = \delta(t)$. More specifically, we ask, "What is the zero-state response of an LTI system described in terms of $h(t) \leftrightarrow H(\omega)$ if the input signal $f(t)$ is an impulse?"

The answer is extremely simple. Clearly, according to the convolution formula for the zero-state response (which follows from the familiar $Y(\omega) = H(\omega)F(\omega)$ relation discussed earlier in the chapter), the answer has to be

$$y(t) = h(t) * f(t) = h(t) * \delta(t) = h(t),$$

since $h(t) * \delta(t) = h(t)$ by the convolution property of the impulse. Thus, $h(t)$, the inverse Fourier transform of the system frequency response $H(\omega)$, is also the zero-state response of the system to an impulse input. For that reason we will refer to $h(t)$ as the *impulse response*.

Impulse response $h(t)$

The concept of impulse response is fundamental and important. The concept will be explored much further in Chapter 10; so, for now, we provide just a single example illustrating its usefulness.

Example 9.20

Suppose that a high-pass filter with frequency response

$$H(\omega) = \frac{j\omega}{1 + j\omega}$$

(see Figure 5.2a) has input

$$f(t) = \text{rect}(t).$$

Determine the zero-state system response $y(t)$, using the system impulse response $h(t)$ and the convolution method.

Solution In Table 7.2 we find no match for $\frac{j\omega}{1+j\omega}$. However,

$$\frac{j\omega}{1 + j\omega} = \frac{j\omega + 1 - 1}{1 + j\omega} = 1 - \frac{1}{1 + j\omega},$$

and, therefore, using the addition rule and the same table, we find that the system impulse response is

$$h(t) = \delta(t) - e^{-t}u(t).$$

Using $y(t) = h(t) * f(t)$ and the convolution property of the impulse, we obtain

$$y(t) = (\delta(t) - e^{-t}u(t)) * \text{rect}(t) = \text{rect}(t) - e^{-t}u(t) * \text{rect}(t)$$

where the last term was calculated earlier in Example 9.13 and plotted in Figure 9.9. By subtracting Figure 9.9 from $\text{rect}(t)$ we obtain Figure 9.14, which is the solution to this problem. Notice that, because the Fourier transform of $f(t)$ is a sinc in the variable ω, this problem would be very difficult to solve by the Fourier technique.

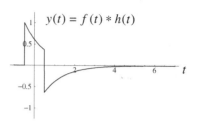

Figure 9.14 The response of high-pass filter $H(\omega) = \frac{j\omega}{1+j\omega}$ to input $f(t) = \text{rect}(t)$.

9.3 Fourier Transform of Distributions and Power Signals

In Chapter 7 we focused primarily on the Fourier transform of signals $f(t)$ having finite signal energy

$$W = \int_{-\infty}^{\infty} |f(t)|^2 dt.$$

After our introduction to the impulse $\delta(t)$, we are ready to explore the Fourier transforms of signals for which the energy W may be infinite. It turns out that the Fourier transform of some such signals $f(t)$ with infinite energy W, but finite instantaneous power $|f(t)|^2$ (e.g., signals like $\cos(\omega_o t)$, $\sin(\omega_o t)$, $u(t)$), can be expressed in terms of the impulse $\delta(\omega)$ in the Fourier domain. Such signals are known as *power signals* (as opposed to energy signals, which have finite W) and appear in the right-hand column of Table 7.2. The same column also includes the Fourier transform of distributions $\delta(t)$ and $\delta(t - t_o)$, discussed in the previous section.

Power signal

In the next set of examples we will verify some of these new Fourier transforms and also illustrate some of their applications.

Example 9.21
Show that

$$1 \leftrightarrow 2\pi\delta(\omega)$$

and

$$e^{j\omega_o t} \leftrightarrow 2\pi\delta(\omega - \omega_o),$$

to confirm entries 15 and 17 in Table 7.2.

Solution First, confirm entry 17 (the second line) by verifying that $e^{j\omega_o t}$ is the inverse Fourier transform

$$e^{j\omega_o t} = \frac{1}{2\pi} \int_{-\infty}^{\infty} 2\pi\delta(\omega - \omega_o)e^{j\omega t} d\omega.$$

This statement is valid, because, by the sifting property, the right-hand side is reduced to

$$\frac{1}{2\pi} 2\pi e^{j\omega_o t} = e^{j\omega_o t},$$

which is the same as the left-hand side. The Fourier pair $1 \leftrightarrow 2\pi\delta(\omega)$ is just a special case of $e^{j\omega_o t} \leftrightarrow 2\pi\delta(\omega - \omega_o)$ for $\omega_o = 0$.

Example 9.22
Show that

$$\cos(\omega_o t) \leftrightarrow \pi[\delta(\omega - \omega_o) + \delta(\omega + \omega_o)]$$

and

$$\sin(\omega_o t) \leftrightarrow j\pi[\delta(\omega + \omega_o) - \delta(\omega - \omega_o)],$$

to confirm entries 18 and 19 in Table 7.2.

Solution   To confirm entry 18, rewrite $\cos(\omega_o t)$, using Euler's formula, and use $e^{j\omega_o t} \leftrightarrow 2\pi\delta(\omega - \omega_o)$ from Example 9.21, giving

$$\cos(\omega_o t) = \frac{1}{2}(e^{j\omega_o t} + e^{-j\omega_o t}) \leftrightarrow \pi[\delta(\omega - \omega_o) + \delta(\omega + \omega_o)].$$

We also can show this same result by using $1 \leftrightarrow 2\pi\delta(\omega)$ and the Fourier modulation property.

To confirm entry 19, rewrite $\sin(\omega_o t)$ by using Euler's formula, and again use $e^{j\omega_o t} \leftrightarrow 2\pi\delta(\omega - \omega_o)$, giving

$$\sin(\omega_o t) = \frac{j}{2}(e^{-j\omega_o t} - e^{j\omega_o t}) \leftrightarrow j\pi[\delta(\omega + \omega_o) - \delta(\omega - \omega_o)].$$

Figures 9.15a through 9.15c show the plots of power signals $\cos(\omega_o t)$, $\sin(\omega_o t)$, and 1 (DC signal), and their Fourier transforms. Notice that all three Fourier transforms are depicted in terms of impulses in the ω-domain. The Fourier transform plots show that frequency-domain contributions to signals $\cos(\omega_o t)$ and $\sin(\omega_o t)$ are confined to frequencies $\omega = \pm\omega_o$. The impulses located at $\omega = \pm\omega_o$ are telling us the obvious: No complex exponentials other than $e^{\pm j\omega_o t}$ are needed to represent

Fourier transform of sine and cosine

$$\cos(\omega_o t) = \frac{e^{j\omega_o t} + e^{-j\omega_o t}}{2}$$

and

$$\sin(\omega_o t) = \frac{e^{j\omega_o t} - e^{-j\omega_o t}}{j2}.$$

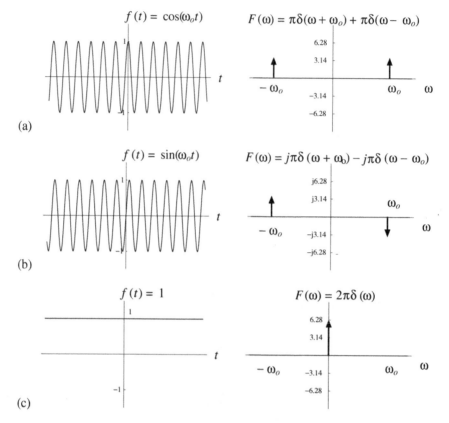

Figure 9.15 Three important power signals and their Fourier transforms: (a)
$f(t) = \cos(\omega_0 t) \leftrightarrow F(\omega) = \pi\delta(\omega - \omega_0) + \pi\delta(\omega + \omega_0)$, (b)
$f(t) = \sin(\omega_0 t) \leftrightarrow F(\omega) = j\pi\delta(\omega + \omega_0) - j\pi\delta(\omega - \omega_0)$, and (c)
$f(t) = 1 \leftrightarrow F(\omega) = 2\pi\delta(\omega)$.

Likewise, DC signal $f(t) = 1$ can be identified with $e^{\pm j0 \cdot t} = 1$; therefore, its Fourier
transform plot is represented by an impulse sitting at $\omega = 0$.

Example 9.23

Given that $f(t) \leftrightarrow F(\omega)$, determine the Fourier transform of $f(t)\sin(\omega_0 t)$
by using the Fourier frequency-convolution property.

Solution   The Fourier frequency-convolution property states that

$$f(t)g(t) \leftrightarrow \frac{1}{2\pi}F(\omega) * G(\omega).$$

Using this property with

$$g(t) = \sin(\omega_0 t)$$

and

$$G(\omega) = j\pi[\delta(\omega + \omega_o) - \delta(\omega - \omega_o)],$$

we obtain

$$f(t)\sin(\omega_o t) \leftrightarrow \frac{1}{2\pi}F(\omega) * j\pi[\delta(\omega + \omega_o) - \delta(\omega - \omega_o)]$$

$$= \frac{j}{2}[F(\omega + \omega_o) - F(\omega - \omega_o)].$$

This is an alternative form of the Fourier modulation property.

Example 9.24
Find the Fourier transform of an arbitrary *periodic* signal

Fourier transform of periodic signals

$$f(t) = \sum_{n=-\infty}^{\infty} F_n e^{jn\omega_o t},$$

with Fourier coefficients F_n.

Solution   Since

$$e^{jn\omega_o t} \leftrightarrow 2\pi\delta(\omega - n\omega_o),$$

application of the Fourier addition property yields

$$f(t) = \sum_{n=-\infty}^{\infty} F_n e^{jn\omega_o t} \leftrightarrow F(\omega) = \sum_{n=-\infty}^{\infty} 2\pi F_n \delta(\omega - n\omega_o).$$

Thus, the Fourier transform $F(\omega)$ of a periodic $f(t)$ with Fourier coefficients F_n is an infinite sum of weighted and shifted frequency-domain impulses $2\pi F_n \delta(\omega - n\omega_o)$, placed at integer multiples of the fundamental frequency $\omega_o = \frac{2\pi}{T}$. The area weights assigned to the impulses are proportional to the Fourier series coefficients.

Example 9.25
In Table 7.2, entry 24,

$$\sum_{n=-\infty}^{\infty} \delta(t - nT) \leftrightarrow \frac{2\pi}{T} \sum_{n=-\infty}^{\infty} \delta(\omega - n\frac{2\pi}{T}),$$

specifies the Fourier transform of the so-called *impulse train* depicted in Figure 9.16 on the left. Verify this Fourier transform pair by first determining the Fourier series of the impulse train.

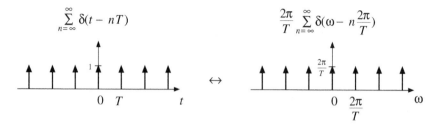

Figure 9.16   The Fourier transform of the impulse train on the left is also an impulse train in the frequency domain, as depicted on the right.

Solution   The impulse train

$$\sum_{n=-\infty}^{\infty} \delta(t - nT),$$

depicted in Figure 9.16, is a periodic signal with period T and fundamental frequency $\frac{2\pi}{T}$. Its exponential Fourier coefficients are

$$F_n = \frac{1}{T} \int_{-T/2}^{T/2} \left\{ \sum_{m=-\infty}^{\infty} \delta(t - mT) \right\} e^{-jn\frac{2\pi}{T}t} \, dt$$

$$= \frac{1}{T} \sum_{m=-\infty}^{\infty} \int_{-T/2}^{T/2} \delta(t - mT) e^{-jn\frac{2\pi}{T}t} \, dt$$

$$= \frac{1}{T} \int_{-T/2}^{T/2} \delta(t) e^{-jn\frac{2\pi}{T}t} \, dt = \frac{1}{T}$$

Impulse train and its Fourier transform

for all n. Hence, equating the impulse train to its exponential Fourier series, we get[5]

$$\sum_{n=-\infty}^{\infty} \delta(t - nT) = \sum_{n=-\infty}^{\infty} \frac{1}{T} e^{jn\frac{2\pi}{T}t},$$

and using the Fourier transform pair (from item 17 from Table 7.2)

$$e^{jn\frac{2\pi}{T}t} \leftrightarrow 2\pi \delta\left(\omega - n\frac{2\pi}{T}\right),$$

---

[5]The meaning of this equality is that when each side is multiplied by a signal $f(t)$ and then the two sides are integrated, the values of the integrals are equal.

we obtain

$$\sum_{n=-\infty}^{\infty} \delta(t - nT) \leftrightarrow \frac{2\pi}{T} \sum_{n=-\infty}^{\infty} \delta(\omega - n\frac{2\pi}{T}),$$

as requested.

Note that entry 25 in Table 7.2 is a straightforward consequence of this result and the frequency-convolution property (item 14 in Table 7.1).

Example 9.26

Suppose that a signal generator produces a periodic signal

$$f(t) = 4\cos(4t) + 2\cos(8t)$$

and a spectrum analyzer is used to examine the frequency-domain composition of $f(t)$. Let us assume that the spectrum analyzer bases its calculations on only a finite-length segment of $f(t)$. In effect, the spectrum analyzer multiplies $f(t)$ with a window function

$$w(t) = \text{rect}(\frac{t}{T_o})$$

and then displays the squared magnitude of the Fourier transform of $f(t)w(t)$ on a screen. What will the screen display look like if $T_o = 10$ s and 20 s?

Solution   Let $g(t) \equiv f(t)w(t)$. Then, according to the Fourier frequency-convolution property,

$$G(\omega) = \frac{1}{2\pi} F(\omega) * W(\omega),$$

where

$$F(\omega) = 4\pi[\delta(\omega - 4) + \delta(\omega + 4)] + 2\pi[\delta(\omega - 8) + \delta(\omega + 8)]$$

and $W(\omega)$ is the Fourier transform of $w(t)$. Substituting the expression for $F(\omega)$ into the convolution, and cancelling the π factors, we have

$$G(\omega) = 2[\delta(\omega - 4) + \delta(\omega + 4)] * W(\omega) + [\delta(\omega - 8) + \delta(\omega + 8)] * W(\omega)$$
$$= 2W(\omega - 4) + 2W(\omega + 4) + W(\omega - 8) + W(\omega + 8),$$

where in the last step we used the convolution property of the shifted impulse. Now,

$$\text{rect}(\frac{t}{T_o}) \leftrightarrow T_o\text{sinc}(\frac{\omega T_o}{2}),$$

and so

$$W(\omega) = T_o \mathrm{sinc}(\frac{\omega T_o}{2}).$$

Figures 9.17a and 9.17b show the plots of $|W(\omega)|^2$ for the cases $T_o = 10$ and 20 s. In both cases, the 90% bandwidth $\frac{2\pi}{T}$ of $w(t)$ is less than the shift frequencies of 4 and 8 rad/s relevant for $G(\omega)$. Thus, the various components of $G(\omega)$ have little overlap, so

$$|G(\omega)|^2 \approx 4|W(\omega - 4)|^2 + 4|W(\omega + 4)|^2 + |W(\omega - 8)|^2 + |W(\omega + 8)|^2.$$

Plots of this approximation for $|G(\omega)|^2$ are shown in Figure 9.17c and 9.17d for the cases $T_o = 10$ and 20 s, respectively. We conclude that the spectrum analyzer display will look like Figures 9.17c and 9.17d (although, typically, only the positive-ω half of the plots would be displayed).

  Notice that a longer analysis window (larger T_o) produces a higher-resolution estimate of the spectrum of $f(t)$, characterized by narrower "spikes."

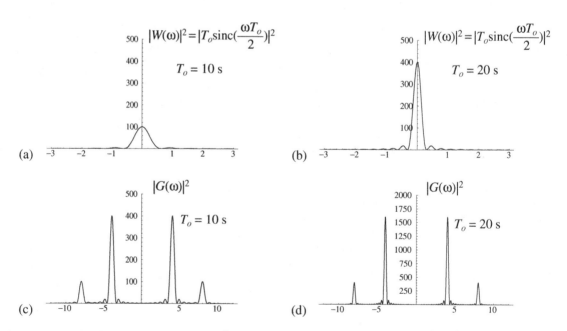

Figure 9.17 The energy spectrum $|W(\omega)|^2$ of the window function $w(t) = \mathrm{rect}(\frac{t}{T_o})$ for (a) $T_o = 10$ s and (b) $T_o = 20$ s. The spectrum $|\frac{1}{2\pi}F(\omega) * W(\omega)|^2$ of $f(t)w(t)$, $f(t) = 4\cos(4t) + 2\cos(8t)$, is shown in (c) and (d) for the two values of T_o. The frequency resolution of the measurement device is set by the window length T_o and can be described as the half-width $\frac{2\pi}{T_o}$ of $|W(\omega)|^2$.

Example 9.27
An incoming AM radio signal

$$y(t) = (f(t) + \alpha)\cos(\omega_c t)$$

is mixed with a signal $\cos(\omega_c t)$, and the result $p(t)$ is filtered with an ideal low-pass filter $H(\omega)$. The filter bandwidth is less than ω_c, but larger than the bandwidth Ω of the low-pass message signal $f(t)$. In addition, $\Omega \ll \omega_c$. What is the output $q(t)$ of the low-pass filter?

Solution   Let

$$p(t) \equiv y(t)\cos(\omega_c t) = (f(t) + \alpha)(\cos(\omega_c t))^2$$

$$= (f(t) + \alpha)\frac{1}{2}(1 + \cos(2\omega_c t)).$$

Using the frequency-convolution property, we find that the Fourier transform of $p(t)$ is

$$P(\omega) = \frac{1}{4\pi}(F(\omega) + \alpha 2\pi\delta(\omega)) * [2\pi\delta(\omega) + \pi\delta(\omega - 2\omega_c) + \pi\delta(\omega + 2\omega_c)]$$

$$= \frac{1}{2}(F(\omega) + \alpha 2\pi\delta(\omega)) + \frac{1}{4}(F(\omega - 2\omega_c) + \alpha 2\pi\delta(\omega - 2\omega_c))$$

$$+ \frac{1}{4}(F(\omega + 2\omega_c) + \alpha 2\pi\delta(\omega + 2\omega_c)).$$

But, only the first term of $P(\omega)$ on the right is within the pass-band of the described low-pass filter $H(\omega)$. Therefore, it follows that

$$Q(\omega) = H(\omega)P(\omega) = \frac{1}{2}(F(\omega) + \alpha 2\pi\delta(\omega)),$$

implying an output

$$q(t) = \frac{1}{2}(f(t) + \alpha),$$

as expected in a successful coherent demodulation of the given AM signal.

Example 9.28
An incoming AM signal

$$y(t) = f(t)\cos(\omega_c t)$$

is mixed with $\sin(\omega_c t)$ and the product signal is filtered with an ideal low-pass filter $H(\omega)$ as in Example 9.27. As before, the filter bandwidth is less than ω_c but larger than the bandwidth Ω of the low-pass $f(t)$, and $\Omega \ll \omega_c$. What is the output $q(t)$ of the filter?

Solution   In this case

$$p(t) = y(t)\sin(\omega_c t) = f(t)\cos(\omega_c t)\sin(\omega_c t).$$

Applying the Fourier modulation property to

$$\sin(\omega_c t) \leftrightarrow j\pi[\delta(\omega + \omega_c) - \delta(\omega - \omega_c)],$$

we first obtain

$$\sin(\omega_c t)\cos(\omega_c t) \leftrightarrow \frac{j\pi}{2}[\delta(\omega) - \delta(\omega - 2\omega_c)] + \frac{j\pi}{2}[\delta(\omega + 2\omega_c) - \delta(\omega)]$$

$$= \frac{j\pi}{2}\delta(\omega + 2\omega_c) - \frac{j\pi}{2}\delta(\omega - 2\omega_c).$$

Therefore, using the frequency-convolution property, the Fourier transform of

$$p(t) = f(t)[\sin(\omega_c t)\cos(\omega_c t)]$$

is

$$P(\omega) = \frac{1}{2\pi}F(\omega) * [\frac{j\pi}{2}\delta(\omega + 2\omega_c) - \frac{j\pi}{2}\delta(\omega - 2\omega_c)]$$

$$= \frac{j}{4}F(\omega + 2\omega_c) - \frac{j}{4}F(\omega - 2\omega_c).$$

Note that $P(\omega)$ contains no term in the passband of the described low-pass filter. Thus,

$$Q(\omega) = H(\omega)P(\omega) = 0,$$

implying that

$$q(t) = 0.$$

Clearly, we cannot demodulate the AM signal

$$y(t) = f(t)\cos(\omega_c t)$$

by mixing it with $\sin(\omega_c t)$, because zero output is obtained from the low-pass filter.

Note that the results of Examples 9.27 and 9.28 suggest that if a signal $g(t)\sin(\omega_c t)$ were added to an AM transmission $f(t)\cos(\omega_c t)$, then it would be possible to recover $f(t)$ and $g(t)$ from the sum unambiguously by mixing the sum with signals $\cos(\omega_c t)$ and $\sin(\omega_c t)$, respectively. This idea is exploited in so-called quadrature amplitude modulation communication systems.

Finding the Fourier transform of power signal $u(t)$, the unit step, is somewhat tricky: First, note that the unit step has an average value of

$$\lim_{T \to \infty} \frac{1}{2T} \int_{-T}^{T} u(t)dt = \frac{1}{2}$$

and it can be expressed as

$$u(t) = \frac{1}{2} + \frac{1}{2}\text{sgn}(t)$$

in terms of its DC component and the signum function $\text{sgn}(t)$ shown in Figure 7.2d. Since

$$1 \leftrightarrow 2\pi\delta(\omega),$$

the Fourier transform of $\frac{1}{2}$ is $\pi\delta(\omega)$; also, from Table 7.2 (and as verified in Example 9.29, next),

$$\text{sgn}(t) \leftrightarrow \frac{2}{j\omega}.$$

Thus, using the addition property of the Fourier transform, we conclude that

$$u(t) = \frac{1}{2} + \frac{1}{2}\text{sgn}(t) \leftrightarrow \pi\delta(\omega) + \frac{1}{j\omega},$$

as indicated in Table 7.2. Note that the term $\pi\delta(\omega)$ in the Fourier transform of $u(t)$ accounts for its DC component equal to $\frac{1}{2}$; the second term $\frac{1}{j\omega}$ results from its AC component $\frac{1}{2}\text{sgn}(t)$.

Example 9.29
Notice that

$$u(t) = \frac{1}{2} + \frac{1}{2}\text{sgn}(t)$$

implies

$$\frac{du}{dt} = \delta(t) = \frac{1}{2}\frac{d}{dt}\text{sgn}(t),$$

or, equivalently,

$$\frac{d}{dt}\text{sgn}(t) = 2\delta(t).$$

Using the Fourier time-derivative property and the fact that $\delta(t) \leftrightarrow 1$, verify

$$\text{sgn}(t) \leftrightarrow \frac{2}{j\omega}.$$

Solution   Let $S(\omega)$ denote the Fourier transform of $\text{sgn}(t)$. Then the Fourier transform of the equation

$$\frac{d}{dt}\text{sgn}(t) = 2\delta(t)$$

is

$$j\omega S(\omega) = 2.$$

Hence,

$$S(\omega) = \frac{2}{j\omega}$$

and, consequently,

$$\text{sgn}(t) \leftrightarrow \frac{2}{j\omega}.$$

Since $\text{sgn}(t)$ is a pure AC signal—that is, $\lim\limits_{T\to\infty} \frac{1}{2T} \int_{-T}^{T} \text{sgn}(t)dt = 0$—its Fourier transform does not contain an impulse term.

We offer one closing comment about the Fourier transform of power signals, such as $u(t)$ and $\cos(\omega_o t)$ where

$$u(t) \leftrightarrow \pi\delta(\omega) + \frac{1}{j\omega}$$

$$\cos(\omega_o t) \leftrightarrow \pi\delta(\omega - \omega_o) + \pi\delta(\omega + \omega_o).$$

It should be apparent that, in the case of these types of signals, the Fourier integral does not converge. Therefore, the corresponding Fourier transform does not exist in the usual sense. This is the reason that the Fourier transforms of signals such as $u(t)$ and $\cos(\omega_o t)$ must be expressed in terms of impulses.

9.4   Sampling and Analog Signal Reconstruction

Now that we know about the impulse train and its Fourier transform (see Example 9.25 in the previous section), we are ready to examine the basic ideas behind *analog-to-digital conversion* and *analog signal reconstruction*. In this section we will learn about the *Nyquist criterion*, which constrains the sampling rates used in CD production, and find out how a CD player may convert data stored on a CD into sound.

Consider a *bandlimited signal*

Bandlimited signal with bandwidth B

$$f(t) \leftrightarrow F(\omega)$$

having *bandwidth B* Hz, so that

$$F(\omega) = 0$$

outside the frequency interval

$$|\omega| \leq 2\pi B \text{ rad/s}.$$

Suppose that only discrete samples of $f(t)$ are available, defined by

$$f_n \equiv f(nT), \quad -\infty < n < \infty,$$

where the samples are equally spaced at times $t = nT$, which are integer multiples of the *sampling interval T*. In the modern digital world, where music and image samples routinely are stored on a computer, we might ask whether it is possible to *reconstruct* the analog signal $f(t)$ from its discrete samples f_n with full fidelity (i.e., identically). It turns out that the answer is *yes* if the sampling interval T is small enough, compared with the reciprocal of the signal bandwidth B. The specific requirement, called the *Nyquist criterion*, is

Samples and sampling interval T

Nyquist criterion

$$T < \frac{1}{2B},$$

or, equivalently,

$$\frac{1}{T} > 2B.$$

Notice that the version of the Nyquist criterion just presented states that the sampling frequency $1/T$ must be larger than twice the highest frequency B (measured in Hertz) in the signal being sampled. That is, each frequency component in $f(t)$ must be sampled at a rate of at least two samples per period. Under this condition, it is theoretically possible for us to exactly *reconstruct* the analog signal $f(t)$ from its discrete samples f_n by using the so-called *reconstruction formula*

Reconstruction formula

$$f(t) = \sum_n f_n \text{sinc}(\frac{\pi}{T}(t - nT)).$$

However, if the Nyquist criterion is violated, then the reconstruction formula becomes invalid in the sense that the sum on the right side of the formula converges, not to the original analog signal $f(t)$ shown on the left, but to another analog signal known as an *aliased* version of $f(t)$. Before we verify the validity of the reconstruction formula and the Nyquist criterion, let us examine the sampling and reconstruction examples shown in Figures 9.18 and 9.19.

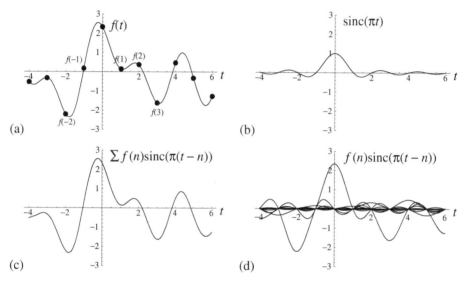

Figure 9.18 (a) An analog signal $f(t)$ and its discrete samples $f(nT)$ (shown by dots) taken at $T = 1$ s intervals, (b) interpolating function $\mathrm{sinc}(\pi \frac{t}{T})$ for the case $T = 1$ s, and (c) reconstructed $f(t)$, using the interpolation formula. (See text.) Panel (d) shows the functions $f(nT)\mathrm{sinc}(\frac{\pi}{T}(t - nT))$ for $T = 1$ s, which are summed up to produce the reconstructed $f(t)$ shown in panel (c). In this example, reconstruction is successful because the Nyquist criterion is satisfied.

In the case illustrated in Figure 9.18, sampling of $f(t)$ is conducted in accordance with the Nyquist criterion; therefore, the reconstructed signal shown in Figure 9.18c is identical to the original $f(t)$. Panel (d) in the figure shows the collection of shifted and amplitude scaled sinc functions (components of the reconstruction formula), the sum of which, shown in panel (c), replicates the original $f(t)$ from panel (a). In the example shown in Figure 9.19, however, the Nyquist criterion is violated and the reconstruction formula fails to reproduce the original $f(t)$. In this case the available samples $f(nT)$ are spaced too far apart. The signal $f(t)$ is said to be *undersampled* and, as a result, the reconstruction is aliased; in this case the original co-sinusoid is "impersonated" by a lower frequency co-sinusoid.

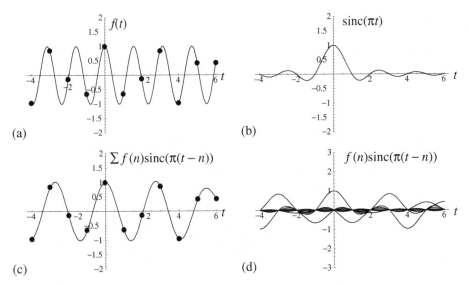

Figure 9.19   (a) $f(t) = \cos(4t)$ and its samples $f(nT)$, $T = 1$ s, (b) interpolating function $\text{sinc}(\pi \frac{t}{T})$ for $T = 1$ s, (c) reconstructed $f(t)$, and (d) $f(nT)\text{sinc}(\frac{\pi}{T}(t - nT))$ for $T = 1$ and all n. Notice that the reconstructed $f(t)$ does not match the original $f(t)$ because of undersampling of $f(t)$ in violation of the Nyquist criterion, which requires $T < \frac{1}{2B}$. (See text.)

Verification of reconstruction formula: The impulse train identity (item 24 in Table 7.2),

$$\sum_{n=-\infty}^{\infty} \delta(t - nT) \leftrightarrow \frac{2\pi}{T} \sum_{n=-\infty}^{\infty} \delta(\omega - n\frac{2\pi}{T}),$$

and the frequency-convolution property of the Fourier transform (item 14 in Table 7.1) imply that the product

$$f(t) \sum_{n=-\infty}^{\infty} \delta(t - nT)$$

has Fourier transform

$$\frac{1}{2\pi} F(\omega) * \frac{2\pi}{T} \sum_{n=-\infty}^{\infty} \delta(\omega - n\frac{2\pi}{T}) = \sum_{n=-\infty}^{\infty} \frac{1}{T} F(\omega - n\frac{2\pi}{T}).$$

Hence,

$$\sum_{n=-\infty}^{\infty} f(t)\delta(t - nT) \leftrightarrow \sum_{n=-\infty}^{\infty} \frac{1}{T} F(\omega - n\frac{2\pi}{T}),$$

as listed in Table 7.2 (item 25), and also

$$\sum_{n=-\infty}^{\infty} f(nT)\delta(t - nT) \leftrightarrow \sum_{n=-\infty}^{\infty} \frac{1}{T}F(\omega - n\frac{2\pi}{T}),$$

in view of the sampling property of the shifted impulse.

The Fourier transform pair just obtained is the key to verification of the reconstruction formula and the Nyquist criterion. Let us first interpret the Fourier transform on the right side, namely

$$F_T(\omega) \equiv \sum_{n=-\infty}^{\infty} \frac{1}{T}F(\omega - n\frac{2\pi}{T}),$$

with the aid of Figure 9.20. Panel (b) in the figure presents a sketch of $F_T(\omega)$ as a collection of amplitude-scaled and frequency-shifted replicas of $F(\omega)$, the Fourier transform of $f(t)$ sketched in panel (a). For the $F_T(\omega)$ shown in panel (b), it is clear that $2\pi B < \frac{\pi}{T}$, or, equivalently, $B < \frac{1}{2T}$, in accordance with the Nyquist criterion. Also, it should be clear that if $F_T(\omega)$ in the same panel were multiplied with

$$H(\omega) = T\text{rect}(\frac{\omega}{2\pi/T}),$$

corresponding to an ideal low-pass filter having bandwidth $\frac{\pi}{T}$ and amplitude T, the result would equal $F(\omega)$, the Fourier transform of $f(t)$. Thus low-pass

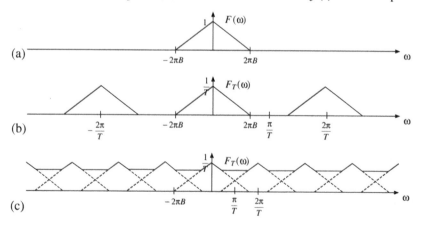

Figure 9.20 (a) The Fourier transform $F(\omega)$ of a band-limited signal $f(t)$ with bandwidth $\Omega = 2\pi B \frac{\text{rad}}{\text{s}}$; (b) and (c) are $F_T(\omega) = \sum_{n=-\infty}^{\infty} \frac{1}{T}F(\omega - \frac{2\pi}{T}n)$ for the same signal, with the sampling interval $T < \frac{1}{2B}$ and $T > \frac{1}{2B}$, respectively. Note that the central feature of $F_T(\omega)$ in panel (b) is identical to $\frac{1}{T}F(\omega)$. Also note that panel (c) contains no isolated replica of $F(\omega)$. Therefore, $F(\omega)$ can be correctly inferred from $F_T(\omega)$ if and only if $T < \frac{1}{2B}$.

filtering of $\sum_{n=-\infty}^{\infty} f(nT)\delta(t-nT)$ (having the Fourier transform $F_T(\omega)$) via a system with impulse response $h(t) = \operatorname{sinc}\left(\frac{\pi}{T}t\right)$ (with Fourier transform $T\operatorname{rect}\left(\frac{\omega}{2\pi/T}\right)$) produces $f(t)$. But, the same convolution operation describing the filter action yields precisely the right side of the reconstruction formula.

The proof of the reconstruction formula just completed is contingent upon satisfying the Nyquist criterion. When the criterion is violated (i.e., when $2\pi B \geq \frac{\pi}{T}$, as illustrated in Figure 9.20c), $F_T(\omega)$ no longer contains an isolated replica of $F(\omega)$ within the passband of $H(\omega)$, and, as a consequence, in such situations the reconstruction formula produces aliased results (as in Figure 9.19).

D/A (digital-to-analog) conversion is a hardware implementation that mimics the reconstruction formula just verified. This is accomplished by utilizing a circuit that creates a weighted pulse train

$$\sum_n f_n p(t-nT),$$

where $p(t)$ generally is a rectangular pulse of width T, and then low-pass filtering the pulse train, using a suitable LTI system

$$h(t) \leftrightarrow H(\omega).$$

The reconstruction of $f(t)$ is nearly ideal if $h(t)$ is designed in such a way that $h(t) * p(t)$ is a good approximation to (a delayed) $\operatorname{sinc}(\frac{\pi t}{T})$. This reconstruction process, which parallels our proof of the reconstruction formula, is illustrated symbolically on the right side of the system shown in Figure 9.21. The left side of the same system is a symbolic representation of an A/D (analog-to-digital) converter, where the input $f(t)$ is sampled every T seconds in order to generate the sequence of discrete samples $f_n = f(nT)$.

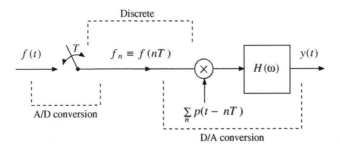

Figure 9.21 A model system that samples an analog input $f(t)$ with a sampling interval T and generates an analog output $y(t)$ by using the samples $f_n = f(nT)$. The input to filter $H(\omega)$ is $\sum_n f_n p(t-nT)$. Mathematically, if $p(t) = \delta(t)$ and $H(\omega) = T\operatorname{rect}(\frac{\omega}{2\pi/T})$, then $y(t) = f(t)$, assuming the Nyquist criterion is satisfied. In real-world systems, $p(t)$ is chosen to be a rectangular pulse of width T, and $H(\omega)$ is chosen so that $|P(\omega)||H(\omega)| \approx T\operatorname{rect}(\frac{\omega}{2\pi/T})$.

The system shown in Figure 9.21 simply regenerates $y(t) = f(t)$ at its output, because the samples f_n of input $f(t)$ are not modified in any way within the system prior to reconstruction. The option of modifying the samples f_n is there, however, and, hence, the endless possibilities of digital signal processing (DSP). DSP can be used to convert analog input signals $f(t)$ into new, desirable analog outputs $y(t)$ by replacing the samples f_n by a newly computed sequence y_n, prior to reconstruction. Examples of digital processing, or manipulating the samples f_n, are

$$y_n = \frac{1}{2}(f_n + f_{n-1}),$$

which is a simple smoothing (averaging) digital low-pass filter, and

$$y_n = \frac{1}{2}(f_n - f_{n-1}),$$

which is a simple high-pass digital filter that emphasizes variations from one sample to the next in f_n. More sophisticated digital filters compute outputs as a more general weighted sum of present and past inputs and, sometimes, past outputs as well. Some other types of digital processing are explored (along with aliasing errors) in one of the labs. (See Appendix B.)

Sound cards, A/D and D/A converters

Sound cards, found nowadays in most PC's, consist of a pair of A/D and D/A converters that work in concert with one another and the rest of the PC—CPU, memory, keyboard, CD player, speaker, etc. These cards can perform a variety of signal processing tasks in the audio frequency range. For instance, the D/A circuitry on the sound card converts samples $f_n = f(nT)$, fetched from a CD (or an MP3 file stored on the hard disk), into a song $f(t)$ by a procedure like the one described here.

44.1 kHz sampling rate

The standard sampling interval used in sound cards is $T = 1/44100$ s. In other words, sound cards sample their input signals at a rate of $1/T = 44100$ samples/s, or, 44.1 kHz, as usually quoted. Since the Nyquist criterion requires that $T < \frac{1}{2B}$ or, equivalently, $B < \frac{1}{2T} = 22.05$ kHz, only signals bandlimited to 22.05 kHz can be processed (without aliasing). Thus, the input stage of a sound card (prior to the A/D) typically incorporates a low-pass filter with bandwidth around 20 kHz (or less) to reduce or prevent aliasing effects. The 44.1 kHz sampling rate of standard sound cards is the same as the sampling rate used in audio CD production. Because human hearing does not exceed 20 kHz, it is possible to low-pass filter analog audio signals to 20kHz prior to sampling, with no audible effect. The choice of sampling rate (44.1 kHz) in excess of twice the filtered sound bandwidth (20 kHz) works quite well and allows for some flexibility in the design of the D/A (i.e., choice of $H(\omega)$ in Figure 9.21). Special cards with wider bandwidths and higher sampling rates are used to digitize and process signals encountered in communications and radar applications. Digital oscilloscopes also use high-bandwidth A/D cards. Some applications use A/D converters operating up into the GHz range.

Returning back to Figure 9.20, recall that

$$F_T(\omega) = \sum_n \frac{1}{T} F(\omega - n\frac{2\pi}{T}),$$

sketched in panel (b), is the Fourier transform of time signal

$$\sum_n f(nT)\delta(t - nT) \equiv f_T(t).$$

Also, recall that if $f(t)$ is bandlimited and T satisfies the Nyquist criterion, then $F_T(\omega) = \frac{1}{T}F(\omega)$ for $|\omega| < \frac{\pi}{T}$. (See Figures 9.20a and 9.20b.) So, if we can find a way of calculating $F_T(\omega)$, we then have a way of calculating $F(\omega)$. This is easily done. Remembering that

$$\delta(t - nT) \leftrightarrow e^{-j\omega nT},$$

and then transforming the preceding expression for $f_T(t)$, term by term, we get

Calculating $F(\omega)$ with discrete samples $f_n = f(nT)$

$$F_T(\omega) = \sum_n f(nT)e^{-j\omega nT},$$

which is an *alternative* formula for $F_T(\omega)$. This formula provides us with a means of computing the Fourier transform $F(\omega)$ of a bandlimited $f(t)$ by using only its sample data $f(nT)$, namely,

$$F(\omega) = T\, F_T(\omega) = T\sum_n f(nT)e^{-j\omega nT}, \quad |\omega| < \frac{\pi}{T},$$

where $\frac{\pi}{T}$ commonly is known as the *Nyquist frequency*.

We have used the approach just described to determine $F(\omega)$ and compute the energy spectra $|F(\omega)|^2$ of the first 6 seconds of Beethoven's 5th symphony and the Beatles' "It's Been a Hard Day's Night," displayed earlier in Figure 7.11.[6]

---

[6] The energy spectrum curves were produced as follows:

(1) The sound card of a laptop computer was controlled to sample the audio input (from a phonograph turntable) at a rate of 44.1 kHz and store the samples as 16-bit integers using the .aiff file format. Approximately 6 s of audio input were sampled to produce a total of $N = 2^{18} = 262144$ data samples $f_n = f(nT)$, where $T = \frac{1}{44100}$ s.

(2) A short *Mathematica* program was written to compute

$$F_m \equiv \sum_{n=0}^{N-1} f_n e^{-j2\pi \frac{mn}{N}} = F_T(\frac{2\pi}{NT}m),$$

$m \in [0, 1, \cdots, N-1]$, using the *Mathematica* "Fourier" function (which implements the popular FFT algorithm). The plotted curves actually correspond to the normalized quantity

$$\frac{|F_m|^2}{\sum_{k=0}^{N-1} |F_k|^2},$$

for $m \in [0, 511, 1023, \cdots, \frac{N}{2} - 1]$, after a 512-point running average operation was applied to $|F_m|^2$. The purpose of the running average was to smooth the spectrum and reduce the number of data points to be plotted.

9.5 Other Uses of the Impulse

The impulse definition and concept play important roles outside of signal processing, for example in various physics problems. Take, for instance, the idealized notion of a *point charge*. Physicists often envision charged particles, such as electrons, as point charges occupying no volume, but carrying some definite amount of charge, say q coulombs. Since the equations of electrodynamics (Newton's laws and Maxwell's equations) are formulated in terms of charge density $\rho(x, y, z, t)$, a physicist studying the motion of a single free electron in some electromagnetic device needs an expression $\rho(x, y, z, t)$ describing the state (the position and motion) of the electron. A model for the charge density of a stationary electron located at the origin and envisioned as a point charge is

$$\rho(x, y, z, t) = q\delta(x)\delta(y)\delta(z) \frac{C}{m^3}$$

in terms of spatial impulses $\delta(x)$, $\delta(y)$, and $\delta(z)$ and electronic charge q. This is a satisfactory representation, because the volume integral of $\rho(x, y, z, t)$ over a small cube centered about coordinates (x_o, y_o, z_o), that is,

$$\int_{\text{cube}} q\delta(x)\delta(y)\delta(z)dxdydz = q \left(\int_{x_o-\varepsilon/2}^{x_o+\varepsilon/2} \delta(x)dx \right)$$
$$\times \left(\int_{y_o-\varepsilon/2}^{y_o+\varepsilon/2} \delta(y)dy \right) \times \left(\int_{z_o-\varepsilon/2}^{z_o+\varepsilon/2} \delta(z)dz \right),$$

equals q for $x_o = y_o = z_o = 0$, for arbitrarily small $\varepsilon > 0$, yielding the total amount of charge contained within the box. Conversely, if the box excludes the origin, then the integration result is zero.

Example 9.30
How would you express the charge density $\rho(x, y, z, t)$ of an oscillating electron with time-dependent coordinates $(0, 0, z_o \cos(\omega t))$?

Solution   In view of the preceding discussion, the answer must be

$$\rho(x, y, z, t) = q\delta(x)\delta(y) \times \delta(z - z_o \cos(\omega t)) \frac{C}{m^3},$$

since the electron always remains on the z-axis and deviates from the origin by an amount $z = z_o \cos(\omega t)$.

Electromagnetic radiation fields of an oscillating electron can be derived by solving Maxwell's equations—a linear set of coupled partial differential equations (PDEs)—with a source function $\rho(x, y, z, t) = q\delta(x)\delta(y)\delta(z - z_o \cos(\omega t)) \frac{C}{m^3}$. The solution, in turn, can be used to describe how radio antennas convert oscillating currents into traveling electromagnetic waves.

Note that dimensional analysis of the foregoing expressions for $\rho(x, y, z, t)$ implies that the units for $\delta(x)$, $\delta(y)$, and $\delta(z)$ must be $\frac{1}{m}$. Likewise, the unit for $\delta(t)$ is $\frac{1}{s}$.

Regarding other application areas, the impulse also can be used as an idealized representation of the force density of a hammer blow on a wall, the current density of a lightning bolt, the acoustic energy density of an explosion, or the light source provided by a star.

EXERCISES

9.1 For the functions $f(t)$ and $g(t)$ sketched as shown,

    (a) Find $x(t) = g(t) * g(t)$ by direct integration and sketch the result.

    (b) Find $y(t) = f(t) * g(t)$, using appropriate properties of convolution and the result of part (a). Sketch the result.

    (c) Find $z(t) = f(t) * f(t - 1)$, using the properties of convolution and sketch the result.

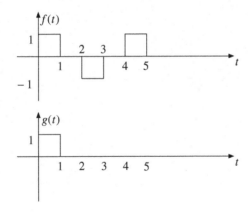

9.2 Given $h(t) = u(t)$ and $f(t) = 2\Delta(\frac{t}{2})$,

    (a) Determine $y(t) = h(t) * f(t)$ and sketch the result.

    (b) Determine $z(t) = h(t) * \frac{df}{dt}$, using the derivative property of convolution, and sketch the result.

    (c) Calculate $z(t) = h(t) * \frac{df}{dt}$, using some alternative method (avoiding direct integration, if you can).

9.3 Given $f(t) = u(t)$, $g(t) = 2tu(t)$, and $q(t) = f(t - 1) * g(t)$, determine $q(4)$.

9.4   Suppose that the convolution of $f(t) = \text{rect}(t)$ with some $h(t)$ produces $y(t) = \triangle(\frac{t-2}{2})$. What is the convolution of $\text{rect}(t) + \text{rect}(t-4)$ with $h(t) - 2h(t-6)$? Sketch the result.

9.5   Determine and plot $c(t) = \text{rect}(\frac{t-5}{2}) * \triangle(\frac{t-8}{4})$.

9.6   Determine and plot $y(t) = h(t) * f(t)$ if

   (a) $h(t) = e^{-2t}u(t)$ and $f(t) = u(t)$.

   (b) $h(t) = \text{rect}(t)$ and $f(t) = u(t)e^{-t} - u(-t)e^{t}$.

9.7   Simplify the following expressions involving the impulse and/or shifted impulse and sketch the results:

   (a) $f(t) = (1 + t^2)(\delta(t) - 2\delta(t - 4))$.

   (b) $g(t) = \cos(2\pi t)(\frac{du}{dt} + \delta(t + 0.5))$.

   (c) $h(t) = \sin(2\pi t)\delta(0.5 - 2t)$.

   (d) $y(t) = \int_{-6}^{\infty}(\tau^2 + 6)\delta(\tau - 2)d\tau$.

   (e) $z(t) = \int_{6}^{\infty}(\tau^2 + 6)\delta(\tau - 2)d\tau$.

   (f) $a(t) = \int_{-\infty}^{t}\delta(\tau + 1)d\tau + \text{rect}(\frac{t}{6})\delta(t - 2)$.

   (g) $b(t) = \delta(t - 3) * u(t)$.

   (h) $c(t) = \triangle(\frac{t}{2}) * (\delta(t) - \delta(t + 2))$.

9.8   For $f(t)$ and $g(t)$ as shown,

   (a)   Determine and sketch $c(t) = f(t) * g(t)$.

   (b)   Given that $p(t) = \text{rect}(t - 1.5)$ and $f(t) = \frac{dp}{dt}$, find $x(t) = p(t) * g(t)$, using the result of part (a). Explain your procedure clearly.

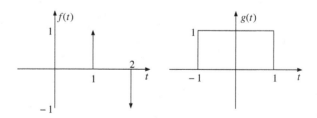

9.9   A system is described by an impulse response $h(t) = \delta(t - 1) - \delta(t + 1)$. Sketch the system response $y(t) = h(t) * f(t)$ to the following inputs:

   (a) $f(t) = u(t)$.

   (b) $f(t) = \text{rect}(t)$.

9.10   For a system with impulse response $h(t)$, the system output $y(t) = h(t) *$
$f(t) = \text{rect}(\frac{t-4}{2})$. Determine and sketch $h(t)$ if

   (a) $f(t) = \text{rect}(\frac{t}{2})$.

   (b) $f(t) = 2u(t)$.

   (c) $f(t) = 4\text{rect}(t)$.

9.11   Determine the Fourier transform of the following signals—simplify the
results as much as you can and sketch the result if it is real valued:

   (a) $f(t) = 5\cos(5t) + 3\sin(15t)$.

   (b) $x(t) = \cos^2(6t)$.

   (c) $y(t) = e^{-t}u(t) * \cos(2t)$.

   (d) $z(t) = (1 + \cos(3t))e^{-t}u(t)$.

9.12   Determine the inverse Fourier transforms of the following:

   (a) $F(\omega) = 2\pi[\delta(\omega - 4) + \delta(\omega + 4)] + 8\pi\delta(\omega)$.

   (b) $A(\omega) = 6\pi\cos(5\omega)$.

   (c) $B(\omega) = \sum_{n=-\infty}^{\infty} 2\pi\frac{1}{1+n^2}\delta(\omega - n2)$.

   (d) $C(\omega) = \frac{8}{j\omega} + 4\pi\delta(\omega)$.

9.13   Signal $f(t) = (5 + \text{rect}(\frac{t}{4}))\cos(60\pi t)$ is mixed with signal $\cos(60\pi t)$ to
produce the signal $y(t)$. Subsequently, $y(t)$ is low-pass filtered with a system
having frequency response $H(\omega) = 4\text{rect}(\frac{\omega}{4\pi})$ to produce $q(t)$. Sketch $F(\omega)$,
$Y(\omega)$, and $Q(\omega)$, and determine $q(t)$.

9.14   If signal $f(t)$ is not bandlimited, would it be possible to reconstruct $f(t)$
exactly from its samples $f(nT)$ taken with some finite sampling interval
$T > 0$? Explain your reasoning.

9.15   The inverse of the sampling interval T—that is, T^{-1}—is known as the
sampling frequency and usually is specified in units of Hz. Determine the
minimum sampling frequencies T^{-1} needed to sample the following analog
signals without causing aliasing error:

   (a) Arbitrary signal $f(t)$ with bandwidth 20 kHz.

   (b) $f(t) = \text{sinc}(4000\pi t)$.

   (c) $f(t) = \text{sinc}(4000\pi t)\cos(20000\pi t)$.

9.16   Using

$$\sum_{n=-\infty}^{\infty} f(t)\delta(t - nT) \leftrightarrow \sum_{n=-\infty}^{\infty} \frac{1}{T} F\left(\omega - n\frac{2\pi}{T}\right)$$

(item 25 in Table 7.2), sketch the Fourier transform $F_T(\omega)$ of signal

$$f_T(t) \equiv \sum_{n=-\infty}^{\infty} f(t)\delta(t - nT)$$

if

$$f(t) = \cos(4\pi t),$$

assuming (a) $T = 1$ s, and (b) $T = 0.2$ s. Which sampling period allows $f(t)$ to be recovered by applying an appropriate low-pass filter to $f_T(t)$?

9.17   Given the identity

$$\int_{-\infty}^{\infty} \delta(t - t_o) f(t)dt = f(t_o),$$

where the function $f(t)$ of time variable t is measured in units of, say, volts (V), what would be the units of the shifted impulse $\delta(t - t_o)$? What would be the units of an impulse $\delta(x)$ if x is a position variable measured in units of meters? What would be the units of a charge distribution specified as $q\delta(x - 4)$ if q is measured in units of coulombs (C)? It is an interesting fact that the impulse $\delta(t)$ has a unit (in the sense of a dimension), but it has no numerical values; only integrals of the impulse have a numerical value.

9.18   To confirm the scaling property of the impulse, show that

$$\delta(a(t - t_o)) * f(t)$$

and

$$\frac{1}{|a|}\delta(t - t_o) * f(t)$$

are identical for $a \neq 0$. Hint: Write the above convolutions explicitly and make use of an appropriate change of variable before applying the sifting property of the impulse.

10

Impulse Response, Stability, Causality, and LTIC Systems

In the last chapter we discovered that the zero-state response $y(t)$ of an LTI system $H(\omega)$ to an arbitrary input can be calculated in the time-domain with the convolution formula

$$y(t) = h(t) * f(t),$$

as shown in Figure 10.1. Here, $h(t)$ is the inverse Fourier transform of the system frequency response $H(\omega)$, or equivalently, the zero-state response of the system to an impulse input.

But what if the Fourier transform of an impulse response $h(t)$ does not exist, as in the case of the impulse response $h(t) = e^t u(t)$? Does this mean that we cannot make use of $h(t) * f(t)$ to calculate the zero-state response in such cases?

To the contrary, as we shall see in Section 10.1, the convolution method $y(t) = h(t) * f(t)$ is always valid for LTI systems, so long as the integral converges, and is, in fact, more fundamental than Fourier inversion of $Y(\omega) = H(\omega)F(\omega)$. It is the latter method that fails in the scenario invoked, when the Fourier transform of $h(t)$ (or of $f(t)$, for that matter) does not exist.

A time-domain perspective: zero-state response $y(t) = h(t) * f(t)$; impulse response and causality; LTIC systems

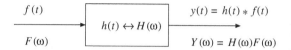

Figure 10.1 The time-domain input–output relation for LTI systems with frequency response $H(\omega)$. The convolution formula $y(t) = h(t) * f(t)$ describes the system *zero-state response* to input $f(t)$, where $h(t)$ is the inverse Fourier transform of the system frequency response $H(\omega)$.

Section 10.1 begins with a discussion of how $h(t)$, the impulse response of an LTI system, can be measured in the lab. The discussion subsequently focuses on the universality of the convolution formula, $h(t) * f(t)$, and the relation between $h(t)$ and $H(\omega)$. In Section 10.2 we examine stability conditions for LTI systems and establish the fact that only those systems with absolutely integrable $h(t)$ are guaranteed to produce bounded outputs when presented with bounded inputs. Next, in Section 10.3 we introduce the concept of *causality* and establish the fact that impulse responses $h(t)$ of causal real-time LTI systems, such as linear circuits built in the lab, vanish for $t < 0$. We refer to causal LTI systems as LTIC systems and consider a sequence of examples that illustrate how to recognize whether a system is causal, linear, and time-invariant. We conclude with short sections that discuss the importance of noncausal models in some settings (Section 10.4) and the modeling of delay lines (Section 10.5).

10.1   Impulse Response $h(t)$ and Zero-State Response $y(t) = h(t) * f(t)$

10.1.1   Measuring $h(t)$ of LTI systems

Measuring the impulse response

So far, we have obtained the impulse response $h(t)$ of LTI systems by inverse Fourier transforming the frequency response $H(\omega)$. Alternatively, the impulse response of a system can be *measured* in the lab with one of the following methods:

(1) Recall that the identity

$$\delta(t) * h(t) = h(t)$$

is a symbolic shorthand for

$$\lim_{\epsilon \to 0}\{p_\epsilon(t) * h(t)\} = h(t),$$

where $p_\epsilon(t)$ is a pulse centered about $t = 0$, having width ϵ and area 1. (See Figure 9.10.) Now, if we were to apply an input $p_\epsilon(t)$ to a system in the lab, having an unknown impulse response $h(t)$, we would measure (or display on a scope) an output $p_\epsilon(t) * h(t)$. Taking a sequence of such measurements with inputs $p_\epsilon(t)$ having decreasing widths ϵ (and increasing heights), we should see the output $p_\epsilon(t) * h(t)$ converge to $h(t)$. We would need to keep reducing ϵ until further changes in the output were imperceptible. If that did not happen and the output kept changing at every step, no matter how small ϵ (see Example 10.1 that follows, for a possible reason), then we instead could use the second method, presented next.

(2) Excite the system with a unit-step input $f(t) = u(t)$ to obtain the *unit-step response*

$$y(t) = h(t) * u(t) \equiv g(t).$$

In symbolic terms,

$$u(t) \longrightarrow \boxed{\text{LTI}} \longrightarrow h(t) * u(t) \equiv g(t).$$

Differentiating $g(t) = h(t) * u(t)$ and using the time-derivative property of convolution, we find

$$\frac{dg}{dt} = h(t) * \frac{du}{dt} = h(t) * \delta(t) = h(t).$$

So, the second method for finding the impulse response $h(t)$ is to differentiate the system unit-step response $g(t)$, which can be measured with a single input.

Unit-step response $g(t)$ and $h(t) = \frac{dg}{dt}$

Example 10.1

Suppose that measurements in the lab indicate that the unit-step response of a certain circuit is

$$g(t) = e^{-t}u(t).$$

What is the system impulse response $h(t)$? Can we measure $h(t)$ by using the first method described above?

Solution We find $h(t)$ by differentiating $g(t)$:

$$h(t) = \frac{dg}{dt} = \frac{d}{dt}(e^{-t}u(t)) = -e^{-t}u(t) + e^{-t}\frac{du}{dt}$$

$$= -e^{-t}u(t) + e^{-t}\delta(t) = \delta(t) - e^{-t}u(t).$$

Notice that we used the sampling property of the impulse to replace $e^{-t}\delta(t)$ with $\delta(t)$.

In implementing the first method, the system response to input $p_\epsilon(t)$ is

$$h(t) * p_\epsilon(t) = (\delta(t) - e^{-t}u(t)) * p_\epsilon(t) = p_\epsilon(t) - e^{-t}u(t) * p_\epsilon(t).$$

As ϵ is reduced, the second term of the output will converge to $e^{-t}u(t)$, because

$$\lim_{\epsilon \to 0}\{e^{-t}u(t) * p_\epsilon(t)\} = e^{-t}u(t) * \delta(t) = e^{-t}u(t).$$

However, the first term $p_\epsilon(t)$ will not converge (i.e., stop changing) as ϵ is reduced. Even if we guess that an impulse is appearing in the output as ϵ is made smaller, it will be difficult to estimate the area of the impulse. Thus, the first method will not be workable in practice. This problem arises because the system impulse response $h(t) = \delta(t) - e^{-t}u(t)$ contains an impulse.

Example 10.2

Measurements in the lab indicate that the unit-step response of a certain circuit is

$$g(t) = te^{-t}u(t).$$

What is the system impulse response $h(t)$? Can we measure $h(t)$ by using the first method described in this section?

Solution   Similar to Example 10.1, the impulse response is

$$h(t) = \frac{dg}{dt} = \frac{d}{dt}(te^{-t}u(t)) = (1 - t)e^{-t}u(t) + te^{-t}\frac{du}{dt}$$

$$= (1 - t)e^{-t}u(t) + te^{-t}\delta(t) = (1 - t)e^{-t}u(t).$$

Because $h(t)$ does not contain an impulse, the first method also will work.

Example 10.3

What is the frequency response of the system described in Example 10.2?

Solution   Given that

$$h(t) = (1 - t)e^{-t}u(t) = e^{-t}u(t) - te^{-t}u(t),$$

the Fourier transform

$$H(\omega) = \frac{1}{1 + j\omega} - \frac{1}{(1 + j\omega)^2} = \frac{j\omega}{(1 + j\omega)^2}$$

must be the corresponding frequency response.

Example 10.4

A system that is known to be LTI responds to the input $u(t + 1)$ with the output $\text{rect}(\frac{t}{2})$, as shown at the top of Figure 10.2. What will be the system response $y(t)$ to the input $f(t) = \text{rect}(t)$? Solve this problem by first finding the system impulse response.

Solution   Since the system is time-invariant, the information

$$u(t + 1) \longrightarrow \boxed{\text{LTI}} \longrightarrow \text{rect}(\frac{t}{2})$$

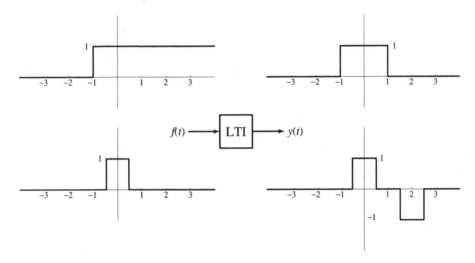

Figure 10.2   An LTI system that responds to an input $u(t + 1)$ with the output $\text{rect}(\frac{t}{2})$ will respond to input $f(t) = \text{rect}(t)$ with output $\text{rect}(t) - \text{rect}(t - 2)$, as shown here. (See Example 10.4.)

implies that

$$u(t) \longrightarrow \boxed{\text{LTI}} \longrightarrow \text{rect}(\frac{t-1}{2}) = u(t) - u(t-2).$$

Thus,

$$g(t) = u(t) - u(t-2)$$

so that

$$h(t) = g'(t) = \delta(t) - \delta(t-2).$$

Consequently, the response to input $f(t) = \text{rect}(t)$ is

$$y(t) = h(t) * f(t) = [\delta(t) - \delta(t-2)] * \text{rect}(t) = \text{rect}(t) - \text{rect}(t-2),$$

as shown at the bottom of Figure 10.2.

We can solve this same problem more directly by using the properties of super-position and time invariance and the fact that the second input can be written as a linear combination of delayed versions of the first input. Working the problem in this way will produce the answer

$$y(t) = \text{rect}(\frac{t - \frac{1}{2}}{2}) - \text{rect}(\frac{t - \frac{3}{2}}{2}),$$

which can be shown to be the same as the former answer. Try it!

10.1.2   General validity of $y(t) = h(t) * f(t)$ for LTI systems

It is surprisingly easy for us to develop the convolution formula, describing the zero-state response of linear, time-invariant (LTI) systems, by working strictly within the time domain, without any need for the frequency response $H(\omega)$ to exist. To do so, we begin by noting that every LTI system in zero state will respond to an impulse input with a specific signal that we denote as $h(t)$ and call the *impulse response*. That is, in symbolic notation,

$$\delta(t) \longrightarrow \boxed{\text{LTI}} \longrightarrow h(t)$$

is true for *any* LTI system as a matter of definition of its impulse response. Now, invoking *time-invariance* of the system—meaning that delayed inputs cause equally delayed outputs—we have

$$\delta(t - \tau) \longrightarrow \boxed{\text{LTI}} \longrightarrow h(t - \tau).$$

Finally, invoking the *linearity* property of the system—wherein a weighted superposition of inputs leads to an equally weighted superposition of outputs—we obtain

$$\int_{-\infty}^{\infty} f(\tau)\delta(t - \tau)d\tau \longrightarrow \boxed{\text{LTI}} \longrightarrow \int_{-\infty}^{\infty} f(\tau)h(t - \tau)d\tau,$$

where $f(\tau)$ is an arbitrary weighting function applied to impulse inputs $\delta(t - \tau)$ with delays τ.[1]

The system input shown in the final statement just asserted is the convolution $f(t) * \delta(t) = f(t)$, and the corresponding system output is $f(t) * h(t) = h(t) * f(t)$. Hence, the final statement is equivalent to

The convolution formula $y(t) = h(t) * f(t)$ is valid for LTI systems, even if $H(\omega)$ does not converge

$$f(t) \longrightarrow \boxed{\text{LTI}} \longrightarrow h(t) * f(t),$$

meaning that

$$y(t) = h(t) * f(t)$$

is the formula for the *zero-state response* of all LTI systems with all possible inputs $f(t)$—the existence of the Fourier transform of the system impulse response $h(t)$ is not necessary.

---

[1]Here, we assume that the LTI system satisfies the superposition property even for an uncountably infinite number of input terms; i.e., for linear combinations of $\delta(t - \tau)$ for all values of τ. This assumption is satisfied by all LTI systems encountered in the lab.

Example 10.5

When applied to a particular LTI system, an impulse input $\delta(t)$ produces the output $h(t) = e^t u(t)$. What is the zero-state response of the same system to the input $f(t) = u(t)$?

Solution

$$y(t) = h(t) * f(t) = e^t u(t) * u(t) = \int_{-\infty}^{t} e^{\tau} u(\tau) d\tau = (e^t - 1)u(t).$$

Notice that we could not have calculated $y(t)$ with the Fourier method, because the Fourier transform of $h(t) = e^t u(t)$ does not exist. Also, notice that the output $y(t)$ is not bounded ($y(t) \to \infty$ as $t \to \infty$), even though the input $f(t) = u(t)$ is a bounded function ($|f(t)| \leq 1$ for all t).

10.1.3 Testing whether a system is (zero-state) LTI

The relationship between the system input $f(t)$ and the system response $y(t)$ always reveals whether a system is LTI.[2] If a system is LTI, then the relationship between $f(t)$ and $y(t)$ can be expressed in the convolution form $y(t) = h(t) * f(t)$, where $h(t)$ does not depend on the choice of $f(t)$. If a system is not LTI, then either the linearity property or the time-invariance property (or possibly, both) is violated. The following examples illustrate how we can recognize whether a given system is LTI.

Example 10.6

For a system with input $f(t)$, the output is given as

$$y(t) = f(t + T).$$

Is this system LTI?

Solution   Because we can write

$$y(t) = f(t + T) = \delta(t + T) * f(t),$$

this system is LTI with impulse response

$$h(t) = \delta(t + T).$$

Thus, the system must satisfy both zero-state linearity and time invariance.

Example 10.7

Suppose a system has input–output relation

$$y(t) = f^2(t + T).$$

Is this system LTI?

---

[2]We assume zero state, which is the condition under which time invariance was defined.

Solution   We can write

$$y(t) = f^2(t + T) = (\delta(t + T) * f(t))^2,$$

which is not in the form $y(t) = h(t) * f(t)$. Thus, the system is not LTI.

Example 10.8
Is the system

$$y(t) = f^2(t + T)$$

time invariant?

Solution   We already know from Example 10.7 that the system is not LTI. But, it still could be time invariant. To test time invariance, we feed the system with a new input

$$f_1(t) = f(t - t_o)$$

and observe that the new output is

$$y_1(t) = f_1^2(t + T) = f^2(t + T - t_o) = f^2((t - t_o) + T) = y(t - t_o).$$

Because the new output is a delayed version of the original output, the system is time invariant. Since the system is time invariant, but not LTI, it must be nonlinear. We easily can confirm this by noting that a doubling of the input does not double the output.

   We also can test zero-state linearity of the system from first principles by checking whether the output formula supports linear superposition. With an input

$$f(t) = f_1(t) + f_2(t),$$

the system output is

$$\begin{aligned} y(t) = f^2(t + T) &= [f_1(t + T) + f_2(t + T)]^2 \\ &= f_1^2(t + T) + f_2^2(t + T) + 2f_1(t + T)f_2(t + T). \end{aligned}$$

Notice that this is not the sum of the responses $f_1^2(t + T)$ and $f_2^2(t + T)$ due to individual inputs $f_1(t)$ and $f_2(t)$. Hence, as expected, the system is not linear.

Example 10.9
A system responds to an unspecified input $f(t)$ with the output $2\text{rect}(t)$. If a delayed version $f(t - 2)$ of the same input is applied, the output is observed to be $4\text{rect}(t - 2)$. Is the system LTI? Time invariant? Zero-state linear?

Solution   The system is not time invariant because, if it were, its response to input $f(t - 2)$ would have been $2\text{rect}(t - 2)$ instead of $4\text{rect}(t - 2)$. Because the system is not time invariant, it cannot be LTI.

Is the system zero-state linear? It could be. There is not enough information provided to test whether the system is zero-state linear. Unless a general input–output formula that relates $f(t)$ and $y(t)$ is given, it is not always possible to test zero-state linearity and time invariance.

10.1.4   Frequency response $H(\omega)$ of LTI systems

In Section 10.1.2 we learned that we can compute the zero-state response of any LTI system to an input $f(t)$ by using $y(t) = h(t) * f(t)$. Thus, for a complex-exponential input $f(t) = e^{j\omega t}$,

$$y(t) = h(t) * e^{j\omega t} = \int_{-\infty}^{\infty} h(\tau)e^{j\omega(t-\tau)}d\tau = e^{j\omega t}\int_{-\infty}^{\infty} h(\tau)e^{-j\omega\tau}d\tau.$$

This fundamental result indicates that, under zero-state conditions,

$$e^{j\omega t} \longrightarrow \boxed{\text{LTI}} \longrightarrow e^{j\omega t} H(\omega),$$

where

$$H(\omega) \equiv \int_{-\infty}^{\infty} h(\tau)e^{-j\omega\tau}d\tau.$$

In other words, given a complex exponential input, the output of an LTI system is the same complex exponential, except scaled by a constant that is the frequency response evaluated at the frequency of the input. As before, the frequency response is the Fourier transform of the impulse response $h(t)$. What we have just shown is consistent with our earlier development of the concept of frequency response, but here we have arrived at the same notion through use of the convolution formula

Notice that this result implicitly assumes that the Fourier transform of $h(t)$ converges so that $H(\omega)$ is well defined (i.e., is finite at every value of ω). If the Fourier integral does not converge (in engineering terms, $H(\omega)$ may be infinite), then our result suggests that the system zero-state response to a bounded input $e^{j\omega t}$ may be unbounded. Such LTI systems with nonconvergent $H(\omega)$ are considered *unstable*, a concept to be examined in more detail in the next section.

Example 10.10
Find the frequency responses $H(\omega)$ of the LTI systems having impulse response functions $h_1(t) = e^{-t}u(t)$, $h_2(t) = tu(t)$, and $h_3(t) = u(t)$.

Solution   The Fourier transform integral of $h_1(t) = e^{-t}u(t)$ equals $H_1(\omega) = \frac{1}{1+j\omega}$, which is the frequency response of the system.

The Fourier transform integral of $h_2(t) = tu(t)$ does not converge, and consequently, the frequency response $H_2(\omega)$ does not exist.

The Fourier transform integral of $h_3(t) = u(t)$ also does not converge. Even though (according to Table 7.2) the Fourier transform of the power signal $u(t)$ is $\pi\delta(\omega) + \frac{1}{j\omega}$, this expression is not a regular function and the frequency response $H_3(\omega)$ of the system does not exist.

A more general frequency-domain description will be discussed in Chapter 11 for systems $h_2(t) = tu(t)$ and $h_3(t) = u(t)$, based on the Laplace transform.

10.2 BIBO Stability

The previous section suggested that if the Fourier transform of an impulse response $h(t)$ does not converge—as in the case with $h(t) = u(t)$—then the zero-state response of the corresponding LTI system to a bounded input may be unbounded. In most cases this behavior of having an unbounded output would be problematic. We generally wish to avoid designing such systems, because as the output signal continues to grow, one of two things can happen: Either the output signal grows large enough that circuit components burn out, or the output saturates at the supply voltage level (a nonlinear effect). In either case the circuit fails to produce the designed response.

Bounded inputs and outputs

Ordinarily, we wish to have systems that produce bounded outputs from bounded inputs. We will use the term *BIBO stable*—an abbreviation for *bounded input, bounded output stable*—to refer to such systems. Systems that are not stable are called *unstable*. The terms "bounded input" and "bounded output" in this context refer to *functions* $f(t)$ and $y(t) = h(t) * f(t)$ having bounded magnitudes. In other words,

$$|f(t)| \leq \alpha < \infty$$

and

$$|y(t)| = |h(t) * f(t)| \leq \beta < \infty$$

for all t, where the bounds α and β are finite constants. For instance, signals $f(t) = e^{-t}u(t)$, $u(t)$, $e^{-|t|}$, $\cos(2t)$, $\text{sgn}[\cos(-t)]$, and e^{j3t} are bounded by $\alpha = 1$ for any value of t. If a signal is not bounded, then it is said to be unbounded; the functions e^{-t}, $tu(t)$, and $e^t u(t)$ are examples of unbounded signals.

If an LTI system is not BIBO stable, this means that there is *at least one* bounded input function $f(t)$ that will cause an unbounded zero-state response $y(t) = h(t) * f(t)$. It does not mean that *all* possible outputs $y(t)$ of an unstable system will be unbounded. For instance, the system $h(t) = u(t)$ is not BIBO stable (e.g., the output $y(t) = u(t) * f(t)$ due to input $f(t) = 1$ is not bounded), but its zero-state response $y(t) = u(t) * \text{rect}(t)$ due to input $f(t) = \text{rect}(t)$ is bounded. In practice, we are concerned if even one bounded input can cause an unbounded output, and hence, we insist on BIBO stability.

We might ask how we can test whether a system is BIBO stable. Certainly, we cannot try *every possible* bounded input and then examine the corresponding outputs to check whether they are bounded. Doing so would require trying an infinite number of inputs, which would require an infinite amount of time. Instead, we need a simpler test. From our earlier discussion, we feel certain that if the Fourier transform of the impulse response $h(t)$ does not converge, then the corresponding LTI system cannot be BIBO stable. So, might this be the test we are looking for? The answer is "not quite," because although convergence of the Fourier transform of $h(t)$ is a *necessary* condition for BIBO stability, it is *not sufficient*. For example, the Fourier transform of sinc(t) converges, and yet a system with impulse response $h(t) = $ sinc(t) is not BIBO stable, as we will see shortly.

The key to BIBO stability turns out to be *absolute integrability* of the impulse response $h(t)$. More specifically, it can be shown that:

An LTI system is BIBO stable *if and only if* its impulse response $h(t)$ is absolutely integrable, satisfying **BIBO stability criterion**

$$\int_{-\infty}^{\infty} |h(t)|dt < \infty.$$

The proof of this BIBO-stability criterion will be given in two parts: First, we will show that absolute integrability of $h(t)$ is *sufficient* for BIBO stability; then we will show that absolute integrability also is *necessary* for BIBO stability.

Proof of sufficiency: With the help of the *triangle inequality*, note that

$$|y(t)| = |h(t) * f(t)| = |\int_{-\infty}^{\infty} h(\tau)f(t-\tau)d\tau|$$

$$\leq \int_{-\infty}^{\infty} |h(\tau)||f(t-\tau)|d\tau \leq |f(t)|_{\max} \int_{-\infty}^{\infty} |h(\tau)|d\tau,$$

where $|f(t)|_{\max}$ is the maximum value of $|f(t)|$ for all t. Now, suppose that $h(t)$ is absolutely integrable, that is.

$$\int_{-\infty}^{\infty} |h(\tau)|d\tau = \gamma < \infty.$$

Then, for any bounded input $f(t)$ such that $|f(t)|_{\max} < \infty$, the corresponding output $y(t) = h(t) * f(t)$ also is bounded, and we have

$$|y(t)|_{max} < \gamma |f(t)|_{\max} < \infty.$$

Hence, absolute integrability of $h(t)$ is a sufficient condition for BIBO stability.

In the second part of the proof we need find only a single example of a bounded input $f(t)$ that causes an unbounded output $h(t) * f(t)$ when $h(t)$ is not absolutely integrable.

Proof of necessity: With an LTI system output expressed as

$$y(t) = \int_{-\infty}^{\infty} h(\tau) f(t - \tau) d\tau,$$

we find that

$$y(0) = \int_{-\infty}^{\infty} h(\tau) f(-\tau) d\tau.$$

Consider now the case of a bounded input signal specified as

$$f(t) = \text{sgn}(h(-t)).$$

In this case $f(-\tau) = \text{sgn}(h(\tau))$ and, consequently,

$$y(0) = \int_{-\infty}^{\infty} h(\tau) \text{sgn}(h(\tau)) d\tau = \int_{-\infty}^{\infty} |h(\tau)| d\tau.$$

Clearly, if $h(t)$ is not absolutely integrable then $y(0)$ will be infinite. Thus, $y(t)$ will not be bounded, even though the input $f(t) = \text{sgn}(h(-t))$ is bounded; the sgn function takes on only ± 1 values. Hence, we have proven the *necessity* of an absolutely integrable $h(t)$ for BIBO stability.

Example 10.11
Determine whether the given systems are BIBO stable. For each system that is unstable, provide an example of a bounded input that will cause an unbounded output.

$$h_1(t) = e^{-t} u(t),$$
$$h_2(t) = e^{-t},$$
$$h_3(t) = 2u(t - 1),$$
$$h_4(t) = 2\delta(t),$$
$$h_5(t) = e^{2t} u(t),$$
$$h_6(t) = e^{-2t} u(t) - e^{-t} u(t),$$
$$h_7(t) = \delta'(t) \equiv \frac{d}{dt} \delta(t),$$
$$h_8(t) = \cos(\omega_o t)\, u(t),$$
$$h_9(t) = \text{rect}(t - 1).$$

Solution   The system with $h_1(t) = e^{-t} u(t)$ is BIBO stable, because

$$\int_{-\infty}^{\infty} |e^{-t} u(t)| dt = \int_{0}^{\infty} e^{-t} dt = 1,$$

showing that $h_1(t)$ is absolutely integrable.

The system with $h_2(t) = e^{-t}$ is not BIBO stable because

$$\int_{-\infty}^{\infty} |e^{-t}| dt = \int_{-\infty}^{\infty} e^{-t} dt,$$

which does not converge. There are many bounded inputs that will cause this system to have an unbounded output. For example, choosing $f(t) = u(t)$, we see that the system response is

$$y(t) = e^{-t} * u(t) = \int_{-\infty}^{t} e^{-\tau} d\tau,$$

which is not bounded.

The system with $h_3(t) = 2u(t-1)$ also is not BIBO stable, because the area under $|h_3(t)| = |2u(t-1)|$ is infinite. Once again, choosing the bounded input $f(t) = u(t)$ will produce an unbounded output, because

$$y(t) = 2u(t-1) * u(t) = 2(t-1)u(t-1)$$

is unbounded.

Our proof that absolute integrability of $h(t)$ is both a necessary and sufficient condition for BIBO stability assumed that $h(t)$ is a function. Because the impulse response for system $h_4(t) = 2\delta(t)$ is not a function (it involves an impulse), we resort to first principles to test whether this system is BIBO stable. In particular, we note that the system zero-state response to an arbitrary input $f(t)$ is

$$y(t) = h_4(t) * f(t) = 2\delta(t) * f(t) = 2f(t).$$

Thus, all bounded inputs $f(t)$ cause bounded outputs $2f(t)$, and so the system must be BIBO stable.

System $h_5(t) = e^{2t}u(t)$ is not BIBO stable, because the area under $|h_5(t)| = e^{2t}u(t)$ is infinite. Many bounded inputs, including $f(t) = u(t)$, will produce an unbounded output.

System $h_6(t) = e^{-2t}u(t) - e^{-t}u(t)$ is BIBO stable because

$$|h_6(t)| = |e^{-2t}u(t) - e^{-t}u(t)| \le e^{-2t}u(t) + e^{-t}u(t)$$

and the area under $e^{-2t}u(t) + e^{-t}u(t)$ is finite.

System $h_7(t) = \delta'(t)$ is not BIBO stable. To see the reason, note that the zero-state response of the system is

$$y(t) = h_7(t) * f(t) = \delta'(t) * f(t) = f'(t).$$

Thus, if $f(t)$ is a bounded input with a discontinuity, the response $f'(t)$ is not bounded (e.g., $f(t) = u(t)$ with $f'(t) = \delta(t)$).

System $h_8(t) = \cos(\omega_o t)u(t)$ is not BIBO stable because $\cos(\omega_o t)u(t)$ is not an absolutely integrable function; the area under $|\cos(\omega_o t)u(t)| = |\cos(\omega_o t)|u(t)$ is infinite. An example of a bounded input that will cause an unbounded output is $f(t) = e^{j\omega_o t}$, which produces

$$y(t) = \cos(\omega_o t)u(t) * e^{j\omega_o t} = e^{j\omega_o t} H(\omega_o),$$

where

$$H(\omega_o) = \int_0^\infty \cos(\omega_o t)e^{-j\omega_o t}dt$$

is non-convergent. Similarly, $\cos(\omega_o t)$ and $\sin(\omega_o t)$ are other bounded inputs that will cause this system to produce an unbounded output.

Finally system $h_9(t) = \text{rect}(t - 1)$ is BIBO stable because the rect function has a finite area under the curve.

Table 10.1 shows plots of some of the impulse responses examined previously, arranged in two columns according to whether each impulse response corresponds to a stable or an unstable system.

We close this section with several observations that reinforce and add to what we have learned about BIBO stability. First, note that the above examples illustrate that the test for stability is the absolute integrability of the impulse response (for $h(t)$ that are functions). The test is *not* the boundedness of the impulse response. This is sometimes a point of confusion. Notice that both the aforementioned $h_3(t)$ and $h_8(t)$ are bounded, and yet they correspond to unstable systems.

Second, it should be clear that BIBO stable systems always have a well-defined frequency response, $H(\omega)$. This follows from Chapter 7, where we learned that absolute integrability is a sufficient condition for the existence of the Fourier transform.

Third, there are some systems with a well defined frequency response that are not BIBO stable. So, the existence of $H(\omega)$ is not sufficient to imply stability. For example, $h(t) = \text{sinc}(t)$ is not absolutely integrable, and so a system having this impulse response is not BIBO stable. Yet, $H(\omega)$ is well defined; indeed, $H(\omega) = \pi \text{rect}(\frac{\omega}{2})$, which represents an ideal low-pass filter.[3]

Fourth, most systems that are not BIBO stable—including very important systems such as ideal integrators ($h(t) = u(t)$) and differentiators ($h(t) = \delta'(t)$)—do not have convergent $H(\omega)$. Unlike in the case of the ideal low-pass filter, Fourier representation of such systems is difficult or impossible. The best tool for frequency-domain analysis of unstable systems is the Laplace transform, which will be introduced in Chapter 11. The Laplace transform also is very convenient for solving zero-input and initial-value problems, as we shall see.

---

[3]The input $f(t) = \text{sgn}(\text{sinc}(t))$ is an example of a bounded input that will cause the ideal low-pass filter to have an unbounded output. Many other bounded inputs, such as $\text{rect}(t)$, $\cos(t)$, etc., will produce a bounded output.

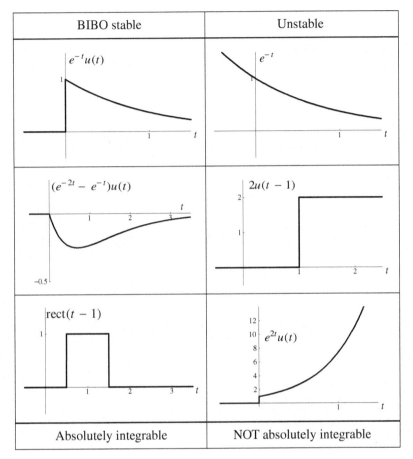

Table 10.1 Some of the impulse response functions $h(t)$ examined in Example 10.11, categorized according to whether they correspond to stable systems ($h(t)$ *absolutely integrable*) or unstable systems.

Finally, some important unstable systems, such as ideal low-pass filters, and ideal integrators and differentiators, can be approximated as closely as we like by stable systems. Chapter 12 introduces this topic of *filter design*.

10.3 Causality and LTIC Systems

An LTI system, stable or not, is said to be *causal* if its zero-state response $h(t) * f(t)$ depends only on past and present, *but not future*, values of the input signal $f(t)$. Systems that are not causal are said to be *noncausal*. All practical analog LTI circuits built in the lab are, of course, causal, because the existence of a noncausal circuit—that is, one that produces an output before its input is applied—would be as impossible as a football flying off the tee before the kicker kicked the ball.[4]

Writing the LTI zero-state response in terms of the convolution formula,

$$y(t) = \int_{-\infty}^{\infty} h(\tau) f(t - \tau) d\tau,$$

we see clearly that $y(t_1)$, the output at some fixed time t_1, can depend on $f(t)$ for $t > t_1$ only if $h(\tau)$ is nonzero for some negative values τ. For example, if $h(-1)$ is nonzero, then we see from the convolution formula that $y(t_1)$ will depend on $f(t_1 - (-1)) = f(t_1 + 1)$, that is, on a *future* value of f. Thus:

Causality criterion for LTI systems

An LTI system with impulse response function $h(t)$ is causal *if and only if* $h(t) = 0$ for $t < 0$.

Clearly, then, the ideal low-pass system $h(t) = \text{sinc}(t)$, just considered in the previous section, is not a causal system (because $\text{sinc}(t)$ is nonzero for $t < 0$) and thus cannot be implemented by an LTI circuit in the lab, no matter how clever the designer. Fortunately, there is no need for an *exact* implementation. In Chapter 12, we will see how to design causal and realizable (as well as BIBO stable) alternatives that will closely approximate the behavior of an ideal low-pass filter.

It is important to note that the *causality criterion* just stated, $h(t) = 0$ for $t < 0$, applies only to systems whose impulse responses can be expressed in terms of functions—for example, $\text{sinc}(t)$, $u(t)$, $e^{-t}u(t)$, etc. If we must write the impulse response in terms of the impulse $\delta(t)$ and related distributions, then we should determine causality by directly examining the dependence of the zero-state output on the system input.

Example 10.12

The zero-state response of an LTI system to arbitrary input $f(t)$ is described by

$$y(t) = f(t - 2).$$

Find the system impulse response $h(t)$ and determine whether the system is causal.

Solution   Since $f(t - 2) = \delta(t - 2) * f(t)$, the input–output formula can be written as

$$y(t) = \delta(t - 2) * f(t).$$

---

[4]As described in Section 10.4, however, noncausal models are highly useful in practice, especially for processing of spatial data (e.g., images) and in digital signal processing, where entire signals can be prestored prior to processing.

Hence, the impulse response of the system is

$$h(t) = \delta(t - 2).$$

Clearly, this system is causal, because the present system output is simply the system input from 2 time units prior. Because the output at any instant does not depend on future values of the input, the system is causal.

Example 10.13
The zero-state response $y(t)$ of an LTI system to a unit-step input $u(t)$ is the function

$$g(t) = \text{rect}(t).$$

Find the system impulse response $h(t)$ and determine whether the system is causal.

Solution   Since

$$\text{rect}(t) = u\left(t + \frac{1}{2}\right) - u\left(t - \frac{1}{2}\right),$$

and because the impulse response $h(t)$ of an LTI system is the derivative of the unit-step response $g(t)$, we have

$$h(t) = \frac{d}{dt}\text{rect}(t) = u'\left(t + \frac{1}{2}\right) - u'\left(t - \frac{1}{2}\right) = \delta\left(t + \frac{1}{2}\right) - \delta\left(t - \frac{1}{2}\right).$$

Thus, the system zero-state response to an arbitrary input $f(t)$ is

$$y(t) = \left[\delta\left(t + \frac{1}{2}\right) - \delta\left(t - \frac{1}{2}\right)\right] * f(t) = f\left(t + \frac{1}{2}\right) - f\left(t - \frac{1}{2}\right).$$

Clearly, this system is noncausal, because the system output $y(t)$ depends on $f(t + \frac{1}{2})$, representing an input half a time unit into the future.

Another way to see that this system is noncausal, without having to find $h(t)$, is to note that the output $\text{rect}(t)$ *turns on* at time $t = -\frac{1}{2}$, which is *earlier* than $t = 0$ when the input $u(t)$ turns on. No practical causal circuit built in the lab can behave this way!

It should be clear from the foregoing examples that LTI systems with impulse responses of the form $\delta(t - t_o)$ are causal when $t_o \geq 0$ and noncausal when $t_o < 0$. In general, we will use the term *causal signal* to refer to signals that could be the impulse response of a causal LTI system, and use the term LTIC to describe LTI systems that are causal. For instance, $\delta(t)$, $u(t - 1)$, $e^{-t}u(t)$, $\delta'(t - 2)$, and $\cos(2\pi t)u(t)$ are examples of causal signals that qualify as impulse responses of possible LTIC systems, whereas signals $\delta(t + 2)$, e^{-t}, $u(t + 1)$, and $\cos(t)$ are noncausal and cannot be impulse responses of LTIC systems. (See Table 10.2.)

LTIC systems and causal signals

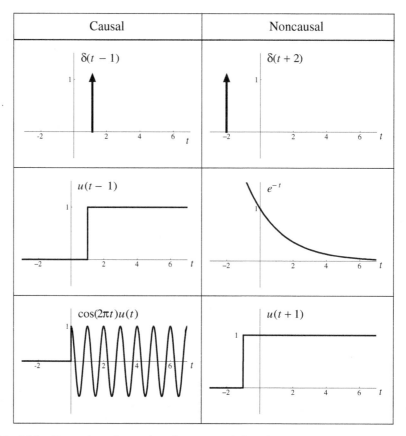

	Causal	Noncausal

Table 10.2   Examples of causal and noncausal signals.

Example 10.14

Determine whether the following signal is causal:

$$h(t) = \delta(t) + u(t + 1).$$

Solution   To solve this problem, we must determine whether the hypothesized $h(t)$ can be the impulse response of a causal LTI system. The output of an LTI system having the given $h(t)$ as its impulse response would be

$$y(t) = (\delta(t) + u(t + 1)) * f(t) = \delta(t) * f(t) + u(t + 1) * f(t)$$

$$= f(t) + f(t + 1) * u(t) = f(t) + \int_{-\infty}^{t} f(\tau + 1)d\tau.$$

Clearly, $y(t)$ depends on values of $f(t)$ up to 1 time unit into the future, and thus the LTI system is not causal. Consequently, the given $h(t)$ is not causal.

A more direct approach to this problem is as follows: $u(t + 1)$ clearly is a noncausal signal since it has nonzero values for $t < 0$. The sum of a noncausal signal with any other signal obviously will be a noncausal signal, unless the two signals cancel one another for $t < 0$. With that logic, the given $h(t)$ clearly is a noncausal signal.

Note that, while all practical LTI systems built in the lab will be causal, not all causal lab systems are LTI. The concept of causality also applies to systems that are nonlinear and/or time-varying. The following examples illustrate some of these possibilities.

Example 10.15
A particular time-varying (and thus not LTI) system is described by the input–output relation

$$y(t) = \cos(t + 5)f(t).$$

Is the system causal or noncausal?

Solution   The system clearly is causal, because the output does not depend on future values of the input $f(t)$.

Example 10.16
A system is described by the input–output relation

$$y(t) = f(t^2).$$

Is the system causal or noncausal?

Solution   The system is noncausal because, for instance,

$$y(-1) = f((-1)^2) = f(1),$$

showing that there are times for which the output depends on future values of the input.

Example 10.17
A particular nonlinear system is described by the input–output relation

$$y(t) = f^2(t + T).$$

Is the system causal or noncausal?

Solution   The answer depends on whether T is negative or positive. If $T \leq 0$, then the output $y(t)$ depends only on the present or a past value of the input $f(t)$, and the system is causal. However, if $T > 0$, then the system is noncausal.

Example 10.18

What type of filter is implemented by an LTI system having the impulse response

$$h(t) = \frac{\Omega}{\pi}\operatorname{sinc}(\Omega t)\cos(\omega_o t),$$

assuming $\omega_o > \Omega$? Discuss why this filter is impossible to build in the lab.

Solution   According to Table 7.2

$$\frac{\Omega}{\pi}\operatorname{sinc}(\Omega t) \leftrightarrow \operatorname{rect}(\frac{\omega}{2\Omega}),$$

so use of the modulation property implies that the frequency response of the given system is

$$H(\omega) = \frac{1}{2}\operatorname{rect}(\frac{\omega - \omega_o}{2\Omega}) + \frac{1}{2}\operatorname{rect}(\frac{\omega + \omega_o}{2\Omega}).$$

Clearly, this frequency response describes an ideal band-pass filter with a center frequency ω_o and bandwidth Ω. The ideal filter is impossible to implement in the lab, because the system impulse response is noncausal.

Example 10.19

The input–output relation of a system is given as

$$y(t) = f(3t).$$

Is the system causal? Is it time invariant? Is it LTIC?

Solution   Since $y(1) = f(3)$, the output at $t = 1$ depends on the input at $t = 3$. So, the system is not causal and, therefore, it cannot be LTIC. However, it still could be time invariant, so let us check.

Time invariance requires that delayed inputs lead to equally delayed, but otherwise unchanged, outputs. Consider a new system input, which is a delayed version of the original, specified as

$$f_1(t) = f(t - t_o).$$

According to the given input–output rule, the corresponding output is

$$y_1(t) = f_1(3t) = f(3t - t_o).$$

Because $y_1(t)$ is different from

$$y(t - t_o) = f(3(t - t_o)),$$

the new output $y_1(t)$ is not a t_o-delayed version of the original output and, therefore, the system is time varying.

10.4   Usefulness of Noncausal System Models

While our focus in the remaining chapters will be on LTIC systems, it should be noted that noncausal LTI system models are important in practice. First, as we saw in the previous section, certain ideal filtering operations (e.g., ideal low-pass) are noncausal. We need our mathematical apparatus to be general enough to model these types of filters, because we often wish to design real filters that approximate such ideal filters. As indicated earlier, Chapter 12 will introduce this *filter design* problem.

A second use for noncausal models arises in the processing of spatial data where one or more of the variables involves position. Such examples abound in signal processing descriptions of imaging systems—for instance, cameras, telescopes, x-ray computer tomography (CT), synthetic aperture radar (SAR), and radio astronomy. In such systems the signal being imaged is a spatial quantity in two or three dimensions. Often, it is convenient to define the origin at the center of the scene, for both the input and output of the system. Doing so generally leads to a noncausal model for the processing. For example, a simple noise-smoothing operation, where each point in the output image is an average of the values surrounding that same point in the input image, can be described by LTI filtering with $\text{rect}(x)$ as the impulse response, which is noncausal. Here, we use the variable x, rather than t, to denote position.

Finally, noncausal models routinely are applied in digital signal processing, where an entire sampled signal may be prestored prior to processing. Depending on how the time origin is specified for the prestored signal samples and the output signal samples, a "current" output sample may depend on "future" input samples. For example,

$$y_n = f_{n+1} + 2f_n + f_{n-1},$$

where y_n depends on the future (already prestored) value f_{n+1}. There even are cases with digital processing where signals are reversed and processed backwards. This is the epitomy of noncausal processing!

10.5   Delay Lines

The system

$$h(t) = K\delta(t - t_o) \leftrightarrow H(\omega) = Ke^{-j\omega t_o}$$

is zero-state linear, time invariant, and BIBO stable. As we have seen in Chapter 8, systems having the frequency response

$$H(\omega) = Ke^{-j\omega t_o}$$

simply delay and amplitude-scale their inputs $f(t)$ to produce outputs

$$y(t) = K\delta(t - t_o) * f(t) = Kf(t - t_o).$$

Clearly, if $t_o \geq 0$, then the output $y(t)$ depends only on past or present values of $f(t)$ and the system is LTIC.

Delay-line systems

The LTIC system just described can be considered a *delay line* having a delay t_o and gain K. A practical example of a delay-line system is the *coaxial cable*. The delay of a coaxial cable or any type of transmission-line cable can be calculated as $t_o = \frac{L}{v}$, where L is the physical cable length and v is the signal propagation speed in the cable, usually a number close to (but less than) the speed of light in free space, $c = 3 \times 10^8$ m/s. If the cable is short, then t_o will be very small and the delay can be neglected in applications where signal durations are larger than t_o. For large L, on the other hand, delay can be an appreciable fraction of signal durations (e.g., $t_o = 0.1$ ms for a cable of length 30 km) and may need to be factored into system calculations. In a lossless cable (with some further conditions satisfied concerning impedance matching), the amplitude scaling constant is $K = 1$. However, in practice, $K < 1$.

Example 10.20

The signal input to a coaxial line is $f(t) = u(t)$. At the far end of the line, an output $y(t) = 0.2u(t - 10)$ is observed. What is the impulse response $h(t)$ of the system?

Solution   From the given information, we deduce that the unit-step response of the system is

$$g(t) = 0.2u(t - 10).$$

Differentiating this equation on both sides, we get

$$h(t) = g'(t) = 0.2\delta(t - 10).$$

Alternatively, the given information suggests that the coax has a 10-second delay and a gain of 0.2. Thus, we can construct the impulse response of the system as

$$h(t) = 0.2\delta(t - 10).$$

Distributed versus lumped-element circuits

Electrical circuits containing finite-length transmission lines are said to be *distributed* circuits, as opposed to *lumped-element* circuits which are composed of discrete elements, such as capacitors and resistors, with insignificantly small (in the sense of associated propagation time delays) physical dimensions. Transmission lines and distributed circuits are studied in detail in courses on electromagnetics and RF circuits. While techniques from elementary circuit analysis are helpful in these studies, transmission lines do not behave like lumped-element circuits.

EXERCISES

10.1   An LTI circuit has the frequency response $H(\omega) = \frac{1}{1+j\omega} + \frac{1}{2+j\omega}$. What is the system impulse response $h(t)$ and what is the system response $y(t) = h(t) * f(t)$ to the input $f(t) = e^{-t}u(t)$?

10.2   Find the impulse responses $h(t)$ of the systems having the following frequency responses:

    (a) $H(\omega) = \frac{1}{3+j\omega}$.

    (b) $H(\omega) = \frac{1}{(4+j\omega)^2}$.

    (c) $H(\omega) = \frac{j\omega}{5+j\omega} = 1 - \frac{5}{5+j\omega}$.

    (d) $H(\omega) = \frac{1}{1+j\omega}e^{-j\omega}$.

10.3   For a system with frequency response $H(\omega) = \frac{1}{1+j\omega}$, plot the system impulse response $h(t)$ and the system output $y(t) = h(t) * f(t)$ for $f(t) = 10\text{rect}(\frac{t}{0.1})$. Explain how the plot of a different output $y(t) = h(t) * p(t)$ would appear for $p(t) = 1000\text{rect}(\frac{t}{0.001})$. You need not do the actual calculation and plotting.

10.4   Determine the zero-state response $y(t) = h(t) * f(t)$ of the following LTI systems to the input $f(t) = u(t) - u(t-2)$:

    (a) $h(t) = u(t)$.

    (b) $h(t) = e^{-2t}u(t)$.

    (c) $h(t) = e^{2t}u(t)$.

10.5   Find the impulse responses $h(t)$ of the LTI systems having the following unit-step responses:

    (a) $g(t) = 5u(t-5)$.

    (b) $g(t) = t^2u(t)$.

    (c) $g(t) = u(t)(2 - e^{-t})$.

10.6   If the unit-step response of an LTI system is $g(t) = 6\text{rect}(\frac{t-6}{3})$, find the system zero-state responses to inputs

    (a) $f(t) = \text{rect}(t)$.

    (b) $f(t) = e^{-4t}u(t)$.

    (c) $f(t) = 2\delta(t)$.

10.7   Each of the given signals represents the impulse response of an LTI system. Determine whether each system is causal and BIBO stable. If a system is not BIBO stable, find an example of a bounded input $f(t)$ that will cause an unbounded response $h(t) * f(t)$.

    (a) $h(t) = e^t$.

    (b) $h(t) = \text{rect}(\frac{t-1}{2})$.

      (c) $h(t) = \text{rect}(t)$.

      (d) $h(t) = \delta(t + 1) - \delta(t - 1)$.

      (e) $h(t) = u(t)e^{-jt}$.

10.8   Determine whether the LTI systems with the following impulse response functions are causal:

      (a) $h(t) = u(t - 1)$.

      (b) $h(t) = u(t + 1)$.

      (c) $h(t) = \delta(t - 2) * u(t + 1)$.

      (d) $h(t) = u(1 - t) - u(t)$.

      (e) $h(t) = u(-t) * u(-t)$.

10.9   Determine whether the following LTIC systems are BIBO stable and explain why or why not:

      (a) $h_1(t) = 5\delta(t) + 2e^{-2t}u(t) + 3te^{-2t}u(t)$.

      (b) $h_2(t) = \delta(t) + u(t)$.

      (c) $h_3(t) = \delta'(t) + e^{-t}u(t)$.

      (d) $h_4(t) = -2\delta(t - 3) - te^{-5t}u(t)$.

10.10   For each unstable system in Problem 10.9, provide an example of a bounded input that will cause an unbounded output.

10.11   Consider the given zero-state input–output relations for a variety of systems. In each case, determine whether the system is zero-state linear, time invariant, and causal.

      (a) $y(t) = f(t - 1) + f(t + 1)$.

      (b) $y(t) = 5f(t) * u(t)$.

      (c) $y(t) = \delta(t - 4) * f(t) - \int_{-\infty}^{t-2} f^2(\tau)d\tau$.

      (d) $y(t) = \int_{-\infty}^{t+2} f(\tau)d\tau$.

      (e) $y(t) = \int_{-\infty}^{t-2} f(\tau^2)d\tau$.

      (f) $y(t) = f^3(t - 1)$.

      (g) $y(t) = f((t - 1)^2)$.

11

Laplace Transform, Transfer Function, and LTIC System Response

Consider applying an exponential input $f(t) = e^{st}$ to an LTIC system having impulse response $h(t)$. Then the zero-state response $y(t) = h(t) * f(t)$ can be calculated as

$$y(t) = h(t) * e^{st} = \int_{-\infty}^{\infty} h(\tau)e^{s(t-\tau)}d\tau = e^{st}\int_{-\infty}^{\infty} h(\tau)e^{-s\tau}d\tau.$$

Since in LTIC systems $h(t)$ is zero for $t < 0$, it should be possible to move the lower integration limit in the above formula to 0; however, in anticipation of a possible $h(t)$ that includes $\delta(t)$ we will move the limit to 0^-. Thus, we obtain a rule

$$e^{st} \longrightarrow \boxed{\text{LTIC}} \longrightarrow \hat{H}(s)e^{st},$$

where

$$\hat{H}(s) \equiv \int_{0^-}^{\infty} h(t)e^{-st}dt$$

is known as both the *Laplace transform* of $h(t)$ and the *transfer function* of the system with impulse response $h(t)$.

The above relations hold whether s is real or complex, so long as the Laplace transform integral defining $\hat{H}(s)$ converges. In general, s is complex, and then $\hat{H}(s)$

Laplace transform and its inverse; partial fraction expansion; transfer function $\hat{H}(s)$; zero-state and general response of LTIC circuits and systems; cascade, parallel and feedback configurations

is a function of a complex variable as described below in Section 11.1.1 and also in Appendix A, Section A.7.

Note that in the special case with $s = j\omega$, the input-output rule above becomes

$$e^{j\omega t} \longrightarrow \boxed{\text{LTIC}} \longrightarrow \hat{H}(j\omega)e^{j\omega t},$$

where

$$\hat{H}(j\omega) = \int_{0^-}^{\infty} h(t)e^{-j\omega t}dt = H(\omega)$$

is both the system *frequency response* and the *Fourier transform* of the impulse response $h(t)$. Clearly, then, the frequency response $H(\omega)$ and transfer function $\hat{H}(s)$ of LTIC systems are related as

$$H(\omega) = \hat{H}(j\omega),$$

assuming that both transforms—Fourier and Laplace—converge.

The existence of the frequency response $H(\omega)$, that is, the convergence of the Fourier transform of $h(t)$, is guaranteed when the system is BIBO stable, (i.e., when the impulse response $h(t)$ is absolutely integrable). If the system is not BIBO stable, then the frequency response $H(\omega)$ usually does not exist (ideal low-pass, band-pass, and high-pass filters are exceptions). Unlike $H(\omega)$, however, the transfer function $\hat{H}(s)$ frequently exists for unstable systems, for some values of s. This follows, since for complex-valued $s \equiv \sigma + j\omega$,

$$\hat{H}(s) = \int_{0^-}^{\infty} h(t)e^{-st}dt = \int_{0^-}^{\infty} h(t)e^{-\sigma t}e^{-j\omega t}dt$$

is convergent for any real σ for which the product $h(t)e^{-\sigma t}$ is absolutely integrable. For instance, $h(t) = e^t u(t)$ represents an unstable system whose frequency response $H(\omega)$ is undefined. But, because $h(t)e^{-\sigma t}$ is absolutely integrable for $\sigma > 1$, a convergent Laplace transform $\hat{H}(s)$ of $h(t) = e^t u(t)$ exists for all $s = \sigma + j\omega$ satisfying $\sigma > 1$.

It should be apparent from the above discussion that the LTIC system transfer function $\hat{H}(s)$ is a *generalization* of the frequency response $H(\omega)$ that remains valid for many unstable systems. In this chapter we will develop an $\hat{H}(s)$-based frequency-domain method applicable to nearly all LTIC systems that we will encounter.

Section 11.1 focuses on the Laplace transform and its basic properties. We shall see that the Laplace transform of the zero-state response $y(t) = h(t) * f(t)$ of LTIC systems can be expressed as

$$\hat{Y}(s) = \hat{H}(s)\hat{F}(s)$$

if $\hat{F}(s)$ denotes the Laplace transform of a causal input signal $f(t)$. Since causal signals and their Laplace transforms form unique pairs (like Fourier transform pairs), a causal zero-state response $y(t)$ can be uniquely inferred from its Laplace transform $\hat{Y}(s)$ as described in Section 11.2. In Section 11.3 we will learn how to determine $\hat{H}(s)$ and $\hat{Y}(s)$ in LTIC circuits using circuit analysis methods similar to the phasor method of Chapters 4 and 5, and then infer $h(t)$ and $y(t)$ from $\hat{H}(s)$ and $\hat{Y}(s)$. Section 11.4 examines the general response of nth-order LTIC circuits and systems in terms of $\hat{H}(s)$, including their zero-input response to initial conditions. Finally, in Section 11.5 we consider systems that are composed of interconnected subsystems, and determine how the transfer function of the overall system is related to the transfer functions of its subsystems.

11.1 Laplace Transform and its Properties

11.1.1 Definition, ROC, and *poles*

The Laplace transform $\hat{H}(s)$ of a signal $h(t)$ is defined[1] as

$$\hat{H}(s) = \int_{0^-}^{\infty} h(t)e^{-st}dt,$$

where

$$s \equiv \sigma + j\omega$$

is a complex variable with real part σ and imaginary part ω. Because a complex variable is a pair of real variables, in this case $s = (\sigma, \omega)$, the Laplace transform $\hat{H}(s)$ is a function of two *real* variables. For conciseness, we choose to write $\hat{H}(s)$ rather than $\hat{H}(\sigma, \omega)$.

Generally, the Laplace transform integral *converges*[2] for some values of s and **Laplace** not for others. The region of the complex number plane containing all **transform**

$$s = (\sigma, \omega) = \sigma + j\omega$$

for which the Laplace transform integral converges is said to be the *region of conver-* **ROC** *gence* (ROC) of the Laplace transform. We refer to the entire complex number **and** plane, containing all possible values of $s = (\sigma, \omega) = \sigma + j\omega$, as the *s*-plane. The **s-plane**

---

[1] $\hat{H}(s)$ defined above also is known as the *one-sided* Laplace transform of $h(t)$ to distinguish it from a *two-sided* transform defined as $\int_{-\infty}^{\infty} h(t)e^{-st}dt$. In these notes, we will use only the one-sided transform. Thus, there will be no occasion for ambiguity when we use the term Laplace transform to refer to its one-sided version.

[2] Meaning that the integral $\int_{0^-}^{T} h(t)e^{-st}dt$ approaches a limit as $T \to \infty$.

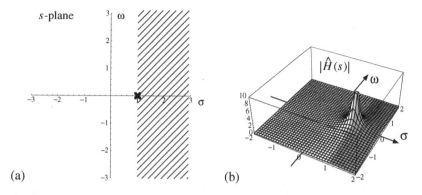

Figure 11.1 (a) The ROC (shaded region) and pole location (×) of Laplace transform expression $\hat{H}(s) = \frac{1}{s-1}$ on the complex s-plane, and (b) surface plot of $|\hat{H}(s)| = |\frac{1}{s-1}| = \frac{1}{|\sigma+j\omega-1|}$ above the s-plane.

horizontal (so-called real) axis of the s-plane is labeled with σ and the vertical (so-called imaginary) axis with ω as shown in Figure 11.1a. It is important to realize that this complex number plane is just the usual two-dimensional plane with a real variable corresponding to each axis. The following example computes a Laplace transform having the ROC shown as the shaded area in Figure 11.1a.

Example 11.1

Determine the Laplace transform $\hat{H}(s)$ of $h(t) = e^t u(t)$ and its ROC.

Solution   The Laplace transform of

$$h(t) = e^t u(t)$$

is

$$\hat{H}(s) = \int_{0^-}^{\infty} e^t u(t) e^{-st} dt = \int_0^{\infty} e^{(1-s)t} dt = \frac{e^{(1-s)t}}{1-s}\Big|_{t=0}$$

for all $s = \sigma + j\omega$ for which the last expression converges. Notice that convergence is possible if and only if

$$e^{(1-s)t} = e^{(1-\sigma-j\omega)t}$$

reaches a limit as $t \to \infty$. This can happen only if $\sigma = \text{Re}\{s\} > 1$ in which case the limit is zero—if $\sigma = 1$, then $e^{(1-\sigma-j\omega)t} = e^{-j\omega t}$, which has no limiting value, and if $\sigma < 1$, then the expression diverges as t is increased. Hence we conclude that we *must* have $\sigma > 1$ for convergence, and thus the ROC for the Laplace transform of $e^t u(t)$ is described by the inequality

$$\sigma = \text{Re}\{s\} > 1.$$

For values of s satisfying this inequality, the Laplace transform is obtained as the algebraic expression

$$\hat{H}(s) = \frac{e^{(1-s)t}}{1-s}\bigg|_{t=0}^{\infty} = \frac{0-1}{1-s} = \frac{1}{s-1}.$$

Figure 11.1 illustrates various aspects of the Laplace transform

$$\hat{H}(s) = \frac{1}{s-1}$$

of signal

$$h(t) = e^t u(t):$$

- The shaded portion of the s-plane depicted in Figure 11.1a corresponds to the ROC: $\{s: \sigma > 1\}$.
- The "×" sign shown in the figure marks the location $s = 1$ in the s-plane where the Laplace transform expression $\frac{1}{s-1}$, diverges.
- The variation of $|\frac{1}{s-1}|$ over the entire s-plane is depicted in Figure 11.1b in the form of a surface plot—note that the surface resembles a circus tent supported by a single *pole* erected above the s-plane at the location × marked in Figure 11.1a.[3]

Locations on the s-plane—or the values of s—where the magnitude of the expression for $\hat{H}(s)$ goes to infinity are called *poles* of the Laplace transform $\hat{H}(s)$.

Poles of the Laplace transform

As we shall discuss below, the ROC always has at least one pole on its boundary.

Example 11.2
Determine the Laplace transform $\hat{F}(s)$ of signal $f(t) = e^{-2t}u(t) - e^{-t}u(t)$.

Solution   Proceeding as in Example 11.1,

$$\hat{F}(s) = \int_{0^-}^{\infty} (e^{-2t} - e^{-t})u(t)e^{-st}dt = \int_{0}^{\infty} (e^{-(2+s)t} - e^{-(1+s)t})dt$$

$$= \frac{e^{-(2+s)t}}{-(2+s)}\bigg|_{t=0}^{\infty} - \frac{e^{-(1+s)t}}{-(1+s)}\bigg|_{t=0}^{\infty} = \frac{1}{s+2} - \frac{1}{s+1} = \frac{-1}{(s+2)(s+1)}$$

under the assumptions $\sigma = \text{Re}\{s\} > -2$ and $\sigma = \text{Re}\{s\} > -1$, dictated by convergence of the two terms above. The first condition is automatically

---

[3]Outside the ROC, $\{s : \sigma > 1\}$, this surface represents the *analytic continuation* of the Laplace transform integral, in analogy with $\frac{1}{1-s}$ representing an extension of the infinite series $1 + s + s^2 + s^3 + \cdots$ beyond its region of convergence, $|s| < 1$, as an analytic function. The concept of analytic continuation arises in the *theory of functions*, where it is shown that analytic functions known over a finite region of the complex plane can be *uniquely* extended across the rest of the plane by a Taylor series expansion process.

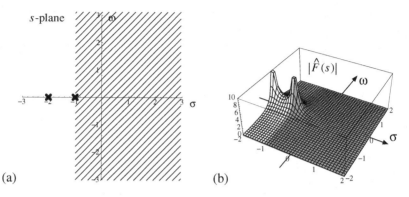

Figure 11.2 (a) The ROC (shaded region) and pole locations ($\times$) of $\hat{F}(s) = \frac{-1}{(s+2)(s+1)}$, and (b) a surface plot of $|\hat{F}(s)|$.

satisfied if the second is satisfied; hence the ROC consists of all complex s such that $\sigma = \text{Re}\{s\} > -1$, which is the cross-hatched region shown in Figure 11.2a.

The pole locations of

$$\hat{F}(s) = \frac{-1}{(s+2)(s+1)}$$

are evident in the surface plot of $|\hat{F}(s)|$ shown in Figure 11.2b, with poles located at $s = -2$ and $s = -1$. Notice that the ROC of $\hat{F}(s)$ lies to the right of the rightmost pole located at $s = -1$.

The ROC and pole locations depicted in Figures 11.1 and 11.2 are consistent with the following general rule:

ROC is all s to the right of the rightmost pole

The ROC of a Laplace transform coincides with the portion of the s-plane to the right of the rightmost pole (not counting a possible pole at $s = \infty$).

As illustrated in Example 11.2, the Laplace transform of a function

$$f(t) = f_1(t) + f_2(t) + \cdots$$

will have the form

$$\hat{F}(s) = \hat{F}_1(s) + \hat{F}_2(s) + \cdots$$

where $\hat{F}_n(s)$ is the Laplace transform of $f_n(t)$, and, furthermore, the ROC of $\hat{F}(s)$ will be the *intersection* of the ROC's of all of the $\hat{F}_n(s)$ components. This is the reason underlying the general rule that the ROC of a Laplace transform $\hat{F}(s)$ is the region to the right of its rightmost pole, not counting a possible pole at $s = \infty$. A pole at infinity arises if $\hat{F}(s)$ contains an additive term proportional to s (or any increasing function of s). But, the ROC of $\hat{F}_n(s) = s$ is the entire s-plane (as shown in Example 11.4 below), and so a pole at $s = \infty$ is not counted in the rule just explained.

Example 11.3
Determine the Laplace transform $\hat{H}(s)$ of signal $h(t) = \delta(t)$.

Solution   In this case, using the sifting property of the impulse,

$$\hat{H}(s) = \int_{0^-}^{\infty} \delta(t)e^{-st}dt = e^{-s\cdot 0} = 1.$$

Because we did not invoke any constraint on s in calculating the Laplace transform, the ROC is the entire s-plane. Notice that in this case the Laplace transform has no pole. Thus the rule that the ROC is the region to the right of the rightmost pole holds here in a trivial way.

Example 11.4
Using the derivative of

$$\delta(t) * e^{st} = e^{st},$$

determine the Laplace transform of $\delta'(t)$, the impulse response of a differentiator.

Solution   Differentiating $\delta(t) * e^{st} = e^{st}$ on both sides (with the help of the derivative property of convolution) we find that

$$\delta'(t) * e^{st} = se^{st},$$

which can be rewritten as

$$\int_{-\infty}^{\infty} \delta'(\tau)e^{s(t-\tau)}d\tau = se^{st}.$$

Evaluating both sides at $t = 0$ we find that

$$\int_{-\infty}^{\infty} \delta'(\tau)e^{-s\tau}d\tau = s,$$

which shows that the Laplace transform of $\delta'(t)$ is simply s (as given in Table 11.1). Since we did not invoke any constraint on s while calculating the Laplace transform, the ROC is the entire s-plane, even though there is a pole at infinity, (i.e., at $s = \infty$).

Example 11.5
Given that the Laplace transform of $h(t) = e^t u(t)$ is $\hat{H}(s) = \frac{1}{s-1}$, show that the Laplace transform of $f(t) = te^t u(t)$ is $\hat{F}(s) = \frac{1}{(s-1)^2}$.

1	$\delta(t) \leftrightarrow 1$	7	$\delta'(t) \leftrightarrow s$
2	$e^{pt}u(t) \leftrightarrow \frac{1}{s-p}$	8	$u(t) \leftrightarrow \frac{1}{s}$
3	$te^{pt}u(t) \leftrightarrow \frac{1}{(s-p)^2}$	9	$tu(t) \leftrightarrow \frac{1}{s^2}$
4	$t^n e^{pt}u(t) \leftrightarrow \frac{n!}{(s-p)^{n+1}}$	10	$t^n u(t) \leftrightarrow \frac{n!}{s^{n+1}}$
5	$\cos(\omega_o t)u(t) \leftrightarrow \frac{s}{s^2+\omega_o^2}$	11	$\sin(\omega_o t)u(t) \leftrightarrow \frac{\omega_o}{s^2+\omega_o^2}$
6	$e^{-\alpha t}\cos(\omega_d t)u(t) \leftrightarrow \frac{s+\alpha}{(s+\alpha)^2+\omega_d^2}$	12	$e^{-\alpha t}\sin(\omega_d t)u(t) \leftrightarrow \frac{\omega_d}{(s+\alpha)^2+\omega_d^2}$

Table 11.1 Laplace transforms pairs $h(t) \leftrightarrow \hat{H}(s)$ involving frequently encountered causal signals—α, ω_o, and ω_d stand for arbitrary real constants, n for nonnegative integers, and p denotes an arbitrary complex constant.

Solution   We are told that

$$\int_0^\infty e^t e^{-st} dt = \frac{1}{s-1},$$

which holds for $\{s : \sigma > 1\}$. Taking the derivative of the above expression with respect to s, we find that

$$\frac{d}{ds}\left(\int_0^\infty e^t e^{-st} dt\right) = -\int_0^\infty te^t e^{-st} dt$$

on the left, and,

$$\frac{d}{ds}\left(\frac{1}{s-1}\right) = -\frac{1}{(s-1)^2}$$

on the right. Thus,

$$\int_0^\infty te^t e^{-st} dt = \frac{1}{(s-1)^2},$$

implying that the Laplace transform of $f(t) = te^t u(t)$ is $\hat{F}(s) = \frac{1}{(s-1)^2}$. The ROC is $\{s : \sigma > 1\}$ since our calculation has not required any additional constraints on s.

The Laplace transforms calculated in the examples so far all are special cases from a list of important *Laplace transform pairs* shown in Table 11.1. Notice that the list contains only causal signals and in each case the ROC can be deduced from corresponding pole locations, which can, in turn, be directly inferred from the form of the Laplace transform. For example,

$$\cos(\omega_o t)u(t) \leftrightarrow \frac{s}{s^2 + \omega_o^2}$$

has a pair of poles at $s = \pm j\omega_o$ on the vertical axis of the s-plane, and, as a consequence, the ROC of this particular Laplace transform coincides with the right half of the s-plane, which we will refer to as the RHP (right half-plane).

Also, notice that the poles of Laplace transforms of *absolutely integrable* signals included in Table 11.1 all are confined to the left half-plane (LHP). For example, the pole $s = p$ for $e^{pt}u(t)$ is in the LHP if and only if $p < 0$ and the signal is absolutely integrable. The same is true with the pole $s = p$ for $te^{pt}u(t)$, etc. This detail is, of course, not a coincidence. It is true more generally, because if a signal $h(t)$ is absolutely integrable and causal, then its Fourier transform integral is guaranteed to *converge* to a bounded $H(\omega) = \hat{H}(j\omega)$—this requires that *all* poles of $\hat{H}(s)$ (if any) be located within the LHP (as in Figure 11.2) so that the vertical axis of the s-plane, where s equals $j\omega$, is contained within the ROC.[4]

Laplace transforms of absolutely integrable signals have only LHP poles

Remembering that BIBO stable systems must have absolutely integrable impulse response functions, it is no surprise that the BIBO stability criterion from Chapter 10 can be restated as:

An LTIC system $h(t) \leftrightarrow \hat{H}(s)$ is BIBO stable *if and only if* its transfer function $\hat{H}(s)$ has all of its poles in the LHP.

LTIC systems are BIBO stable if and only if all poles in LHP

This alternative test for BIBO stability holds for any LTIC system with an $\hat{H}(s)$ that is a *rational function* written in *minimal form* (polynomial in s divided by a polynomial in s, where all possible cancellations of terms between the numerator and denominator have been performed), giving rise to poles at distinct locations, as in the examples above. A proof of this alternative stability test is accomplished by simply writing $\hat{H}(s)$ in a partial fraction expansion via the method described in Section 11.2.

Example 11.6

Using Table 11.1 and the version of the BIBO stability criterion stated above, determine whether the LTIC systems with the following impulse response functions are BIBO stable:

$$h_a(t) = u(t)$$
$$h_b(t) = e^{-t}u(t) + e^{2t}u(t)$$
$$h_c(t) = e^{-t}\cos(t)u(t)$$
$$h_d(t) = \sin(2t)u(t)$$
$$h_e(t) = e^{-t}u(t) + h_c(t)$$
$$h_f(t) = \delta'(t)$$

---

[4]A pole at $s = \infty$ is no exception to this rule, because the corresponding $\hat{H}(s) = s$ does not lead to a bounded $H(\omega) = \hat{H}(j\omega)$ as required by an absolutely integrable $h(t)$.

Solution   Using Table 11.1, we find that the transfer function that corresponds to impulse response

$$h_a(t) = u(t)$$

is

$$\hat{H}_a(s) = \frac{1}{s}.$$

This transfer function has a pole at $s = 0$, which is just outside the LHP. Thus, the system is not BIBO stable, which of course is consistent with the fact that $h_a(t) = u(t)$ is not absolutely integrable.

For

$$h_b(t) = e^{-t}u(t) + e^{2t}u(t),$$

we have

$$\hat{H}_b(s) = \frac{1}{s+1} + \frac{1}{s-2} = \frac{2s-1}{(s+1)(s-2)}.$$

This transfer function has two poles, one at $s = -1$ within the LHP, and another at $s = 2$ outside the LHP. Therefore, the system is not BIBO stable.

The system

$$h_c(t) = e^{-t}\cos(t)u(t)$$

is BIBO stable because the poles of its transfer function

$$\hat{H}_c(s) = \frac{s+1}{(s+1)^2 + 1} = \frac{s+1}{(s+1+j)(s+1-j)},$$

located at $s = -1 \pm j$, are within the LHP. The pole locations and surface plot of $|\hat{H}_c(s)|$ for this BIBO stable system are shown in Figures 11.3a and 11.3b. We refer to locations where $\hat{H}_c(s) = 0$ as "zeros" of the transfer function; Figure 11.3a shows an "O" that marks the location of the zero of $\hat{H}_c(s)$. Finally, Figure 11.3c shows a plot of $|\hat{H}_c(j\omega)| = |H_c(\omega)|$ as a function of ω, which is the magnitude of the frequency response of the system.

"Zero" of a transfer function

The poles of

$$\hat{H}_d(s) = \frac{2}{s^2 + 4} = \frac{2}{(s+j2)(s-j2)}$$

are at $s = \pm j2$, just outside the LHP; therefore the system is not BIBO stable.

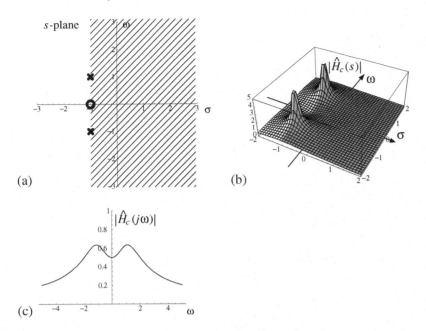

Figure 11.3 (a) The ROC (shaded region), pole locations ($\times$), and the zero location (O) of $\hat{H}_c(s) = \frac{s+1}{(s+1+j)(s+1-j)}$, (b) a surface plot of $|\hat{H}_c(s)|$, and (c) line plot of $|\hat{H}_c(j\omega)|$ versus ω representing the magnitude of the frequency response of system $\hat{H}_c(s)$.

All three poles of

$$\hat{H}_e(s) = \frac{1}{s+1} + \hat{H}_c(s),$$

at $s = -1$ and $s = -1 \pm j$ are within the LHP. Therefore the system is BIBO stable.

Finally, for

$$h_f(t) = \delta'(t)$$

the system transfer function is

$$\hat{H}_f(s) = s$$

(see Example 11.4). Since $|\hat{H}_f(s)| \to \infty$ as $|s| \to \infty$, this transfer function has poles at infinities to the right and left of the s-plane. A pole at $s = +\infty$ obviously is outside the LHP, and hence the system should not be BIBO stable. This conclusion checks with our earlier observation that a differentiator described by an impulse response $\delta'(t)$ cannot be BIBO stable.

Hidden poles

Poles at infinities as seen in the previous and upcoming examples are known as *hidden poles*.

Example 11.7
Determine the Laplace transform $\hat{Q}(s)$ of causal signal

$$q(t) = \text{rect}(t - \frac{1}{2}) = u(t) - u(t - 1).$$

Solution

$$\hat{Q}(s) = \int_{0^-}^{\infty} q(t)e^{-st}dt = \int_0^1 e^{-st}dt = \frac{e^{-st}}{-s}\bigg|_{t=0}^1$$

$$= \frac{e^{-s\cdot 1} - e^{-s\cdot 0}}{-s} = \frac{1 - e^{-s}}{s},$$

except in the limit as $\sigma = \text{Re}\{s\} \to -\infty$. Hence, in this case the ROC is described by $\{s : \sigma = \text{Re}\{s\} > -\infty\}$.

Note that $\hat{Q}(s)$ has a hidden pole at $s = -\infty$ in the LHP. Also note that $s = 0$ is not a pole, since, using l'Hospital's rule,

$$\lim_{s \to 0} \hat{Q}(s) = \lim_{s \to 0} \frac{\frac{d}{ds}(1 - e^{-s})}{\frac{ds}{ds}} = \lim_{s \to 0} \frac{e^{-s}}{1} = 1$$

is finite.

11.1.2   Properties of the Laplace transform

Since the Laplace transform operation on a signal $f(t)$ ignores the portion of $f(t)$ for $t < 0$, only causal signals $f(t)$ can be expected to form unique transform pair relations with $\hat{F}(s)$, such as those listed in Table 11.1. In fact, if $f(t)$ is not causal, then its Laplace transform will match the Laplace transform of the causal signal $f(t)u(t)$. For example, the noncausal signal

$$f(t) = e^t u(t + 2),$$

shown in Figure 11.4a, shares the same Laplace transform, $\frac{1}{s-1}$, with causal signal

$$g(t) = e^t u(t) = f(t)u(t)$$

shown in Figure 11.4b. Likewise, the Laplace transform of noncausal $f(t) = \delta(t + 1) + 2\delta(t)$ is the same as the Laplace transform of causal $g(t) = 2\delta(t)$, namely 2.

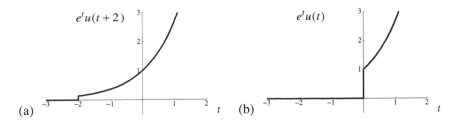

Figure 11.4   (a) Noncausal $f(t) = e^t u(t+2)$ and (b) causal $g(t) = f(t)u(t)$.

We shall indicate Laplace transform pair relationships involving causal signals using *double-headed* arrows, as in

$$e^t u(t) \leftrightarrow \frac{1}{s-1},$$

and use single-headed arrows to indicate the Laplace transform of noncausal signals, as in

$$e^t u(t+1) \rightarrow \frac{1}{s-1}.$$

Laplace transform of noncausal signals

Keep in mind the distinction between $\rightarrow$ and $\leftrightarrow$ in reading and interpreting Table 11.2, which lists some of the general properties of the Laplace transform. Properties given in terms of $\rightarrow$ are applicable to both causal and noncausal signals, while those given in terms of $\leftrightarrow$ apply only to causal signals. For instance, the time-delay property (item 4 in Table 11.2) is stated using $\leftrightarrow$, and therefore it applies only to causal $f(t)$. The next example illustrates use of the time-delay property when calculating the Laplace transform of a delayed causal signal.

Example 11.8
Given that (according to Table 11.1)

$$f(t) = tu(t) \leftrightarrow \hat{F}(s) = \frac{1}{s^2},$$

find the Laplace transform $\hat{P}(s)$ of

$$p(t) = (t-1)u(t-1).$$

Signals $f(t)$ and $p(t)$ are shown in Figures 11.5a and 11.5b, respectively.

Solution   We first note that $p(t) = f(t-1)$ is a delayed version of the causal $f(t)$, with a positive delay of $t_o = 1$ s (as seen in Figure 11.5). Thus, the time-delay property given in Table 11.2 is applicable, and

$$\hat{P}(s) = \hat{F}(s)e^{-st_o} = \frac{e^{-s}}{s^2}.$$

	Name	Condition	Property
1	Multiplication	$f(t) \to \hat{F}(s)$, constant K	$K f(t) \to K \hat{F}(s)$
2	Addition	$f(t) \to \hat{F}(s)$, $g(t) \to \hat{G}(s) \cdots$	$f(t) + g(t) + \cdots \to \hat{F}(s) + \hat{G}(s) + \cdots$
3	Time scaling	$f(t) \to \hat{F}(s)$, real $a > 0$	$f(at) \to \frac{1}{a}\hat{F}(\frac{s}{a})$
4*	Time delay	$f(t) \leftrightarrow \hat{F}(s), t_o \geq 0$	$f(t - t_o) \leftrightarrow \hat{F}(s)e^{-st_o}$
5	Frequency shift	$f(t) \to \hat{F}(s)$	$f(t)e^{s_o t} \to \hat{F}(s - s_o)$
6	Time derivative	Differentiable $f(t) \to \hat{F}(s)$	$f'(t) \to s\hat{F}(s) - f(0^-)$ $f''(t) \to s^2\hat{F}(s) - sf(0^-) - f'(0^-)$ $\cdots$ $f^{(n)}(t) \to s^n\hat{F}(s) - \cdots - f^{(n-1)}(0^-)$
7	Time integration	$f(t) \to \hat{F}(s)$	$\int_{0^-}^{t} f(\tau)d\tau \to \frac{1}{s}\hat{F}(s)$
8	Frequency derivative	$f(t) \to \hat{F}(s)$	$-tf(t) \to \frac{d}{ds}\hat{F}(s)$
9*	Time convolution	$h(t) \leftrightarrow \hat{H}(s)$, $f(t) \leftrightarrow \hat{F}(s)$	$h(t) * f(t) \leftrightarrow \hat{H}(s)\hat{F}(s)$
10	Frequency convolution	$f(t) \to \hat{F}(s)$, $g(t) \to \hat{G}(s)$	$f(t)g(t) \to \frac{1}{2\pi j}\hat{F}(s) * \hat{G}(s)$
11	Poles	$f(t) \to \hat{F}(s)$	Values of s such that $|\hat{F}(s)| = \infty$
12	ROC	$f(t) \to \hat{F}(s)$	Portion of s − plane to the right of rightmost pole $\neq \infty$
13*	Fourier transform	$f(t) \leftrightarrow \hat{F}(s)$	$F(\omega) = \hat{F}(j\omega)$ if and only if ROC includes $s = j\omega$
14	Final value	Poles of $s\hat{F}(s)$ in LHP	$f(\infty) = \lim_{s \to 0} s\hat{F}(s)$
15	Initial value	Existence of the limit	$f(0^+) = \lim_{s \to \infty} s\hat{F}(s)$

Table 11.2 Important properties and definitions for the one-sided Laplace transform. Properties marked by * in the first column hold only for causal signals.

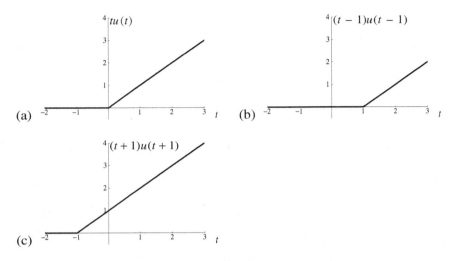

Figure 11.5 (a) Causal ramp function $f(t) = tu(t)$, (b) delayed ramp $p(t) = (t - 1)u(t - 1)$, and (c) noncausal $q(t) = (t + 1)u(t + 1)$.

Example 11.9
Determine the Laplace transform $\hat{Q}(s)$ of

$$q(t) = (t + 1)u(t + 1)$$

plotted in Figure 11.5c.

Solution   Clearly, the signal $q(t) = f(t + 1)$ shown in Figure 11.5c is an advanced, rather than a delayed, version of $f(t)$, and so we cannot proceed as in Example 11.8. Instead, we note that the Laplace transform of the noncausal $q(t)$ must be the same as the transform of the causal $q(t)u(t)$, which can be written as

$$q(t)u(t) = (t + 1)u(t + 1)u(t) = (t + 1)u(t) = tu(t) + u(t).$$

Using Table 11.1 as well as the addition property from Table 11.2, we find that

$$q(t)u(t) \leftrightarrow \hat{Q}(s) = \frac{1}{s^2} + \frac{1}{s}.$$

The previous two examples illustrated how the properties in Table 11.2 can be used to deduce new Laplace transforms from ones already known. Here is another example of a similar type.

Example 11.10

Using Table 11.2 confirm that

$$f(t) = e^{-\alpha t}\cos(\omega_d t)u(t) \leftrightarrow \hat{F}(s) = \frac{s + \alpha}{(s + \alpha)^2 + \omega_d^2}.$$

Solution   We first note that

$$f(t) = e^{-\alpha t}\cos(\omega_d t)u(t) = e^{-\alpha t}\frac{e^{j\omega_d t} + e^{-j\omega_d t}}{2}u(t)$$

$$= \frac{e^{-(\alpha - j\omega_d)t}}{2}u(t) + \frac{e^{-(\alpha + j\omega_d)t}}{2}u(t).$$

Therefore, using the multiplication and addition properties from Table 11.2, as well as

$$e^{pt}u(t) \leftrightarrow \frac{1}{s - p}$$

from Table 11.1 (with $p = -(\alpha \pm j\omega_d)$), we find

$$\hat{F}(s) = \frac{1}{2}\frac{1}{s + (\alpha - j\omega_d)} + \frac{1}{2}\frac{1}{s + (\alpha + j\omega_d)}$$

$$= \frac{1}{2}\{\frac{1}{(s + \alpha) - j\omega_d} + \frac{1}{(s + \alpha) + j\omega_d}\}$$

$$= \frac{1}{2}\{\frac{(s + \alpha) + j\omega_d + (s + \alpha) - j\omega_d}{(s + \alpha)^2 + \omega_d^2}\}$$

$$= \frac{s + \alpha}{(s + \alpha)^2 + \omega_d^2}$$

as desired, which is entry 6 in Table 11.1.

Verification of the time-delay property: Assume $f(t)$ is causal. Then, by definition, the Laplace transform of $f(t - t_o)$, $t_o \geq 0$, is

$$\int_{t=0^-}^{\infty} f(t - t_o)e^{-st}dt = \int_{t=t_o}^{\infty} f(t - t_o)e^{-st}dt = \int_{\tau=0^-}^{\infty} f(\tau)e^{-s(\tau + t_o)}d\tau$$

$$= e^{-st_o}\int_{\tau=0^-}^{\infty} f(\tau)e^{-s\tau}d\tau = e^{-st_o}\hat{F}(s).$$

Notice that we used a change of variables $\tau = t - t_o$ with $d\tau = dt$ for a fixed t_o in order to obtain the result. Also note that exchange of the lower integration limit from 0^- to t_o^- is justified only for $t_o \geq 0$ and for causal $f(t)$. Therefore, the time-delay property is guaranteed to hold only for causal $f(t)$ and $t_o \geq 0$, as noted in Table 11.2.

The time-delay property is guaranteed to work only for causal signals and for positive delays $t_o \geq 0$.

Example 11.11

Using the time-delay property, determine the Laplace transform of

$$\text{rect}(t - \frac{1}{2}).$$

Solution   Since

$$\text{rect}(t - \frac{1}{2}) = u(t) - u(t - 1),$$

where the second term is a delayed version of the causal first term, we have

$$u(t) - u(t - 1) \leftrightarrow \frac{1}{s} - \frac{1}{s}e^{-s}.$$

Thus, it follows that, in agreement with Example 11.7,

$$\text{rect}(t - \frac{1}{2}) \leftrightarrow \frac{1}{s}(1 - e^{-s}).$$

The frequency-derivative property in Table 11.2 can be verified using a procedure similar to that used to prove the frequency-derivative property of the Fourier transform. We leave this to the reader. The next example shows how to apply the property.

Example 11.12

Using the frequency-derivative property, show that

$$e^{pt}u(t) \leftrightarrow \frac{1}{s - p}$$

(from Table 11.1) implies

$$te^{pt}u(t) \leftrightarrow \frac{1}{(s - p)^2}.$$

Solution   According to the frequency-derivative property, the Laplace transform of $te^{pt}u(t)$ must be minus the derivative with respect to s of the Laplace transform $\frac{1}{s-p}$ of $e^{pt}u(t)$, giving

$$-\frac{d}{ds}\frac{1}{s - p} = -\frac{d}{ds}(s - p)^{-1} = \frac{1}{(s - p)^2}.$$

Hence,

$$te^{pt}u(t) \leftrightarrow \frac{1}{(s - p)^2}.$$

**Time-derivative
property**

Verification of the time-derivative property: By definition, the Laplace transform of $\frac{df}{dt}$ is the integral $\int_{0^-}^{\infty} \frac{df}{dt} e^{-st} dt$, for all values of s for which the integral converges (i.e. within its ROC). Using integration by parts,

$$\int_{0^-}^{\infty} \frac{df}{dt} e^{-st} dt = f(t)e^{-st}\Big|_{0^-}^{\infty} - \int_{0^-}^{\infty} f(t)\frac{d}{dt}(e^{-st}) dt$$

$$= \lim_{t\to\infty} f(t)e^{-st} - f(0^-) + s\int_{0^-}^{\infty} f(t)e^{-st} dt.$$

Since $\int_{0^-}^{\infty} f(t)e^{-st} dt \equiv \hat{F}(s)$, we have

$$\frac{df}{dt} \equiv f'(t) \rightarrow s\hat{F}(s) - f(0^-) + \lim_{t\to\infty} f(t)e^{-st}.$$

Now,

$$\lim_{t\to\infty} f(t)e^{-st} = 0$$

for all values of s in the ROC for $\hat{F}(s)$; otherwise the Laplace integral $\hat{F}(s)$ cannot converge. Hence, for s in the ROC of $\hat{F}(s)$,

$$f'(t) \rightarrow s\hat{F}(s) - f(0^-),$$

and by induction

$$f^{(n)}(t) \rightarrow s^n \hat{F}(s) - s^{(n-1)} f(0^-) - \cdots s f^{(n-2)}(0^-) - f^{(n-1)}(0^-).$$

**For
causal $f(t)$,
$f^{(n)}(t) \leftrightarrow s^n \hat{F}(s)$.**

Note that for causal $f(t)$, we have the simpler time-derivative rule

$$f^{(n)}(t) \rightarrow s^n \hat{F}(s) \text{ for all } n \geq 0.$$

Example 11.13
Given that

$$f(t) = e^{2t} \rightarrow \hat{F}(s) = \frac{1}{s-2},$$

use the time-derivative property to determine the Laplace transforms of

$$\frac{df}{dt} = 2e^{2t}$$

and

$$\frac{d^2 f}{dt^2} = 4e^{2t}.$$

Note that $f(t)$ is a noncausal signal.

Solution   Since

$$\hat{F}(s) = \frac{1}{s-2}$$

and

$$f(0^-) = e^{2 \cdot 0^-} = e^0 = 1,$$

we have, using the formula $s\hat{F}(s) - f(0^-)$ for the Laplace transform of the derivative of $f(t)$,

$$\frac{d}{dt}e^{2t} = 2e^{2t} \rightarrow s\frac{1}{s-2} - 1 = \frac{s-(s-2)}{s-2} = \frac{2}{s-2}.$$

Next, we notice that

$$f'(0^-) = 2e^{2t}\big|_{t=0^-} = 2.$$

Hence, using the formula $s^2\hat{F}(s) - sf(0^-) - f'(0^-)$ for the Laplace transform of the second derivative of $f(t)$, we have

$$\frac{d^2}{dt^2}e^{2t} = 4e^{2t} \rightarrow s^2\frac{1}{s-2} - s \cdot 1 - 2 = \frac{s^2 - s(s-2) - 2(s-2)}{s-2} = \frac{4}{s-2}.$$

Of course, the results $2e^{2t} \rightarrow \frac{2}{s-2}$ and $4e^{2t} \rightarrow \frac{4}{s-2}$ could have been obtained by inspection using the multiplication rule and the fact that $e^{2t} \rightarrow \frac{1}{s-2}$.

Verification of the convolution property: Table 11.2 states that $h(t) * f(t) \leftrightarrow \hat{H}(s)\hat{F}(s)$, for causal $h(t)$ and $f(t)$. To prove this, write the Laplace transform of the convolution of $h(t)$ and $f(t)$ as

$$\int_{0^-}^{\infty} \{h(t) * f(t)\}e^{-st}dt = \int_{t=0^-}^{\infty} \{\int_{\tau=-\infty}^{\infty} h(\tau)f(t-\tau)d\tau\}e^{-st}dt$$

$$= \int_{\tau=-\infty}^{\infty} h(\tau)\{\int_{t=0^-}^{\infty} f(t-\tau)e^{-st}dt\}d\tau$$

$$= \int_{\tau=0^-}^{\infty} h(\tau)\{\int_{t=0^-}^{\infty} f(t-\tau)e^{-st}dt\}d\tau$$

$$= \int_{\tau=0^-}^{\infty} h(\tau)e^{-s\tau}\hat{F}(s)d\tau$$

$$= \hat{F}(s)\int_{\tau=0^-}^{\infty} h(\tau)e^{-s\tau}d\tau = \hat{F}(s)\hat{H}(s),$$

The convolution property applies to causal signals

which proves the result. Notice that the change of the bottom integration limit in line 3 from $-\infty$ to 0^- requires that $h(t)$ be causal. Also, use of the time-shift property between lines 3 and 4 requires that $f(t)$ be causal and $\tau \geq 0$.

Example 11.14

Given that

$$f(t) = e^{-t}u(t),$$

determine $y(t) \equiv f(t) * f(t)$ using the time-convolution property.

Solution   We have

$$f(t) = e^{-t}u(t) \leftrightarrow \hat{F}(s) = \frac{1}{s+1}.$$

Because $f(t)$ is causal, we can apply the time-convolution property to yield

$$\hat{Y}(s) = \hat{F}(s)\hat{F}(s) = \frac{1}{(s+1)^2}.$$

But, according to Table 11.1

$$te^{-t}u(t) \leftrightarrow \frac{1}{(s+1)^2},$$

and so we deduce that

$$y(t) = te^{-t}u(t).$$

Example 11.15

If $f(t) = e^{-t}$, can we take advantage of the time-convolution property to calculate $y(t) \equiv f(t) * f(t)$?

Solution   Because the given $f(t)$ is not causal, the answer is "no"—in particular,

$$e^{-t} * e^{-t} \neq te^{-t}.$$

Try computing the convolution $e^{-t} * e^{-t}$ directly in the time domain to see that it does not equal either te^{-t} or $te^{-t}u(t)$.

11.2   Inverse Laplace Transform and PFE

Causal signals and their Laplace transforms form unique pairs, just like Fourier transform pairs familiar from earlier chapters. Hence, we can associate with each Laplace transform $\hat{H}(s)$ a unique causal signal $h(t)$ such that

$$h(t) \leftrightarrow \hat{H}(s),$$

where $h(t)$ is said to be the inverse Laplace transform of $\hat{H}(s)$. For instance, the inverse Laplace transform[5] of $\hat{H}(s) = \frac{1}{(s+1)^2}$ is the causal signal $h(t) = te^{-t}u(t)$. Inverse Laplace transforms for elementary cases, such as s^n, $\frac{1}{s-p}$, $\frac{1}{(s-p)^2}$, etc., can be directly identified by using the pairs in Table 11.1. More complicated cases call for the use of systematic algebraic procedures discussed in this section, as well as properties in Table 11.2.

The most important class of Laplace transforms encountered in LTIC system theory is the ratio of polynomials. In fact, as we will see in Section 11.3, transfer functions of all lumped-element LTIC circuits have this form, namely

$$\hat{H}(s) = \frac{B(s)}{P(s)},$$

where both $B(s)$ and $P(s)$ denote polynomials in s. As indicated earlier, such transfer functions and/or Laplace transforms are said to be *rational functions* or to have **Rational** *rational form*. If the denominator polynomial $P(s)$ has degree $n \geq 1$ then the rational **form** function can be written as

$$\hat{H}(s) = \frac{B(s)}{(s - p_1)(s - p_2) \cdots (s - p_n)}.$$

---

[5]**Uniqueness of and a formula for the inverse Laplace transform for causal signals:** The Laplace transform $\hat{H}(s) = \hat{H}(\sigma + j\omega)$ of a causal signal $h(t)$ is also the Fourier transform of a causal signal $e^{-\sigma t}h(t)$ with $\sigma = \text{Re}\{s\}$ selected so that s is in the ROC of $\hat{H}(s)$. Thus, the uniqueness of the inverse Fourier transform

$$e^{-\sigma t}h(t) = \frac{1}{2\pi}\int_{-\infty}^{\infty} \hat{H}(\sigma + j\omega)e^{j\omega t}\,d\omega$$

implies the uniqueness of the *inverse Laplace transform*

$$h(t) \equiv \frac{e^{\sigma t}}{2\pi}\int_{-\infty}^{\infty} \hat{H}(\sigma + j\omega)e^{j\omega t}\,d\omega.$$

The inverse Laplace transform formula can be expressed more compactly as

$$h(t) = \frac{1}{2\pi j}\int_{\sigma - j\infty}^{\sigma + j\infty} \hat{H}(s)e^{st}\,ds,$$

which is a line integral in the complex plane, within the ROC, along a vertical line intersecting the horizontal axis at $\text{Re}\{s\} = \sigma$. Although this formula always is available to us, we generally will strive for simpler methods of computing inverse Laplace transforms.

	Rational $\hat{H}(s)$	Expanded form $\hat{H}(s)$	$h(t)$
Distinct real poles	$\frac{1}{s^2+3s+2} = \frac{1}{(s+1)(s+2)}$	$\frac{1}{s+1} + \frac{-1}{s+2}$	$(e^{-t} - e^{-2t})u(t)$
Repeated poles	$\frac{s}{(s+1)^2}$	$\frac{1}{s+1} + \frac{-1}{(s+1)^2}$	$(e^{-t} - te^{-t})u(t)$
Complex conjugate pair	$\frac{s}{s^2+1} = \frac{s}{(s+j)(s-j)}$	$\frac{1/2}{s+j} + \frac{1/2}{s-j}$	$\frac{1}{2}(e^{-jt} + e^{jt})u(t)$
Mixed	$\frac{s+1}{s^3-s^2} = \frac{s+1}{s^2(s-1)}$	$\frac{-2}{s} + \frac{-1}{s^2} + \frac{2}{s-1}$	$(-2 - t + 2e^t)u(t)$

Table 11.3 Four examples of *proper rational* expressions with different types of poles are shown in the second column. In proper rational form, the degree of the denominator polynomial exceeds the degree of the numerator polynomial. The third column of the table shows the same proper rational-form expressions on the left rearranged as a weighted sum of elementary terms, known as a partial fraction expansion (PFE). Finally, the last column shows the inverse Laplace transform in each case.

Proper
rational
form

The second column in Table 11.3 lists a few examples of such rational Laplace transforms where the poles p_1, p_2, $\cdots$, p_n correspond to the roots of the denominator polynomial $P(s)$. In the following discussion we will assume that the degree of polynomial $P(s)$ is larger than the degree of $B(s)$—just like in the examples shown in Table 11.3—so that there are no poles[6] of $\hat{H}(s)$ at infinity. When that condition holds, a rational form is said to be *proper*.

Our strategy for calculating inverse Laplace transforms of proper rational functions will be to rewrite the Laplace transform as a sum of simple terms, called a *partial fraction expansion* (PFE), and to then identify the inverse Laplace transform of each simple term by inspection. Table 11.3 illustrates the strategy, where the third column shows the Laplace transforms given on the left rearranged as PFEs (i.e., as weighted sums of elementary terms $\frac{1}{(s-p)}$, $\frac{1}{(s-p)^2}$). It then becomes an easy matter to write $h(t)$ in the last column. The subsections below explain how to rewrite a rational function as a PFE.

11.2.1   PFE for distinct poles

When a Laplace transform in proper rational form has distinct poles, as in the first row of Table 11.3, then finding the PFE consists of finding coefficients K_1, K_2, $\cdots$, K_n such that

$$\frac{B(s)}{(s-p_1)(s-p_2)\cdots(s-p_n)} = \frac{K_1}{s-p_1} + \frac{K_2}{s-p_2} + \cdots + \frac{K_n}{s-p_n}$$

is true for all s.

---

[6]We will assume here that $\hat{H}(s)$ is written in minimal form, so that the factored $B(s) = b_o(s - z_1)(s - z_2)\cdots(s - z_m)$ does not contain a factor that is identical to some $s - p_k$ from the denominator.

This can be done most easily by repeating the following procedure for each unknown coefficient on the right: To determine K_1, for instance, we multiply both sides of the expression above by $s - p_1$, yielding

$$\frac{B(s)}{(s - p_2) \cdots (s - p_n)} = K_1 + \frac{K_2(s - p_1)}{s - p_2} + \cdots + \frac{K_n(s - p_1)}{s - p_n},$$

and then evaluate this expression at $s = p_1$. The right-hand side then becomes just K_1, matching, on the left, *effectively* what you would see after "covering up" the original $s - p_1$ factor in the denominator (e.g., with your THUMB) evaluated at $s = p_1$; i.e.,

$$\left. \frac{B(s)}{\text{THUMB}(s - p_2) \cdots (s - p_n)} \right|_{s=p_1} = K_1.$$

For example, given

$$\frac{1}{(s + 1)(s + 2)} = \frac{K_1}{s + 1} + \frac{K_2}{s + 2},$$

use of the *cover-up method* explained above leads to

$$K_1 = \left. \frac{1}{\text{THUMB}(s + 2)} \right|_{s=-1} = 1$$

and

$$K_2 = \left. \frac{1}{(s + 1)\text{THUMB}} \right|_{s=-2} = -1,$$

confirming the PFE in the first row of Table 11.3.

The cover-up method can be used in exactly the same way when the poles are *complex valued*, so long as they are distinct. This is illustrated in the next example, which works out the PFE pertinent to the third row of Table 11.3.   **Cover-up method**

Example 11.16

Using the cover-up method, find the PFE of the transfer function

$$\hat{H}(s) = \frac{s}{s^2 + 1}$$

and then find $h(t)$.

Solution   First, we factor the denominator of $\hat{H}(s)$ to write

$$\hat{H}(s) = \frac{s}{(s + j)(s - j)}$$

so that the poles are visible. Since the poles (at $s = \pm j$) are distinct, the PFE has the form:

$$\frac{s}{(s+j)(s-j)} = \frac{K_1}{s+j} + \frac{K_2}{s-j}.$$

Now, using the cover-up idea,

$$K_1 = \left.\frac{s}{\text{THUMB}(s-j)}\right|_{s=-j} = \frac{-j}{-2j} = \frac{1}{2}$$

and

$$K_2 = \left.\frac{s}{(s+j)\text{THUMB}}\right|_{s=j} = \frac{j}{2j} = \frac{1}{2}.$$

Thus,

$$\hat{H}(s) = \frac{1/2}{s+j} + \frac{1/2}{s-j}$$

as shown in the third row of Table 11.3 and, consequently,

$$h(t) = \frac{1}{2}(e^{-jt} + e^{jt})u(t) = \cos(t)u(t).$$

With a little practice, one can apply the cover-up method "in place," as illustrated in the following example.

Example 11.17
We write

$$\hat{H}(s) = \frac{s+1}{s(s-1)} = \frac{\overset{0+1}{\overline{\text{THUMB}(0-1)}}}{s} + \frac{\overset{1+1}{\overline{1\text{THUMB}}}}{s-1}.$$

Thus,

$$h(t) = -u(t) + 2e^t u(t).$$

11.2.2   PFE with repeated poles

If a pole of a proper rational function is repeated, as in

$$\frac{b_0 s^m + \cdots + b_m}{(s-p_1)(s-p_1)\cdots(s-p_n)},$$

then there is no choice of PFE coefficients K_i that will satisfy

$$\frac{b_0 s^m + \cdots + b_m}{(s - p_1)(s - p_1) \cdots (s - p_n)} = \frac{K_1}{s - p_1} + \frac{K_2}{s - p_1} + \cdots + \frac{K_n}{s - p_n}.$$

For example, there are no K_1, K_2, and K_3 such that

$$\frac{1}{(s + 1)(s + 1)(s + 2)} = \frac{K_1}{s + 1} + \frac{K_2}{s + 1} + \frac{K_3}{s + 2} = \frac{K_1 + K_2}{s + 1} + \frac{K_3}{s + 2}.$$

However, by modifying the form of the PFE, we still can express a rational function as a sum of simple terms. In the above example, we can write

$$\frac{1}{(s + 1)(s + 1)(s + 2)} = \frac{K_1}{(s + 1)^2} + \frac{K_2}{s + 1} + \frac{K_3}{s + 2}$$

with $K_1 = 1$, $K_2 = -1$, and $K_3 = 1$. But, what is a simple way to find the K_i for the repeated-pole case?

In general, the PFE of a proper-form rational expression with repeated poles will contain terms such as

$$\frac{K_i}{(s - p_i)^r} + \frac{K_{i+1}}{(s - p_i)^{r-1}} + \cdots + \frac{K_{i+r-1}}{(s - p_i)}$$

associated with poles p_i that repeat r times—or have *multiplicity r*. *Simple poles*, i.e., nonrepeating poles with multiplicity $r = 1$, contribute to the expansion in the same way as before. The weighting coefficients for terms involving simple poles, as well as the coefficient of the leading term for each repeated pole, (i.e., K_i above), can be determined using the cover-up method. The remaining coefficients are obtained using a strategy illustrated in the next set of examples.

Simple poles versus repeated poles with multiplicity $r > 1$

Example 11.18
Find the PFE of

$$\hat{F}(s) = \frac{s}{(s + 1)^2}.$$

Solution   Since the pole at $s = -1$ has a multiplicity of $r = 2$, we choose the form of the PFE to be

$$\frac{s}{(s + 1)^2} = \frac{K_1}{(s + 1)^2} + \frac{K_2}{s + 1}.$$

Next, using the cover-up method (where the entire $(s + 1)^2$ factor is covered up on the left), we determine K_1 as

$$K_1 = \frac{s}{\text{THUMB}}\bigg|_{s=-1} = -1.$$

Thus, the problem is now reduced to finding the weight K_2 so that so that

$$\frac{s}{(s+1)^2} = \frac{-1}{(s+1)^2} + \frac{K_2}{s+1}$$

is true for all s. This is accomplished by evaluating the above expression with any convenient value of s and then solving for K_2. For example, using $s = 0$ we find

$$\frac{0}{(0+1)^2} = \frac{-1}{(0+1)^2} + \frac{K_2}{0+1},$$

which yields $K_2 = 1$. Thus, we have

$$\frac{s}{(s+1)^2} = \frac{-1}{(s+1)^2} + \frac{1}{s+1},$$

confirming the example in the second row of Table 11.3.

Example 11.19

Determine the inverse Laplace transform $h(t)$ of

$$\hat{H}(s) = \frac{s+1}{s^2(s-1)}.$$

Solution   We have a repeated pole at $s = 0$, so we begin with

$$\frac{s+1}{s^2(s-1)} = \frac{\frac{1+1}{1^2\text{THUMB}}}{s-1} + \frac{\frac{0+1}{\text{THUMB}(0-1)}}{s^2} + \frac{K}{s},$$

giving

$$\frac{s+1}{s^2(s-1)} = \frac{2}{s-1} + \frac{-1}{s^2} + \frac{K}{s}.$$

Next, evaluating this expression at $s = -1$, we find

$$0 = \frac{2}{-1-1} + \frac{-1}{(-1)^2} + \frac{K}{-1}$$

yielding

$$K = \frac{2}{-1-1} + \frac{-1}{(-1)^2} = -2.$$

Thus,

$$\hat{H}(s) = \frac{2}{s-1} + \frac{-1}{s^2} + \frac{-2}{s}$$

and

$$h(t) = (2e^t - t - 2)u(t)$$

as in the bottom row of Table 11.3.

11.2.3   Inversion of improper rational expressions

The inversion of *improper* rational Laplace transforms can be handled in a number of ways. We will illustrate two approaches to the problem, treating the very simple case of

$$\hat{H}(s) = \frac{s}{s+1}.$$

This function is improper because the degree of the denominator is not larger than the degree of the numerator.

Option 1:   Think of $\hat{H}(s) = \frac{s}{s+1}$ as

$$\hat{H}(s) = s\frac{1}{s+1} = s\hat{G}(s),$$

where

$$g(t) = e^{-t}u(t) \leftrightarrow \hat{G}(s) = \frac{1}{s+1}.$$

Then the derivative property of the Laplace transform for causal signals (so that $g(0^-) = 0$) implies that

$$h(t) = \frac{dg}{dt} = \frac{d}{dt}(e^{-t}u(t)) = -e^{-t}u(t) + e^{-t}\delta(t) = \delta(t) - e^{-t}u(t).$$

Option 2:   Think of $\hat{H}(s) = \frac{s}{s+1}$ as

$$\hat{H}(s) = \frac{s+1-1}{s+1} = 1 - \frac{1}{s+1}.$$

Since $\delta(t) \leftrightarrow 1$ and $e^{-t}u(t) \leftrightarrow \frac{1}{s+1}$, the expression above implies that

$$h(t) = \delta(t) - e^{-t}u(t),$$

the same result as obtained above.

In both strategies, we first rewrite the improper-form expression in terms of a proper rational expression and then take advantage of the methods specific for those two options. The following two examples illustrate further details of the inversion of improper rational Laplace transforms.

Example 11.20

Find the inverse Laplace transform of the improper rational expression

$$\hat{F}(s) = \frac{s^2 + 1}{(s + 1)(s + 2)}.$$

Solution   Write

$$\hat{F}(s) = \frac{s^2}{(s + 1)(s + 2)} + \frac{1}{(s + 1)(s + 2)}.$$

So, letting

$$g(t) \leftrightarrow \hat{G}(s) = \frac{1}{(s + 1)(s + 2)}$$

we have

$$f(t) = \frac{d^2 g}{dt^2} + g(t),$$

where

$$g(t) \leftrightarrow \frac{\frac{1}{\text{THUMB}(-1+2)}}{s + 1} + \frac{\frac{1}{(-2+1)\text{THUMB}}}{s + 2}.$$

Thus,

$$g(t) = e^{-t}u(t) - e^{-2t}u(t), \quad \frac{dg}{dt} = -e^{-t}u(t) + 2e^{-2t}u(t),$$

and

$$\frac{d^2 g}{dt^2} = e^{-t}u(t) - 4e^{-2t}u(t) + \delta(t).$$

Therefore,

$$f(t) = \delta(t) + 2e^{-t}u(t) - 5e^{-2t}u(t).$$

Alternatively, we could perform a long division operation with the original form of $\hat{F}(s)$ to obtain

$$\hat{F}(s) = \frac{s^2 + 1}{s^2 + 3s + 2} = \frac{s^2 + 3s + 2 - (3s + 1)}{s^2 + 3s + 2} = 1 - \frac{3s + 1}{(s + 1)(s + 2)}.$$

Substituting the PFE of the proper second term in $\hat{F}(s)$ gives

$$\hat{F}(s) = 1 - \frac{\frac{-3+1}{-1+2}}{s+1} - \frac{\frac{-6+1}{-2+1}}{s+2} = 1 - \frac{-2}{s+1} - \frac{5}{s+2}.$$

Thus, we obtain

$$f(t) = \delta(t) + 2e^{-t}u(t) - 5e^{-2t}u(t)$$

as before.

Example 11.21

Determine the inverse Laplace transform of

$$\hat{G}(s) = \frac{s^3}{s^2+1}.$$

Solution   We can attack this problem as follows: First we note

$$\hat{G}(s) = s\frac{s^2}{s^2+1} = s(1 - \frac{1}{s^2+1}).$$

Since

$$\delta(t) - \sin(t)u(t) \leftrightarrow 1 - \frac{1}{s^2+1},$$

implying

$$\frac{d}{dt}[\delta(t) - \sin(t)u(t)] \leftrightarrow s(1 - \frac{1}{s^2+1}),$$

we find that

$$g(t) = \delta'(t) - \cos(t)u(t).$$

Note that in applying the derivative property, above, we used the fact that $\delta(t) - \sin(t)u(t)$ is a causal signal.

11.3 s-Domain Circuit Analysis

Consider an LTIC circuit in zero-state that is excited by some causal input

$$f(t) \leftrightarrow \hat{F}(s).$$

Then all of the voltage and current signals $v(t)$ and $i(t)$ excited in the circuit are necessarily causal and satisfy the initial conditions

$$v(0^-) = i(0^-) = 0.$$

Thus, the Laplace transform of the v-i relation

$$v(t) = L\frac{di}{dt}$$

for an inductor in the circuit is

$$\hat{V}(s) = Ls\hat{I}(s),$$

where

$$v(t) \leftrightarrow \hat{V}(s) \text{ and } i(t) \leftrightarrow \hat{I}(s).$$

Likewise, the Laplace transform of the v-i relation

$$i(t) = C\frac{dv}{dt}$$

for a capacitor is

$$\hat{I}(s) = Cs\hat{V}(s),$$

implying

$$\hat{V}(s) = \frac{1}{sC}\hat{I}(s).$$

Finally, the Laplace transform of the v-i relation

$$v(t) = Ri(t)$$

for a resistor is simply

$$\hat{V}(s) = R\hat{I}(s).$$

These results imply a general *s-domain* relation between $\hat{V}(s)$ and $\hat{I}(s)$ having the form

$$\hat{V}(s) = Z\hat{I}(s),$$

s-domain impedance

where

$$Z \equiv \begin{cases} sL & \text{for an inductor } L \\ \frac{1}{sC} & \text{for a capacitor } C \\ R & \text{for a resistor } R. \end{cases}$$

Here Z denotes an *s-domain impedance*, a generalization of the familiar phasor-domain impedance. These algebraic $\hat{V}(s)$-$\hat{I}(s)$ relations, along with Laplace transform versions of KVL and KCL, namely

$$\left\{\sum \hat{V}(s)_{\text{drop}} = \sum \hat{V}(s)_{\text{rise}}\right\}_{\text{loop}}$$

and

$$\left\{\sum \hat{I}(s)_{\text{in}} = \sum \hat{I}(s)_{\text{out}}\right\}_{\text{node}}$$

can be used to perform *s*-domain circuit analysis, analogous to phasor circuit calculations familiar from earlier chapters. Because KVL and KCL hold in the *s*-domain, techniques such as voltage division and current division do as well.

Assuming that the LTIC circuit shown in Figure 11.6a has zero initial state and a causal input $f(t)$, Figure 11.6b shows the equivalent *s*-domain circuit. We can use the *s*-domain equivalent and simple voltage division to calculate the Laplace transform $\hat{Y}(s)$ of the zero-state response

$$y(t) = h(t) * f(t)$$

of the circuit to a causal input. The result will have the form (because of the convolution property of Laplace transforms)

$$\hat{Y}(s) = \hat{H}(s)\hat{F}(s),$$

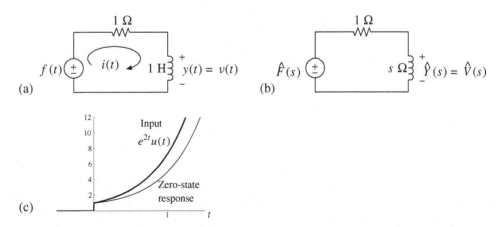

Figure 11.6 (a) An LTIC circuit with zero initial state, (b) its *s*-domain equivalent assuming a causal $f(t) \leftrightarrow \hat{F}(s)$, (c) plots of a causal input $f(t) = e^{2t}u(t)$ (heavy curve) and zero-state response (light curve) calculated in Example 11.22.

allowing us to identify the transfer function

$$\hat{H}(s) = \frac{\hat{Y}(s)}{\hat{F}(s)}$$

Zero-state circuit response to causal inputs

of the circuit, as well as finding the zero-state response $y(t)$ by computing the inverse Laplace transform of $\hat{Y}(s)$. The following set of examples illustrate such s-domain circuit calculations.

Example 11.22

In the LTIC circuit shown in Figure 11.6a, determine the transfer function $\hat{H}(s) = \hat{Y}(s)/\hat{F}(s)$ and also find the zero-state response $y(t)$, assuming the circuit input signal is

$$f(t) = e^{2t}u(t) \leftrightarrow \hat{F}(s) = \frac{1}{s-2}.$$

Finding the transfer function

Solution   Figure 11.6b shows the equivalent s-domain circuit where the 1 H inductor has been replaced by an $s\,\Omega$ impedance as a consequence of $Z = sL$, pertinent for inductors. This equivalent circuit has been obtained by assuming that the circuit input $f(t)$ is a causal signal having some Laplace transform $\hat{F}(s)$.

Now, using voltage division in Figure 11.6b (similar to phasor analysis, but with s-domain impedances), we have

$$\hat{Y}(s) = \hat{F}(s)\frac{s}{1+s}.$$

Hence the system transfer function is

$$\hat{H}(s) = \frac{\hat{Y}(s)}{\hat{F}(s)} = \frac{s}{s+1}.$$

Given the causal input

$$f(t) = e^{2t}u(t) \leftrightarrow \hat{F}(s) = \frac{1}{s-2},$$

the Laplace transform of the zero-state response of the circuit is

$$\hat{Y}(s) = \frac{s}{s+1}\hat{F}(s) = \frac{s}{s+1}\frac{1}{s-2} = \frac{s}{(s+1)(s-2)} = \frac{1/3}{s+1} + \frac{2/3}{s-2}.$$

Hence, the system zero-state response to input $f(t) = e^{2t}u(t)$ is the inverse Laplace transform

$$y(t) = \frac{1}{3}(e^{-t} + 2e^{2t})u(t).$$

Figure 11.6c shows plots of the causal input $f(t) = e^{2t}u(t)$ and the causal zero-state response $y(t)$ derived above.

Example 11.23
Determine the transfer function and the impulse response of the LTIC circuit shown in Figure 11.7a.

Solution   Transfer functions and impulse responses are defined under zero-state conditions. Figure 11.7b shows the s-domain equivalent zero-state circuit, where $\hat{F}(s)$ denotes the Laplace transform of a causal source $f(t)$. Using KVL in the s-domain, we find that

$$\hat{F}(s) = (2 + s + \frac{1}{s})\hat{Y}(s),$$

giving

$$\hat{Y}(s) = \frac{\hat{F}(s)}{2 + s + \frac{1}{s}} = \frac{s}{s^2 + 2s + 1}\hat{F}(s) = \frac{s}{(s+1)^2}\hat{F}(s).$$

Hence, the system transfer function is

$$\hat{H}(s) = \frac{\hat{Y}(s)}{\hat{F}(s)} = \frac{s}{(s+1)^2}$$

with a double pole at $s = -1$. Expanding $\hat{H}(s)$ in a PFE, we have

$$\hat{H}(s) = \frac{s}{(s+1)^2} = \frac{-1}{(s+1)^2} + \frac{K}{s+1}.$$

Finding the impulse response

Evaluating at $s = 0$, we find that $\hat{H}(0) = 0 = -1 + K$, implying that $K = 1$. Hence,

$$\hat{H}(s) = \frac{-1}{(s+1)^2} + \frac{1}{s+1}$$

yielding

$$h(t) = -te^{-t}u(t) + e^{-t}u(t) = (1 - t)e^{-t}u(t).$$

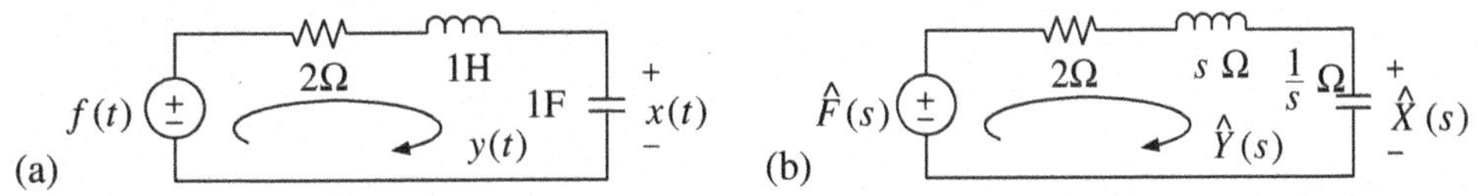

(a)
(b)

Figure 11.7   (a) An RLC circuit, and (b) its s-domain equivalent.

Notice how the result of Example 11.23 indicates an effective jump in the inductor current $y(t) = h(t)$ at $t = 0$ in response to an impulse input $f(t) = \delta(t)$. As pointed out in Chapter 3, inductor currents and capacitor voltages cannot jump in response to *practical* sources. However, when *theoretical* sources such as $\delta(t)$ are involved, jumps can occur, as in Example 11.23.

Example 11.24

Determine the transfer function $\hat{H}(s)$ and impulse response $h(t)$ of the circuit described in Figure 11.8.

Solution   Applying ideal op-amp approximations, and writing a KCL equation at the inverting node of the op amp,

$$\frac{\hat{F}(s) - 0}{2} = \frac{0 - \hat{Y}(s)}{1 + \frac{1}{s}}.$$

Therefore, the transfer function is

$$\hat{H}(s) = \frac{\hat{Y}(s)}{\hat{F}(s)} = -\frac{1}{2}(1 + \frac{1}{s}).$$

Taking the inverse Laplace transform yields the impulse response,

$$h(t) = -\frac{1}{2}[\delta(t) + u(t)].$$

Notice that the circuit in Example 11.24 is not BIBO stable—the transfer function has a pole at $s = 0$, which is outside the LHP, and the impulse response is not absolutely integrable. Evidently, unlike the phasor method and Fourier analysis, the s-domain method can work perfectly well for the analysis of unstable circuits. The next example further illustrates this point.

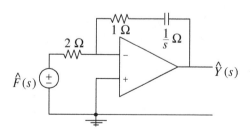

Figure 11.8   *s*-domain equivalent of an op-amp circuit with a 1 F capacitor.

Example 11.25
Find the transfer function of the LTIC circuit described in Figure 11.9 and determine the range of values of the real parameter A for which the circuit is BIBO stable.

**Circuit
stability
analysis**

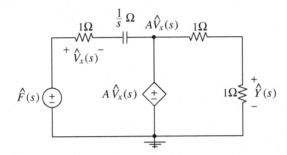

Figure 11.9 *s*-domain equivalent of a circuit with a single independent input $f(t)$ and a voltage-controlled voltage source $Av_x(t)$, where A is a real gain constant.

Solution   Note that $\hat{Y}(s)$ is simply one half of $A\hat{V}_x(s)$, so we begin by finding $\hat{V}_x(s)$ in terms of $\hat{F}(s)$. Applying voltage division, we note that

$$\hat{V}_x(s) = (\hat{F}(s) - A\hat{V}_x(s))\frac{1}{1 + \frac{1}{s}},$$

from which we obtain

$$\hat{V}_x(s) = \frac{s}{s(1 + A) + 1}\hat{F}(s).$$

Thus,

$$\hat{Y}(s) = \frac{A\hat{V}_x(s)}{2} = \frac{A}{2}\frac{s}{s(1 + A) + 1}\hat{F}(s) = (\frac{A}{2(1 + A)})\frac{s}{s + \frac{1}{1+A}}\hat{F}(s),$$

and the circuit transfer function is

$$\hat{H}(s) = (\frac{A}{2(1 + A)})\frac{s}{s + \frac{1}{1+A}}.$$

The circuit is BIBO stable if and only if the single pole of the system at

$$s = -\frac{1}{1 + A}$$

is within the LHP. This is will be the case only if

$$-\frac{1}{1+A} < 0,$$

which implies

$$1 + A > 0,$$

or

$$A > -1.$$

For $A \leq -1$ the circuit is unstable.

11.4   General Response of LTIC Circuits and Systems

As we have seen in Section 11.3, the impulse response $h(t)$ of a lumped-element LTIC circuit can be determined as the inverse Laplace transform of a rational transfer function

$$\hat{H}(s) = \frac{B(s)}{P(s)}.$$

Here, $B(s)$ and $P(s)$ are mth- and nth-order polynomials

$$B(s) = b_0 s^m + b_1 s^{m-1} + \cdots + b_m$$

and

$$P(s) = s^n + a_1 s^{n-1} + \cdots + a_n = (s - p_1)(s - p_2) \cdots (s - p_n),$$

respectively. Since

$$\hat{H}(s) = \frac{\hat{Y}(s)}{\hat{F}(s)},$$

where $\hat{Y}(s)$ is the Laplace transform of a causal zero-state response to a causal input $f(t) \leftrightarrow \hat{F}(s)$, it follows that for an LTIC circuit,

$$\frac{\hat{Y}(s)}{\hat{F}(s)} = \frac{B(s)}{P(s)}$$

or, equivalently,

$$P(s)\hat{Y}(s) = B(s)\hat{F}(s).$$

Thus, for an arbitrary LTIC circuit we can write

$$s^n \hat{Y}(s) + a_1 s^{n-1} \hat{Y}(s) + \cdots + a_n \hat{Y}(s) = b_0 s^m \hat{F}(s) + b_1 s^{m-1} \hat{F}(s) + \cdots + b_m \hat{F}(s),$$

which has the inverse Laplace transform

$$\frac{d^n y}{dt^n} + a_1 \frac{d^{n-1} y}{dt^{n-1}} + \cdots + a_n y(t) = b_0 \frac{d^m f}{dt^m} + b_1 \frac{d^{m-1} f}{dt^{m-1}} \cdots + b_m f(t).$$

*n*th-order
LTIC
circuits are
described
by *n*th-order
linear
ODEs with
constant
coefficients

This expression is a linear constant-coefficient ODE relating the circuit input and output. This development implies that all LTIC circuits and systems with rational transfer functions are described by such ODE's.

Example 11.26

Determine an ODE that describes the LTIC circuit shown in Figure 11.10a. What are the polynomials $B(s)$ and $P(s)$ for the circuit? What is the order n of the circuit? What homogeneous ODE is satisfied by the zero-input response of the circuit?

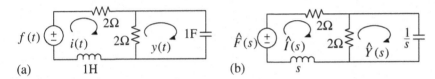

(a)   1H   (b)   s

Figure 11.10 (a) A 2nd-order LTIC circuit and (b) its *s*-domain equivalent.

Solution   Using the *s*-domain equivalent of the circuit, shown in Figure 11.10b, and writing a KVL equation for each loop, we have

$$\hat{F}(s) = 2\hat{I}(s) + 2(\hat{I}(s) - \hat{Y}(s)) + s\hat{I}(s)$$

and

$$2(\hat{I}(s) - \hat{Y}(s)) = \frac{1}{s}\hat{Y}(s).$$

Solving the second equation for $\hat{I}(s)$ gives

$$\hat{I}(s) = (1 + \frac{1}{2s})\hat{Y}(s).$$

Substituting for $\hat{I}(s)$ in the first equation then yields

$$\hat{F}(s) = (s + \frac{5}{2} + \frac{2}{s})\hat{Y}(s).$$

Hence, the transfer function is

$$\hat{H}(s) = \frac{1}{s + 2.5 + \frac{2}{s}} = \frac{s}{s^2 + 2.5s + 2}.$$

Therefore,

$$B(s) = s,$$

and

$$P(s) = s^2 + 2.5s + 2.$$

Furthermore, because

$$\frac{\hat{Y}(s)}{\hat{F}(s)} = \frac{B(s)}{P(s)},$$

we have

$$s^2\hat{Y}(s) + 2.5s\hat{Y}(s) + 2\hat{Y}(s) = s\hat{F}(s),$$

so that an ODE describing the circuit is

$$\frac{d^2y}{dt^2} + 2.5\frac{dy}{dt} + 2y(t) = \frac{df}{dt}$$

and the circuit order is $n = 2$. The zero-input response of the circuit must satisfy the homogeneous ODE

$$\frac{d^2y}{dt^2} + 2.5\frac{dy}{dt} + 2y(t) = 0,$$

which is obtained by setting the input $f(t)$ to zero in the ODE for the circuit.

Beginning with Chapter 4, we have emphasized the *zero-state response* of circuits and systems. The fact that LTIC circuits are described by ODEs is a reminder that the output of a circuit is, in general, a sum of two components. The zero-state response is $h(t) * f(t)$, where $h(t) \leftrightarrow \hat{H}(s)$. When initial conditions are zero, the zero-state response is the entire response. However, if initial conditions are not zero, then the output includes an additional signal $y_o(t)$ that satisfies the homogeneous form of the

General response

ODE. Thus, in general, for n^{th}-order LTIC circuits and systems with rational transfer functions, we have

$$f(t) \rightarrow \boxed{\text{LTIC}} \rightarrow y(t) = y_o(t) + h(t) * f(t).$$

Zero-input response and homogeneous ODE

The output component $y_o(t)$ is the *zero-input response*, the solution of some

$$\frac{d^ny}{dt^n} + a_1\frac{d^{n-1}y}{dt^{n-1}} + \cdots + a_ny(t) = 0$$

satisfying a set of initial conditions related to energy storage within the system.

We already know how to determine the zero-state response $h(t) * f(t)$ for LTIC circuits and systems using various methods—direct convolution, the Fourier method, or the Laplace method if $f(t)$ is causal—so we next focus our attention on the zero-input response.

11.4.1 Zero-input response and asymptotic stability in LTIC systems

For LTIC systems with rational transfer functions described by

$$\hat{H}(s) = \frac{B(s)}{P(s)},$$

with

$$B(s) = b_0(s - z_1)(s - z_2) \cdots (s - z_m)$$

and

$$P(s) = (s - p_1)(s - p_2) \cdots (s - p_n),$$

the denominator $P(s)$ is known as the *characteristic polynomial*. $P(s)$ is the same characteristic polynomial that is associated with the differential equation describing the system. The *roots*, $p_1, p_2, \cdots, p_n$, are called *characteristic poles*.[7] The number n of characteristic poles is the system order and coincides with the number of distinct energy storage elements (e.g., capacitors and inductors) in the system.

Characteristic poles

  In this section we will see that the zero-input response of LTIC circuits and systems, with rational transfer functions, always can be expressed as a weighted sum of functions

$$c_l(t) = t^k e^{p_i t}, \ \ 0 \le k < r_i,$$

known as *characteristic modes*. As explained below, characteristic modes of a system can be identified unambiguously once the characteristic poles p_i and their multiplicities r_i are known.

Characteristic modes

  For example, the characteristic modes of a circuit with simple (nonrepeated) characteristic poles

$$p_1 = -1$$
$$p_2 = -2$$

are

$$c_1(t) = e^{-t}$$
$$c_2(t) = e^{-2t},$$

and the zero-input response of the circuit would have the form

$$y(t) = Ae^{-t} + Be^{-2t}.$$

If the initial conditions imposed on $y(t)$ were $y(0) = 2$ and $y'(0) = -3$, then $A = B = 1$.

---

[7]The set of poles of $\hat{H}(s)$ defined earlier is the subset of characteristic poles that are not cancelled by any term $z - z_i$ in the numerator of $\hat{H}(s)$. In most cases, the set of poles and the set of characteristic poles are the same.

For another 2nd-order system with transfer function

$$\hat{H}(s) = \frac{1}{(s+1)^2}$$

there is a single characteristic pole at $s = -1$ with multiplicity 2. In this case, the characteristic modes are

$$c_1(t) = te^{-t}$$
$$c_2(t) = e^{-t},$$

and the zero-input response would be

$$y(t) = Ate^{-t} + Be^{-t}$$

where the constants A and B would depend on the initial conditions for $y(t)$.

Explanation: The zero-input response of an LTIC circuit is the solution of an nth-order linear and homogeneous ODE

$$\frac{d^n y}{dt^n} + a_1 \frac{d^{n-1} y}{dt^{n-1}} + \cdots + a_n y(t) = 0.$$

Assume that the ODE and its solution $y(t)$ are valid for all t, positive and negative. Then you may recall from a course on differential equations that a time-domain solution of this homogeneous ODE is a weighted sum of the above characteristic modes. Alternatively, the Laplace transform of $y(t)$ and its derivatives result in

$$y(t) \rightarrow \hat{Y}(s),$$
$$y'(t) \rightarrow s\hat{Y}(s) - y(0^-),$$
$$y''(t) \rightarrow s^2\hat{Y}(s) - sy(0^-) - y'(0^-),$$

etc. Hence, the Laplace transform of the homogeneous ODE above can be expressed as

$$(s^n + a_1 s^{n-1} + \cdots + a_n)\hat{Y}(s) = C(s),$$

where $C(s)$ is an $(n-1)^{\text{th}}$-order polynomial in s of the form

$$y(0^-)(s^{n-1} + a_1 s^{n-2} + \cdots) + y'(0^-)(s^{n-2} + a_1 s^{n-3} + \cdots) + \cdots + y^{(n-1)}(0^-).$$

Consequently, the Laplace transform of the system zero-input response is

$$\hat{Y}(s) = \frac{C(s)}{(s - p_1)(s - p_2) \cdots (s - p_n)}.$$

Note that $\hat{Y}(s)$ is a *proper* rational expression with poles p_i that coincide with the characteristic poles of the system $\hat{H}(s)$. Also note that $\hat{Y}(s)$ is *linearly* dependent, through the polynomial $C(s)$, on the initial conditions $y(0^-)$, $y'(0^-)$, etc., representing the *state* of the system at $t = 0^-$. Therefore, the system zero-input response—the inverse Laplace transform of $\hat{Y}(s)$ above, which can be calculated using a PFE—turns out to be a weighted sum of the characteristic modes of the system proportional to $e^{p_i t}$ (or $t^k e^{p_i t}$ if p_i is repeated), with weighting coefficients that depend linearly on the system's initial state (as expected).

In the following examples, we will identify the characteristic poles and modes of various LTIC systems and then calculate zero-input responses.

Example 11.27

Consider an LTIC circuit described by the ODE

$$\frac{d^2 y}{dt^2} - 4y(t) = f(t)$$

for $-\infty < t < \infty$. Determine the characteristic polynomial $P(s)$, characteristic poles p_i, and characteristic modes $c_i(t)$ of the system. Also, find the zero-input response $y(t)$ assuming that $y(0^-) = 1$ and $y'(0^-) = -1$.

Solution   Noting that the characteristic polynomial is the denominator of the transfer function $\hat{H}(s)$, we take the Laplace transform of the ODE under the assumption of causal $y(t)$ and $f(t)$ to obtain

$$(s^2 - 4)\hat{Y}(s) = \hat{F}(s).$$

Clearly, the characteristic polynomial of the system is

$$P(s) = s^2 - 4 = (s - 2)(s + 2).$$

Consequently, the characteristic poles are $p_1 = 2$, $p_2 = -2$, and the characteristic modes are $c_1(t) = e^{2t}$ and $c_2(t) = e^{-2t}$. We, therefore, can express the zero-input response of the system as

$$y(t) = Ae^{2t} + Be^{-2t},$$

which is valid for all t, just like the homogeneous ODE it satisfies.

To determine A and B, we equate $y(0)$ and $y'(0)$ directly to the specified initial conditions, $y(0^-) = 1$ and $y'(0^-) = -1$, and solve for A and B. Doing so gives

$$y(0) = A + B = y(0^-) = 1,$$
$$y'(0) = 2A - 2B = y'(0^-) = -1,$$

yielding

$$A = \frac{1}{4} \text{ and } B = \frac{3}{4}.$$

Hence, the zero-input response of the circuit with the given initial conditions is

$$y(t) = 0.25e^{2t} + 0.75e^{-2t}.$$

Notice that the zero-input response examined in this example is nontransient because of the contribution of a nontransient characteristic mode e^{2t} due to the characteristic pole at $s = p_1 = 2$ in the RHP.

Example 11.28
Determine the characteristic modes and the form of the zero-input response of the circuit shown in Figure 11.7a.

Solution   From Example 11.23 in Section 11.3, the system transfer function is known to be

$$\hat{H}(s) = \frac{s}{(s + 1)^2}.$$

Therefore, the characteristic polynomial is $P(s) = (s + 1)^2$ and the characteristic modes are $c_1(t) = te^{-t}$ and $c_2(t) = e^{-t}$. Hence, the zero-input solution for the circuit is

$$y(t) = Ate^{-t} + Be^{-t},$$

where A and B depend on initial conditions.

Example 11.29
Evaluate the constants A and B of Example 11.28 if the zero-input response $y(t)$ is known to satisfy $y(0) = 1$ and $y(-1) = 0$.

Solution   Evaluating $y(t) = Ate^{-t} + Be^{-t}$ at $t = 0$ and $t = -1$ s, we find that

$$y(0) = B = 1,$$
$$y(-1) = -Ae^{1} + Be^{1} = 0.$$

Thus,

$$A = B = 1$$

and

$$y(t) = te^{-t} + e^{-t}.$$

Example 11.30
Applying the Laplace transform to the ODE

$$\frac{dy}{dt} + 3y(t) = f(t),$$

determine the zero-input solution corresponding to $y(0^-) = 2$.

Solution   Transforming the ODE gives

$$s\hat{Y}(s) - y(0^-) + 3\hat{Y}(s) = \hat{F}(s).$$

Because we are interested in the zero-input solution, we set $\hat{F}(s) = 0$ and obtain

$$\hat{Y}(s) = \frac{y(0^-)}{s+3}.$$

Therefore, we find

$$y(t) = y(0^-)e^{-3t} = 2e^{-3t},$$

valid for $t \geq 0$ (and also before $t = 0$ if the ODE is valid for all times t).

Example 11.31
Suppose that a 3rd-order LTIC system is described by the ODE

$$\frac{d^3y}{dt^2} - 2\frac{d^2y}{dt^2} - 6\frac{dy}{dt} - 8y(t) = \frac{df}{dt} - 4f(t).$$

Determine the system transfer function $\hat{H}(s)$ and find the characteristic poles and characteristic modes of the system. Find also the poles of the transfer function $\hat{H}(s)$ (distinct from characteristic poles as will be seen in the solution). Is the system BIBO stable? What is the zero-input response if $y(0^-) = 2$, $y'(0^-) = 3$, $y''(0^-) = 16$?

Solution   To determine the transfer function, we take the Laplace transform of the ODE under the assumption of a causal zero-state response $y(t)$ to a causal input $f(t)$. The transform is

$$s^3\hat{Y}(s) - 2s^2\hat{Y}(s) - 6s\hat{Y}(s) - 8\hat{Y}(s) = s\hat{F}(s) - 4\hat{F}(s).$$

Hence, the system transfer function is

$$\hat{H}(s) = \frac{\hat{Y}(s)}{\hat{F}(s)} = \frac{s-4}{s^3 - 2s^2 - 6s - 8} = \frac{s-4}{(s-4)(s+1+j)(s+1-j)}.$$

The characteristic polynomial of the system is

$$P(s) = (s - 4)(s + 1 + j)(s + 1 - j),$$

and the characteristic poles are at

$$s = 4, \quad -1 - j, \quad \text{and} \quad -1 + j.$$

Since all of the characteristic poles are simple (i.e., non-repeated), the characteristic modes are

$$c_1(t) = e^{4t}, \quad c_2(t) = e^{(-1-j)t}, \quad c_3(t) = e^{(-1+j)t}.$$

Because $\hat{H}(s)$ has a pole-zero cancellation at $s = 4$, the transfer function can be simplified to

$$\hat{H}(s) = \frac{1}{(s + 1 + j)(s + 1 - j)}.$$

Thus, the poles of the transfer function are at $s = -1 - j$ and $-1 + j$. Because both poles are in the LHP, the system is BIBO stable.

Finally, we will find the zero-input response to the stated initial conditions. The result may be surprising!

The zero-input response is a linear combination of the characteristic modes,

$$y(t) = Ae^{4t} + Be^{(-1-j)t} + Ce^{(-1+j)t}.$$

Applying the initial conditions produces the set of equations

$$y(0^-) = 2 = A + B + C,$$

$$y'(0^-) = 3 = 4A + (-1 - j)B + (-1 + j)C,$$

$$y''(0^-) = 16 = 16A + (-1 - j)^2 B + (-1 + j)^2 C,$$

yielding

$$A = 1 \quad \text{and} \quad B = C = \frac{1}{2},$$

so that

$$y(t) = e^{4t} + e^{-t} \cos(t).$$

This is surprising indeed! The zero-input response is unbounded even though the system is BIBO stable.

The reason for the behavior encountered in the above example is that the system analyzed contains a characteristic pole at $s = 4$ that is cancelled in the transfer function and yet it affects the zero-input response. It is easy to imagine how this might happen. For example, a circuit may consist of two cascaded parts, with transfer functions

$\hat{H}_1(s)$ and $\hat{H}_2(s)$, and the overall transfer function $\hat{H}_1(s)\hat{H}_2(s)$. Part 2 of the circuit may have an unstable pole that is cancelled by a zero in the transfer function of Part 1. This can lead to a situation where the overall system is BIBO stable (all poles of $\hat{H}_1(s)\hat{H}_2(s)$ are in the LHP) and yet nonzero initial conditions in Part 2 of the circuit may yield a zero-input response that is unbounded, because $\hat{H}_2(s)$ has a pole located outside the LHP.

Because we do not wish to have systems whose zero-input responses can be unbounded, we frequently require that systems exhibit a *transient* zero-input response. Such systems with transient zero-input responses are called *asymptotically stable*:

> An LTIC circuit with a rational transfer function is asymptotically stable if and only if all of its characteristic poles are in the LHP.

Asymptotic stability criterion

For LTIC systems with proper rational transfer functions, BIBO stability and asymptotic stability are equivalent unless there is cancellation of an unstable characteristic pole in the transfer function. The BIBO stable system in Example 11.31 is not asymptotically stable, because of the unstable characteristic pole at $s = 4$. Note that the output of a circuit that is not asymptotically stable may be dominated by (or at least include) components of the zero-input response. Likewise, any thermal noise injected within such a circuit may create an unbounded signal either interior to or at the output of the system. Thus, circuits that are not asymptotically stable (a stronger condition than BIBO stability) must be used with care. One possible use of such systems is exemplified in the following section.

11.4.2 Marginal stability and resonance

Systems with simple (i.e., nonrepeated) characteristic poles on the imaginary axis deserve further discussion. Such systems are neither BIBO stable nor asymptotically stable; yet, in the absence of RHP poles, they display bounded, but nontransient, zero-input responses—such systems are said to be *marginally stable*.

The 1st-order system

$$\hat{H}(s) = \frac{1}{s}$$

with a simple pole at $s = 0$ on the ω-axis, and no RHP poles, is marginally stable because its zero-input response

$$y(t) = y(0^-)$$

is bounded. Likewise, the 2nd-order system

$$\hat{H}(s) = \frac{1}{s^2 + \omega_o^2} = \frac{1}{(s - j\omega_o)(s + j\omega_o)},$$

with $\omega_o^2 > 0$, is marginally stable. The system has simple poles at $s = \pm j\omega_o$. Under initial conditions $y(0^-) = 1$ and $y'(0^-) = 0$, the system exhibits the bounded and sinusoidal steady-state zero-input response

$$y(t) = \frac{e^{j\omega_o t} + e^{-j\omega_o t}}{2} = \cos(\omega_o t).$$

Reiterating, an

Marginal stability criterion

LTIC system is marginally stable *if and only if* it has simple characteristic poles on the imaginary axis and no characteristic poles in the RHP.

Example 11.32

Determine the zero-input response of the nth-order system

$$\hat{H}(s) = \frac{1}{s^n}, \quad n \geq 1.$$

Show that the system has a bounded zero-input response and is marginally stable only for $n = 1$.

Solution   With characteristic poles at $s = 0$, the zero-input response of the system is

$$y(t) = K_1 t^{n-1} + K_2 t^{n-2} + \cdots + K_n,$$

where the constants $K_1, K_2, \cdots, K_n$ depend on initial conditions. Clearly, given arbitrary initial conditions, $y(t)$ is unbounded unless $n = 1$, which corresponds to a single pole on the imaginary axis and marginal stability.

The zero-input response $y(t) = \cos(\omega_o t)$ of the marginally stable 2nd-order system

$$\hat{H}(s) = \frac{1}{s^2 + \omega_o^2},$$

$\omega_o^2 > 0$, indicates that the system can be used as an "oscillator," or as a co-sinusoidal signal source. Because the oscillations $\cos(\omega_o t)$ are produced in the absence of an external input, the system is *resonant* (see Chapter 4), which in turn implies that no net energy dissipation takes place within the system.[8] Clearly, resonance and marginal stability are related concepts—all marginally stable systems also are resonant. For instance, the 4th-order marginally stable system

$$\hat{H}(s) = \frac{1}{(s^2 + 4)(s^2 + 9)}$$

---

[8]Such systems can be built with dissipative components so long as the systems also include active elements that can be modeled by negative resistances.

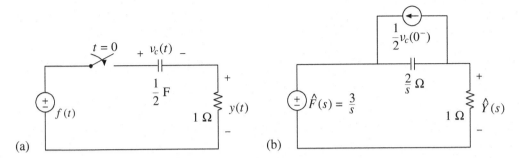

Figure 11.11   (a) An LTIC circuit with a switch, and (b) its s-domain equivalent including an initial-value source.

can sustain unforced and steady-state co-sinusoidal oscillations at two resonant frequencies 2 and 3 rad/s. On the other hand, another 4th-order system

$$\hat{H}(s) = \frac{1}{(s^2 + 4)(s^2 + 4)}$$

with the resonant frequency 2 rad/s is not marginally stable because the poles at $s = \pm j2$ are not simple.

11.4.3   Circuit initial-value problems

We will describe in this section an extension of the s-domain technique of Section 11.3 that can be used in LTIC circuit problems where the zero initial-state condition may not be applicable. See, for instance, the circuit in Figure 11.11a, where a solution for $y(t)$ is sought for $t > 0$, assuming an arbitrary $v_c(0^-)$ just before the switch is closed. Although finding $y(t)$, $t > 0$, in the circuit is just a matter of superposing zero-state and zero-input responses (determined using methods of previous sections), an *alternate* approach is to apply s-domain analysis to an equivalent circuit shown in Figure 11.11b, as worked out in Example 11.33 below. The procedure by which the equivalent circuit can be obtained—where a *current source* $\frac{1}{2}v_c(0^-)$ accounts for the nonzero initial state—will be described after the example.

Example 11.33
The source in the circuit shown in Figure 11.11a is specified as $f(t) = 3$ V. Determine the system response $y(t)$ for $t > 0$, for an arbitrary initial state $v_c(0^-)$, by using the s-domain equivalent circuit shown in Figure 11.11b.

Solution   Note that the source $f(t) = 3$ V appears in the equivalent circuit as the Laplace transform $\hat{F}(s) = \frac{3}{s}$. The impedance $\frac{2}{s}$ Ω is the counterpart of the $\frac{1}{2}$ F capacitor (as in Section 11.3), but it is in parallel with a current source $\frac{1}{2}v_c(0^-)$ for reasons to be explained after this example. Focusing, for now, on the analysis of the circuit in Figure 11.11b, we note that we may use superposition to determine $\hat{Y}(s)$. First, suppressing the current source $\frac{1}{2}v_c(0^-)$ and applying voltage division, we find

$$\hat{Y}(s) = \hat{F}(s)\frac{1}{1+\frac{2}{s}} = \frac{3}{s}\frac{1}{1+\frac{2}{s}} = \frac{3}{s+2}.$$

Next, suppressing $\hat{F}(s)$ and applying the current $\frac{1}{2}v_c(0^-)$ through the parallel combination of impedances of 1 and $\frac{2}{s}$ Ω, we obtain

$$\hat{Y}(s) = -\frac{1}{2}v_c(0^-)\frac{1\cdot\frac{2}{s}}{1+\frac{2}{s}} = -\frac{v_c(0^-)}{s+2}.$$

Superposing the above contributions, the overall response is

$$\hat{Y}(s) = \frac{3}{s+2} - \frac{v_c(0^-)}{s+2}.$$

Taking the inverse Laplace transform of this result, we find that for $t > 0$

$$y(t) = 3e^{-2t} - v_c(0^-)e^{-2t}.$$

The first term of the solution is the zero-state component (driven by the 3 V source) while the second term, proportional to $v_c(0^-)$, is the zero-input response.

Figures 11.12a and 11.12b, below, depict the s-domain equivalents of an inductor L and capacitor C, with *nonzero* initial states $i(0^-)$ and $v(0^-)$, respectively. In the equivalent networks, the s-domain impedances sL and $1/sC$ Ω are accompanied by series and parallel voltage and current sources, $Li(0^-)$ and $Cv(0^-)$, having polarity and flow direction opposite to those associated with $v(t)$ and $i(t)$. For instance, in Figure 11.12a the *voltage* $Li(0^-)$ is a drop in the direction of voltage rise $v(t)$, while in Figure 11.12b the *current* $Cv(0^-)$ flows counter to $i(t)$. The equivalence depicted in Figure 11.12b is the transformation rule used to obtain the s-domain circuit shown above in Figure 11.11b.

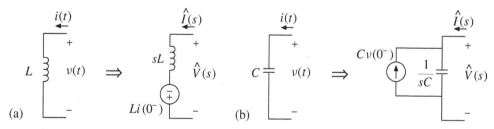

Figure 11.12 Transformation rule for obtaining s-domain equivalents of (a) an inductor L and (b) a capacitor C, with initial states $i(0^-)$ and $v(0^-)$, respectively (see text for explanation).

To validate the transformation depicted in Figure 11.12a, we note that the Laplace transform of the inductor v-i relation

$$v(t) = L\frac{di}{dt}$$

is

$$\hat{V}(s) = sL\hat{I}(s) - Li(0^-),$$

an equality that can be seen to satisfy the s-domain circuit constraint at the network terminals shown at the right side of the figure. Likewise, the s-domain constraint at the network terminals at the right side of Figure 11.12b is

$$\hat{I}(s) = sC\hat{V}(s) - Cv(0^-),$$

which is in agreement with the Laplace transform of the capacitor v-i relation

$$i(t) = C\frac{dv}{dt}.$$

The Laplace transforms in both cases above result from applying the Laplace transform derivative property (item 6 in Table 11.2) to account for the initial-value source terms included in the above models. Note that by applying *source transfor-* **Initial-value** *mations* to the s-domain equivalents shown in Figures 11.12a and 11.12b we can **sources** generate additional forms (Norton and Thevenin types, respectively) of the equivalents. That will not be necessary for solving the example problems given below, but remembering the equivalences shown in Figure 11.12, and in particular the proper initial-value source directions, is important.

Example 11.34
In the circuit shown in Figure 11.13a, the initial capacitor voltage is $v(0^-) = 0$ and the initial inductor current is $i(0^-) = 2$ A. Determine $v(t)$ for $t > 0$ if $f(t) = 1$ for $t > 0$.

Solution   Figure 11.13b shows the equivalent s-domain circuit, where an initial-value voltage source, due to the nonzero initial inductor current $i(0^-) = 2$ A, has been introduced. Since the initial capacitor voltage is zero,

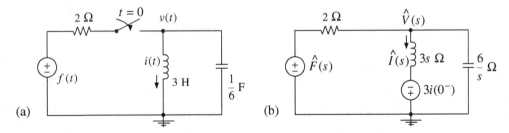

(a)   (b)

Figure 11.13   (a) An LTIC circuit with a switch and two energy storage elements, and (b) its s-domain equivalent including an initial-value source.

the capacitor has been transformed to the s-domain without an accompanying initial-value source. Applying KCL in the s-domain to the equivalent circuit, we have

$$\frac{\hat{V}(s) - \hat{F}(s)}{2} + \frac{\hat{V}(s) + 3i(0^-)}{3s} + \frac{\hat{V}(s)}{6/s} = 0,$$

leading to

$$\left(\frac{1}{2} + \frac{1}{3s} + \frac{s}{6}\right)\hat{V}(s) = \frac{\hat{F}(s)}{2} - \frac{i(0^-)}{s}.$$

Solving for $\hat{V}(s)$ gives

$$\hat{V}(s) = \frac{3s\hat{F}(s)}{s^2 + 3s + 2} - \frac{6i(0^-)}{s^2 + 3s + 2}.$$

Thus, with $f(t) = 1 \rightarrow \hat{F}(s) = \frac{1}{s}$ and $i(0^-) = 2$ A, we have

$$\hat{V}(s) = \frac{3 - 12}{(s + 1)(s + 2)} = \frac{-9}{s + 1} + \frac{9}{s + 2},$$

which implies a total response, for $t > 0$, of

$$v(t) = -9e^{-t} + 9e^{-2t}.$$

Example 11.35
In Figure 11.14a $f_1(t) = 3t$ V for $t > 0$ and $f_2(t) = 1$ A. Assuming that $v(0^-) = 1$ V, and using s-domain techniques, determine the zero-state and zero-input components of the current $i(t)$ for $t > 0$.

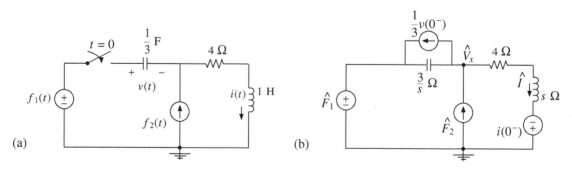

Figure 11.14   (a) A circuit with two independent sources, and (b) its s-domain equivalent including initial-value sources.

Solution   Figure 11.14b shows the s-domain equivalent of Figure 11.14a for $t > 0$, where $i(0^-) = 1$ A in order to satisfy KCL in the circuit prior to the switching action.

To determine the zero-state response of the circuit, we analyze the s-domain equivalent with suppressed initial-value sources. In that case writing a KCL equation at the top node, with $\hat{F}_1(s) = \frac{3}{s^2}$ and $\hat{F}_2(s) = \frac{1}{s}$, yields

$$\frac{\hat{V}_x - \frac{3}{s^2}}{\frac{3}{s}} - \frac{1}{s} + \frac{\hat{V}_x}{4+s} = 0,$$

from which

$$\hat{V}_x = \frac{\frac{2}{s}}{\frac{s}{3} + \frac{1}{4+s}}.$$

Thus,

$$\hat{I}(s) = \frac{\hat{V}_x}{4+s} = \frac{6}{s^2(s+4) + 3s} = \frac{6}{s(s+1)(s+3)} = \frac{2}{s} + \frac{-3}{s+1} + \frac{1}{s+3}.$$

Taking the inverse Laplace transform, we find that the zero-state response for $t > 0$ must be

$$i(t) = 2 - 3e^{-t} + e^{-3t} \text{ A}.$$

Now, for the zero-input response we suppress input sources $\hat{F}_1(s)$ and $\hat{F}_2(s)$, and then use the superposition method once again to calculate the circuit response to initial value-sources $\frac{1}{3}v(0^-)$ and $i(0^-)$. First, due to the current source $\frac{1}{3}v(0^-)$, we have, using current division,

$$\hat{I}(s) = -\frac{1}{3}v(0^-)\frac{\frac{3}{s}}{\frac{3}{s} + (s+4)} = \frac{-v(0^-)}{3 + s(s+4)}.$$

Next, due to the voltage source $i(0^-)$, we have, dividing by the total impedance around the loop,

$$\hat{I}(s) = \frac{i(0^-)}{\frac{3}{s} + (s+4)} = \frac{si(0^-)}{3 + s(s+4)}.$$

Thus, the total circuit response to initial-value sources is

$$\hat{I}(s) = \frac{si(0^-) - v(0^-)}{s^2 + 4s + 3} = \frac{s-1}{(s+1)(s+3)} = \frac{-1}{s+1} + \frac{2}{s+3}$$

using $v(0^-) = 1$ V and $i(0^-) = f_2(0^-) = 1$ A. Taking the inverse Laplace transform, we find that the zero-input response for $t > 0$ must be

$$i(t) = -e^{-t} + 2e^{-3t} \text{ A.}$$

Naturally, the zero-state and zero-input responses calculated above can be summed to obtain the expression for the total response of the system for $t > 0$.

11.5 LTIC System Combinations

We close this chapter with a discussion of higher-order LTIC systems composed of interconnections of lower-order subsystems. In particular, we consider cascade, parallel, and feedback configurations. For each of these configurations, we determine how the transfer function of the overall system is related to the transfer functions of the subsystems. Cascade and parallel configurations are used for the implementation of higher-order filters, whose design is the subject of Chapter 12. The feedback configuration is of fundamental importance in the field of control systems.

11.5.1 Cascade configuration

Higher-order systems frequently are composed of cascaded subsystems. Figure 11.15 shows a cascade of k LTIC systems with transfer functions $\hat{H}_i(s)$, $1 \le i \le k$. Given a causal input $f(t) \leftrightarrow \hat{F}(s)$, the first-stage output $h_1(t) * f(t) \leftrightarrow \hat{H}_1(s)\hat{F}(s)$ is also the second-stage input. Therefore, the second-stage output is $h_2(t) * (h_1(t) * f(t)) \leftrightarrow \hat{H}_2(s)\hat{H}_1(s)\hat{F}(s)$. Through similar reasoning, the overall system output is described by

$$y(t) = h(t) * f(t) \leftrightarrow \hat{Y}(s) = \hat{H}(s)\hat{F}(s)$$

with

$$h(t) \equiv h_1(t) * h_2(t) * \cdots * h_k(t) \leftrightarrow \hat{H}(s) \equiv \hat{H}_1(s)\hat{H}_2(s)\cdots\hat{H}_k(s).$$

The order of the cascade system $h(t) \leftrightarrow \hat{H}(s)$ is less than or equal to the sum of the orders of its k components—the order can be less if there are pole-zero cancellations in the product of the $\hat{H}_i(s)$ that forms $\hat{H}(s)$.

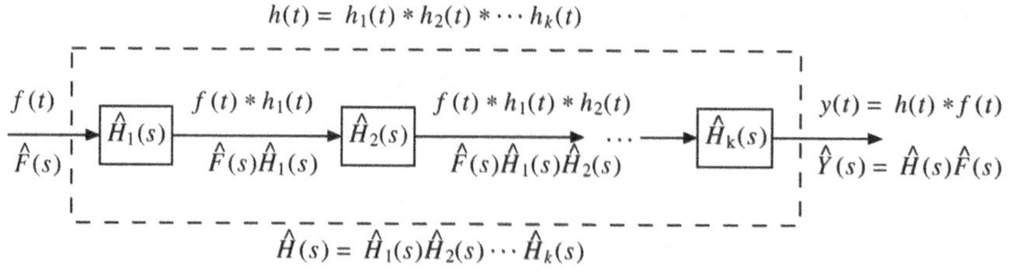

$$h(t) = h_1(t) * h_2(t) * \cdots h_k(t)$$

Figure 11.15 A cascade system configuration.

Example 11.36

The 3rd-order system

$$\hat{H}(s) = \frac{1}{(s+1)(s+1-j)(s+1+j)}$$

is to be realized by cascading systems $\hat{H}_1(s)$ and $\hat{H}_2(s)$. Determine $h_1(t) \leftrightarrow \hat{H}_1(s)$ and $h_2(t) \leftrightarrow \hat{H}_2(s)$ in such a way that the transfer function of each subsystem has real-valued coefficients.

Solution   Since

$$\hat{H}(s) = \frac{1}{s+1} \cdot \frac{1}{(s+1-j)(s+1+j)},$$

the system can be realized by cascading

$$h_1(t) = e^{-t}u(t) \leftrightarrow \hat{H}_1(s) = \frac{1}{s+1}$$

(e.g., an RC circuit) with a 2nd-order system

$$h_2(t) = e^{-t}\sin(t)u(t) \leftrightarrow \hat{H}_2(s) = \frac{1}{(s+1-j)(s+1+j)}$$

$$= \frac{1}{s^2+2s+2}.$$

Example 11.37

An LTIC system $\hat{H}_1(s) = \frac{s}{s+1}$ is cascaded with the LTIC system $h_2(t) = \delta(t) + e^{-2t}u(t)$ to implement a system $\hat{H}(s) = \hat{H}_1(s)\hat{H}_2(s)$. Discuss the stability of system $\hat{H}(s)$ and examine $H(\omega) = \hat{H}(j\omega)$ to determine the kind of filter implemented by $\hat{H}(s)$.

Solution   Since

$$h_2(t) = \delta(t) + e^{-2t}u(t) \leftrightarrow \hat{H}_2(s) = 1 + \frac{1}{s+2} = \frac{s+3}{s+2},$$

it follows that

$$\hat{H}(s) = \hat{H}_1(s)\hat{H}_2(s) = \frac{s}{s+1} \cdot \frac{s+3}{s+2} = \frac{s(s+3)}{(s+1)(s+2)}.$$

Because both poles of $\hat{H}(s)$ are in the LHP, the system is BIBO stable. Furthermore, both components of the system have only LHP system poles and therefore they are asymptotically stable. The system frequency response

$$\hat{H}(j\omega) = \frac{j\omega(j\omega+3)}{(j\omega+1)(j\omega+2)}$$

vanishes as $\omega \to 0$ and approaches unity as $\omega \to \infty$. Hence, the system implements a type of high-pass filter.

Cascading of subsystems is a straightforward idea, but in practical implementations we must be sure that stage $i + 1$ does not affect the properties of (change the transfer function of) the preceding stage i. For instance, cascading two RC circuits does not produce a system having a transfer function that is the product of the two original transfer functions, unless a buffer stage is inserted between the two to prevent the loading of (drawing current from) the front-end circuit by the second stage. On the other hand, active filters that use op amps readily can be cascaded, because an op amp in a later stage draws essentially no current from the preceding stages and, therefore, does not alter the transfer functions of those stages.

11.5.2 Parallel configuration

Figure 11.16 shows a parallel combination of k LTIC subsystems with transfer functions $\hat{H}_i(s)$, $1 \le i \le k$. In this configuration, a single input $f(t) \leftrightarrow \hat{F}(s)$ is applied in parallel to the input of each subsystem (represented by the fork on the left), and the outputs $h_i(t) * f(t) \leftrightarrow \hat{H}_i(s)\hat{F}(s)$ of the individual stages are summed to obtain an overall system output of

$$y(t) = h(t) * f(t) \leftrightarrow \hat{Y}(s) = \hat{H}(s)\hat{F}(s)$$

with

$$h(t) \equiv h_1(t) + \cdots + h_k(t) \leftrightarrow \hat{H}(s) \equiv \hat{H}_1(s) + \cdots + \hat{H}_k(s).$$

The order of the parallel system $h(t) \leftrightarrow \hat{H}(s)$ is less than or equal to the sum of the orders of its k subcomponents. (The order can be less if pole-zero cancellations occur in the sum of the $\hat{H}_i(s)$ that forms $\hat{H}(s)$.)

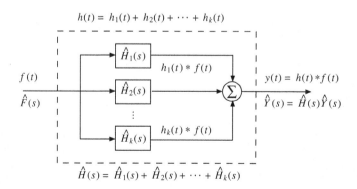

Figure 11.16 A parallel system configuration.

Example 11.38

A 3rd-order LTIC system

$$\hat{H}(s) = \frac{s(s+1)}{(s+3)(s^2+s+1)}$$

is to be constructed using a parallel configuration with a 1st-order system, $\hat{H}_1(s)$, and a 2nd-order system, $\hat{H}_2(s)$. Identify possible choices for $\hat{H}_1(s)$ and $\hat{H}_2(s)$ and discuss their properties.

Solution   Writing $\hat{H}(s)$ in a PFE, and then recombining the two terms involving complex poles, we have

$$\hat{H}(s) = \frac{s(s+1)}{(s+3)(s^2+s+1)} = \frac{1}{7}\frac{6}{s+3} + \frac{1}{7}\frac{s-2}{s^2+s+1}.$$

Hence, a possible parallel configuration has

$$\hat{H}_1(s) = \frac{6/7}{s+3},$$

and

$$\hat{H}_2(s) = \frac{1}{7}\frac{s-2}{s^2+s+1}.$$

$\hat{H}_1(s)$ is a low-pass system. $\hat{H}_2(s)$ is another low-pass system and is 2nd-order. However, the overall system behavior is bandpass since the DC outputs of $\hat{H}_1(s)$ and $\hat{H}_2(s)$ are equal in magnitude but opposite in sign ($\hat{H}_1(0) = \frac{2}{7}$ and $\hat{H}_2(0) = -\frac{2}{7}$).

In this example, we also could have expressed $\hat{H}(s)$ in the more standard PFE, as a sum of three 1st-order terms. In that case the transfer functions of two of the subsystems would have complex coefficients. Such filter sections can be implemented (see Exercise Problem 11.28), but the implementation is more complicated than for transfer functions having real-valued coefficients.

As just one example of a real-world parallel system, high-quality loudspeakers are configured and implemented as parallel systems with woofer, mid-range, and tweeter subcomponents, responding to the low-, mid-, and high-frequency bands of the same audio signal input, $f(t)$, respectively.

11.5.3 Feedback configuration

Figure 11.17 shows a feedback system with subcomponents $h_1(t) \leftrightarrow \hat{H}_1(s)$ and $h_2(t) \leftrightarrow \hat{H}_2(s)$. The input of system $h_1(t) \leftrightarrow \hat{H}_1(s)$ is

$$f(t) + h_2(t) * y(t) \leftrightarrow \hat{F}(s) + \hat{H}_2(s)\hat{Y}(s),$$

and therefore its output is

$$y(t) = h_1(t) * (f(t) + h_2(t) * y(t)) \leftrightarrow \hat{Y}(s) = \hat{H}_1(s)(\hat{F}(s) + \hat{H}_2(s)\hat{Y}(s)).$$

Moving all terms involving $\hat{Y}(s)$ to the left-hand side gives

$$(1 - \hat{H}_1(s)\hat{H}_2(s))\hat{Y}(s) = \hat{H}_1(s)\hat{F}(s).$$

Thus, the transfer function of the feedback configuration is obtained as

$$\hat{H}(s) = \frac{\hat{Y}(s)}{\hat{F}(s)} = \frac{\hat{H}_1(s)}{1 - \hat{H}_1(s)\hat{H}_2(s)}.$$

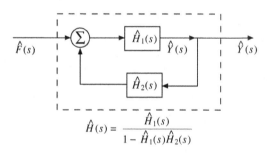

$$\hat{H}(s) = \frac{\hat{H}_1(s)}{1 - \hat{H}_1(s)\hat{H}_2(s)}$$

Figure 11.17 A feedback system

Example 11.39
Consider a LTIC feedback system with $\hat{H}_1(s) = \frac{1}{s}$ and $\hat{H}_2(s) = A$, where A is a real constant. Determine the conditions under which the feedback system is BIBO stable.

Solution

$$\hat{H}(s) = \frac{\hat{H}_1(s)}{1 - \hat{H}_1(s)\hat{H}_2(s)} = \frac{\frac{1}{s}}{1 - \frac{1}{s}A} = \frac{1}{s - A}.$$

The pole of $\hat{H}(s)$, at $s = A$, is in the left-half plane only for $A < 0$; therefore the system is BIBO stable if and only if $A < 0$.

Example 11.39 illustrates that a marginally stable system $\hat{H}_1(s) = \frac{1}{s}$ (which is not BIBO stable) can be stabilized by "feeding back" an amount $Ay(t)$ of the system output, $y(t)$, to the system input. This is an example of "negative feedback" stabilization, since for stability $A < 0$ is required. Positive feedback ($A > 0$) leads to an unstable configuration. Notice that for A very small and negative, the transfer functions $\hat{H}(s)$ and $\hat{H}_1(s)$ can be nearly identical, and yet the former is stable and has a frequency response, whereas the latter does not.

Example 11.40

Consider a LTIC feedback system with $\hat{H}_1(s) = \frac{1}{s+1}$ and $\hat{H}_2(s) = A$, where A is a real constant. Note that both $\hat{H}_1(s)$ and $\hat{H}_2(s)$ are stable systems. Determine the values of A for which the feedback system itself is unstable.

Solution

$$\hat{H}(s) = \frac{\hat{H}_1(s)}{1 - \hat{H}_1(s)\hat{H}_2(s)} = \frac{\frac{1}{s+1}}{1 - \frac{1}{s+1}A} = \frac{1}{s+1-A}.$$

The pole at $s = A - 1$ is in the left-half plane if and only if $A - 1 < 0$ (i.e., $A < 1$). Hence, for $A \geq 1$ the feedback system is unstable, even though its components are not.

Notice that instability in Example 11.40 occurs once again with positive feedback ($A \geq 1$). However, in Example 11.40 negative feedback is not necessary for system stability (e.g., with $A = 0.5$, the system is stable) because $\hat{H}_1(s)$ is stable to begin with.

Example 11.41

For a LTIC feedback system with

$$\hat{H}_1(s) = \frac{1}{s + 1}$$

and

$$\hat{H}_2(s) = \frac{1}{s + K},$$

where K is real-valued, determine the range of K for which the system is BIBO stable.

Solution

$$\hat{H}(s) = \frac{\hat{H}_1(s)}{1 - \hat{H}_1(s)\hat{H}_2(s)} = \frac{\frac{1}{s+1}}{1 - \frac{1}{s+1}\frac{1}{s+K}} = \frac{s + K}{(s + 1)(s + K) - 1}$$

$$= \frac{s + K}{s^2 + (1 + K)s + (K - 1)}.$$

The poles of $\hat{H}(s)$ are at

$$s = \frac{-(1+K) \pm \sqrt{(1+K)^2 - 4(K-1)}}{2} = \frac{-(1+K) \pm \sqrt{K^2 - 2K + 5}}{2}.$$

It is easy to verify that $K^2 - 2K + 5 > 0$, for all K, which ensures that both poles are real-valued. For BIBO stability, both poles must lie in the LHP, requiring that

$$-(1+K) \pm \sqrt{K^2 - 2K + 5} < 0$$

or

$$\pm\sqrt{K^2 - 2K + 5} < 1 + K,$$

implying

$$K^2 - 2K + 5 < K^2 + 2K + 1.$$

Thus,

$$K > 1$$

for stability.

Example 11.42

Determine the transfer function $\hat{H}(s)$ of the system shown in Figure 11.18.

Solution   Labeling the output of the bottom adder as $\hat{W}(s)$, we have

$$\hat{W}(s) = \hat{H}_3(s)\hat{Y}(s) + \hat{H}_4(s)\hat{F}(s).$$

The system output then can be written as

$$\hat{Y}(s) = \hat{H}_1(s)[\hat{F}(s) + \hat{H}_2(s)\hat{W}(s)],$$

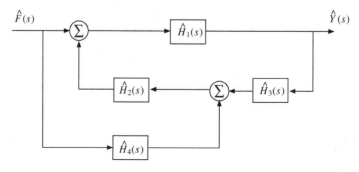

Figure 11.18   A system with parallel and feedback connections.

so that substituting for $\hat{W}(s)$ gives

$$\hat{Y}(s) = \hat{H}_1(s)[\hat{F}(s) + \hat{H}_2(s)(\hat{H}_3(s)\hat{Y}(s) + \hat{H}_4(s)\hat{F}(s))].$$

Thus

$$\hat{Y}(s) - \hat{H}_1(s)\hat{H}_2(s)\hat{H}_3(s)\hat{Y}(s) = \hat{H}_1(s)\hat{F}(s) + \hat{H}_1(s)\hat{H}_2(s)\hat{H}_4(s)\hat{F}(s),$$

and the system transfer function is

$$\hat{H}(s) = \frac{\hat{Y}(s)}{\hat{F}(s)} = \frac{\hat{H}_1(s)(1 + \hat{H}_2(s)\hat{H}_4(s))}{1 - \hat{H}_1(s)\hat{H}_2(s)\hat{H}_3(s)}.$$

EXERCISES

11.1 Determine the Laplace transform $\hat{F}(s)$, and the ROC, for the following signals $f(t)$. In each case identify the corresponding pole locations where $|\hat{F}(s)|$ is not finite.

  (a) $f(t) = u(t) - u(t - 8)$.

  (b) $f(t) = u(t) - u(t + 8)$.

  (c) $f(t) = u(t + 8)$.

  (d) $f(t) = 6$.

  (e) $f(t) = \text{rect}(\frac{t-4}{2})$.

  (f) $f(t) = \text{rect}(\frac{t+8}{3})$.

  (g) $f(t) = te^{2t}u(t)$.

  (h) $f(t) = te^{2t}u(t - 2)$.

  (i) $f(t) = 2te^{2t}$.

  (j) $f(t) = te^{-4t} + \delta(t) + u(t - 2)$.

  (k) $f(t) = e^{2t}\cos(t)u(t)$.

11.2 For each of the following Laplace transforms $\hat{F}(s)$, determine the inverse Laplace transform $f(t)$.

  (a) $\hat{F}(s) = \frac{s+3}{(s+2)(s+4)}$.

  (b) $\hat{F}(s) = \frac{s^2}{(s+2)(s+4)}$.

  (c) $\hat{F}(s) = \frac{1}{s(s-5)^2}$.

  (d) $\hat{F}(s) = \frac{s^2+2s+1}{(s+1)(s+2)}$.

  (e) $\hat{F}(s) = \frac{s}{s^2+2s+5}$.

  (f) $\hat{F}(s) = \frac{s^3}{s^2+4}$.

11.3   Sketch the amplitude response $|H(\omega)|$ and determine the impulse response $h(t)$ of the LTIC systems having the following transfer functions:

(a) $\hat{H}(s) = \frac{s}{s+10}$.

(b) $\hat{H}(s) = \frac{10}{s+1}$.

(c) $\hat{H}(s) = \frac{s}{s^2+3s+2}$.

(d) $\hat{H}(s) = \frac{1}{s+1}e^{-s}$.

11.4   Determine the zero-state responses of the systems defined in Problem 11.3 to a causal input $f(t) = u(t)$. Use $y(t) = h(t) * f(t)$ or find the inverse Laplace transform of $\hat{Y}(s) = \hat{H}(s)\hat{F}(s)$, whichever is more convenient.

11.5   Repeat Problem 11.4 with $f(t) = e^{-t}u(t)$.

11.6   Given the frequency response $H(\omega)$, below, determine the system transfer function $\hat{H}(s)$ and impulse response $h(t)$.

(a) $H(\omega) = \frac{j\omega}{(1+j\omega)(2+j\omega)}$.

(b) $H(\omega) = \frac{j\omega}{1-\omega^2+j\omega}$.

11.7   Determine whether the LTIC systems with the following transfer functions are BIBO stable and explain why or why not.

(a) $\hat{H}_1(s) = \frac{s^3+1}{(s+2)(s+4)}$.

(b) $\hat{H}_2(s) = 2 + \frac{s}{(s+1)(s-2)}$.

(c) $\hat{H}_3(s) = \frac{s^2+4s+6}{(s+1+j6)(s+1-j6)}$.

(d) $\hat{H}_4(s) = \frac{1}{s^2+16}$.

(e) $\hat{H}_5(s) = \frac{s-2}{s^2-4}$.

11.8   For each unstable system in Problem 11.7 give an example of a bounded input that causes an unbounded output.

11.9   Given

$$s\hat{F}(s) - f(0^-) = \int_{0^-}^{\infty} \frac{df}{dt}e^{-st}dt = f(0^+) - f(0^-) + \int_{0^+}^{\infty} \frac{df}{dt}e^{-st}dt,$$

and assuming that Laplace transforms of $f(t)$ and $f'(t)$ exist, show that

(a) $\lim_{s \to 0} s\hat{F}(s) = f(\infty)$.

(b) $\lim_{s \to \infty} s\hat{F}(s) = f(0^+)$.

11.10   Consider the LTIC circuit shown in Figure 11.7a. What is the zero-state response $x(t)$ if the input is $f(t) = u(t)$? Hint: Use s-domain voltage division to relate $\hat{X}(s)$ to $\hat{F}(s)$ in Figure 11.7b.

11.11   Repeat Problem 11.10 for (a) $f(t) = \delta(t)$, and (b) $f(t) = tu(t)$.

11.12   Consider the following circuit with $C > 0$:

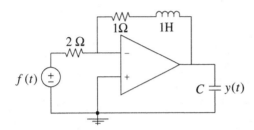

(a)   Determine the zero-state response $y(t)$ if $f(t) = te^{-t}u(t)$.

(b)   Determine the zero-state response $y(t)$ if $f(t) = tu(t)$.

11.13   Determine the transfer functions $\hat{H}(s)$ and the zero-state responses for LTIC systems described by the following ODEs:

(a) $\frac{d^2y}{dt^2} + 3\frac{dy}{dt} + 2y(t) = e^{3t}u(t)$.

(b) $\frac{d^2y}{dt^2} + y(t) = \cos(2t)u(t)$.

(c) $\frac{d^2y}{dt^2} + y(t) = \cos(t)u(t)$.

11.14   If an LTIC system has the transfer function $\hat{H}(s) = \frac{\hat{Y}(s)}{\hat{F}(s)} = \frac{s+1}{(s+2)^2}$, determine a linear ODE that describes the relationship between the system input $f(t)$ and the output $y(t)$.

11.15   Determine the characteristic polynomial $P(s)$, characteristic poles, characteristic modes, and the zero-input solution for each of the LTIC systems described below.

(a) $\frac{d^2y}{dt^2} + 2\frac{dy}{dt} - 8y(t) = 6f(t)$, $y(0^-) = 0$, $y'(0^-) = 1$.

(b) $\frac{d^3y}{dt^3} + 2\frac{d^2y}{dt^2} - \frac{dy}{dt} - 2y(t) = f(t)$, $y(0^-) = 1$, $y'(0^-) = 1$, $y''(0^-) = 0$.

(c) $\frac{d^2y}{dt^2} + 2\frac{dy}{dt} + y(t) = 2f(t)$, $y(0^-) = 1$, $y'(0^-) = 1$.

11.16   (a)   Take the Laplace transform of the following ODE to determine $\hat{Y}(s)$ assuming $f(t) = u(t)$, $y(0^-) = 1$, and $y'(0^-) = 0$. Determine $y(t)$ for $t > 0$ by taking the inverse Laplace transform of $\hat{Y}(s)$.

$$\frac{d^2y}{dt^2} + 5\frac{dy}{dt} + 4y(t) = \frac{df}{dt} + 2f(t).$$

(b) Repeat (a) for $y(0^-) = 0$ and

$$3\frac{dy}{dt} + 6y(t) = \delta(t).$$

(c) Repeat (a) for $y(0^-) = 0$ and

$$\frac{dy}{dt} - y(t) = e^{-t}u(t).$$

11.17 The transfer function of a particular LTIC system is $\hat{H}(s) = \frac{\hat{Y}(s)}{\hat{F}(s)} = \frac{4}{s-2}$. Is the system asymptotically stable? Explain. Is the system BIBO stable? Explain.

11.18 What are the resonance frequencies in a system with the transfer function

$$\hat{H}(s) = \frac{s}{(s+1)(s^2+4)(s^2+25)}?$$

Is the system marginally stable? BIBO-stable? Explain.

11.19 Determine the zero-state response $y(t) = h(t) * f(t)$ of the marginally stable system

$$\hat{H}(s) = \frac{1}{(s^2+4)(s^2+9)}$$

to an input $f(t) = \cos(2t)u(t)$.

11.20 **(a)** Determine the transfer function and characteristic modes of the circuit shown below assuming that $C_a = C_b = 1/2\,\text{F}$, $R = 1.5\,\Omega$, and $L = 1/2\,\text{H}$.

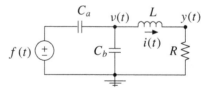

(b) Given that $v(0^-) = 1\,\text{V}$ and $i(0^-) = 0.5\,\text{A}$, and using the element values given in part (a), determine $y(t)$ for $t > 0$ in the circuit:

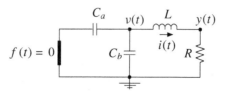

11.21 Consider the following circuit, which is in DC steady-state until the switch is opened at $t = 0$:

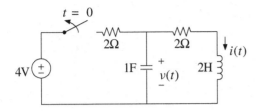

(a) Determine $i(0^-)$ and $v(0^-)$.

(b) Determine the characteristic modes of the circuit to the right of the switch for $t > 0$.

(c) Determine $i(t)$ for $t > 0$.

11.22 In the circuit:

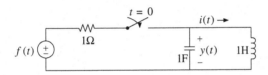

(a) Determine the transfer function $\hat{H}(s) = \frac{\hat{Y}(s)}{\hat{F}(s)}$ for $t > 0$.

(b) Determine the zero-state response for $t > 0$ if $f(t) = e^{3t}$.

(c) Determine the zero-input response for $t > 0$ if $y(0^-) = 1$ V and $i(0^-) = 0$.

11.23 Consider the circuit:

(a) Show that the transfer function of the circuit for $t > 0$ is $\hat{H}(s) = \frac{\hat{Y}(s)}{\hat{F}(s)} = \frac{s}{4s^2+5s+2}$.

(b) What are the characteristic modes of the circuit?

(c) Determine $y(t)$ for $t > 0$ if $f(t) = 1$ V, $y(0^-) = 1$ V, and $i(0^-) = 0$.

11.24   The system shown below can be implemented as a cascade of two 1st-order systems $\hat{H}_1(s)$ and $\hat{H}_2(s)$. Identify the possible forms of $\hat{H}_1(s)$ and $\hat{H}_2(s)$.

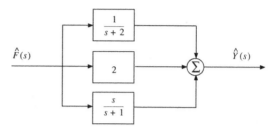

11.25   Determine the impulse response $h(t)$ of the system shown below. Also determine whether the system is BIBO stable.

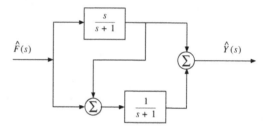

11.26   Determine the transfer function $\hat{H}(s)$ of the system shown below. Also determine whether the system is BIBO stable.

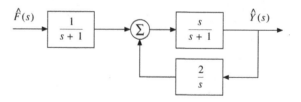

11.27   Determine the transfer function $\hat{H}(s)$ of the system shown below and determine for which K the system is BIBO stable if

(a) $\hat{H}_f(s) = \frac{1}{s+K}$.

(b) $\hat{H}_f(s) = s + K$.

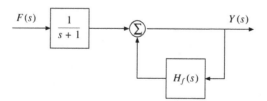

11.28 Consider a system with transfer function $\hat{H}(s) = \frac{2}{s+1-j3}$. Draw a block diagram that implements this transfer function. Individual blocks in the diagram may denote addition of real-valued signals, amplification by real values, and integration of real-valued signals. Your diagram should contain no complex numbers. Hints: 1) The transfer function $\hat{H}(s) = \frac{a}{s+b}$ corresponds in the time domain to the equation $\frac{dy}{dt} + by(t) = af(t)$ or, equivalently, $y(t) = -b \int y(\tau)d\tau + a \int f(\tau)d\tau$. So, it is easy to draw a block diagram of this system. 2) In this homework exercise, you may assume that $f(t)$ is real-valued, but $y(t)$ is complex-valued, which means that $y(t)$ is a *pair of real-valued* signals. Your diagram will need to show $y_R(t)$ and $y_I(t)$, the real and imaginary parts of $y(t)$, separately. 3) Multiplication of a complex signal by a complex number involves four real multiplications. Addition of complex signals requires two real additions.

12

Analog Filters and Low-Pass Filter Design

Ideal filters; practical low-order filters; 2nd-order filters and quality factor Q; Butterworth filter design

In earlier chapters we focused our efforts on the *analysis* of LTIC circuits and systems. For example, given a circuit, we learned how to find its transfer function and its corresponding frequency response. In this final chapter, we consider the other side of the coin, the problem of *design*. For instance, suppose we wish to create a filter whose frequency response approximates that of an ideal low-pass shape with a particular cutoff frequency. How can this be accomplished? The solution lies in answering two questions. First, how do we choose the coefficients in the transfer function? And, second, given the coefficients of a transfer function, how do we build a circuit having this transfer function (and, therefore, approximating the desired frequency response)?

We will tackle these questions in reverse order, but before doing so, in Section 12.1 we first describe the desired frequency response characteristics for ideal filters. Section 12.2 then examines how closely the ideal characteristics can be approximated by 1st- and 2nd-order filters and describes how to implement such low-order filters using op-amps. This section also includes discussion of a system design parameter known as Q and its relation to frequency response and pole locations of 2nd-order systems. Section 12.3 tackles the important problem of designing the filter transfer function for higher-order filters, describing a method for designing one common class of filters called Butterworth. The higher-order filters then can be implemented as cascade or parallel op-amp first- and second-order circuits.

12.1   Ideal Filters: Distortionless and Nondispersive

Many real-world applications require filters that pass signal content in one frequency band, with no distortion (other than a possible amplitude scaling and signal delay), and that attenuate (reduce in amplitude) all signal content outside of this frequency band. The sets of frequencies where signal content is passed and attenuated are called the *passband* and *stopband*, respectively.

Figures 12.1a through 12.1c depict the magnitude and angle variations of three such ideal frequency response functions $H(\omega) = \hat{H}(j\omega)$. Linear systems with the frequency response curves shown in Figures 12.1a through 12.1c are known as *distortionless* low-pass, band-pass, and high-pass filters, respectively.

To understand this terminology, consider the low-pass filter with frequency response

$$H(\omega) = K\operatorname{rect}(\frac{\omega}{2\Omega})e^{-j\omega t_o},$$

represented in Figure 12.1a. This filter responds to a low-pass input $f(t) \leftrightarrow F(\omega)$ of bandwidth less than Ω with an output $y(t) = Kf(t - t_o)$, which is a time-delayed and amplitude-scaled but *undistorted* replica of the input. In fact, all three filters shown in Figures 12.1a through 12.1c produce delayed and scaled, but undistorted, copies of their inputs within their respective passbands because of the filters' flat-amplitude responses and linear-phase characteristics.

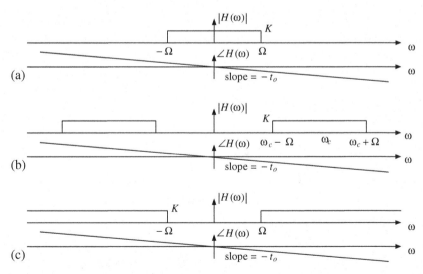

Figure 12.1   Sketches of the amplitude and phase responses of *distortionless* filters: (a) low-pass, (b) band-pass, and (c) high-pass. Note that each filter has a flat amplitude $|H(\omega)|$ within its passband (frequency band with nonzero $|H(\omega)|$) and makes step transitions from passband to stopband (where $|H(\omega)| = 0$).

It so happens that none of these distortionless filters can be exactly implemented as circuits, because each of the filters is associated with a noncausal impulse response $h(t) \leftrightarrow H(\omega)$. For instance, the impulse response corresponding to the distortionless low-pass filter is

$$h(t) = \frac{K\Omega}{\pi}\text{sinc}(\Omega(t - t_o)) \leftrightarrow H(\omega) = K\text{rect}(\frac{\omega}{2\Omega})e^{-j\omega t_o}.$$

No physical circuit can have such an impulse response that begins prior to $t = 0$.

Distortionless filters are noncausal

Given that distortionless filters are unrealizable, we must accept some amount of signal distortion in real-world filtering operations. Fortunately, practical analog filters can be designed to approximate the distortionless filter characteristics of Figure 12.1 as closely as we like, and such filters can be built using analog circuits. Practical filter design involves determining a realizable and stable (with only LHP poles) transfer function $\hat{H}(s)$ that provides

Practical filter design criteria

(1) As flat an amplitude response $|\hat{H}(j\omega)|$ as "needed" within a desired passband,
(2) As fast a "decay" of $|\hat{H}(j\omega)|$ as needed outside the passband, and
(3) A phase response $\angle\hat{H}(j\omega)$ that is as linear as needed within the desired passband.

The "needs" may vary from application to application, and filter design may require compromises, because the three design goals above generally are in conflict to some degree. Tighter specifications related to the design goals generally lead to the requirement of a filter transfer function $\hat{H}(s)$ having higher order, which in turn requires a more complex circuit implementation.

Phase linearity

As indicated above, we usually seek phase linearity (corresponding to signal delay, but no distortion) only within the passband. This is sufficient, because high-quality low-pass, band-pass, and high-pass filters pass very little signal energy within the stopband, and therefore distortion of signal components lying in the stopband is of little concern. Moreover, as we shall see below, we need not require that the linear phase characteristic pass through the origin (i.e., $\angle H(\omega) = -\omega t_o$).

Overall, there are two types of phase linearity that we will find acceptable. The first is true linear phase, where $\angle H(\omega) = -\omega t_o$, for $-\infty < \omega < \infty$, corresponding to a distortionless filter introducing signal delay t_o. The second form of linear phase requires that $\angle H(\omega)$ be linear only for ω within the passband of the filter (and the phase characteristic need not pass through the origin). Such filters are called *nondispersive* (rather than distortionless). To illustrate the difference, let

$$h_1(t) \leftrightarrow H_1(\omega) = \text{rect}(\frac{\omega}{2\Omega})e^{-j\omega t_o}$$

denote the impulse response of the distortionless low-pass filter having magnitude and phase responses depicted in Figure 12.1a ($K = 1$ case) and repeated in Figure 12.2a. Let

$$h_2(t) = h_1(t)\cos(\omega_c t) \leftrightarrow H_2(\omega) = \frac{1}{2}H_1(\omega - \omega_c) + \frac{1}{2}H_1(\omega + \omega_c)$$

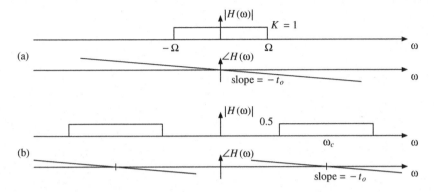

Figure 12.2 (a) A distortionless low-pass filter, and (b) a *nondispersive* band-pass filter.

(remember the Fourier modulation property) denote a band-pass filter having magnitude and phase responses depicted in Figure 12.2b. Clearly, $H_2(\omega)$ is a band-pass filter with a linear phase variation within its passband, but the phase curve in Figure 12.2b is different from the phase curve of the distortionless band-pass filter shown in Figure 12.1b. We refer to the filter in Figure 12.2b as nondispersive. Now, how is the output signal of a nondispersive filter different from the output of a distortionless filter?

Nondispersive filter

To answer this question, consider the response of the nondispersive band-pass filter

$$H(\omega) = \text{rect}(\frac{\omega - \omega_c}{2\Omega})e^{-j(\omega-\omega_c)t_o} + \text{rect}(\frac{\omega + \omega_c}{2\Omega})e^{-j(\omega+\omega_c)t_o}$$

depicted in Figure 12.3 to an AM signal

$$f(t) = m(t)\cos(\omega_c t) \leftrightarrow F(\omega) = \frac{1}{2}M(\omega - \omega_c) + \frac{1}{2}M(\omega + \omega_c),$$

where $m(t) \leftrightarrow M(\omega)$ is a low-pass signal with the triangular Fourier transform shown in the same figure. Also shown in the figure is the Fourier transform of the filter output,

$$Y(\omega) = H(\omega)F(\omega) = \frac{1}{2}e^{-j(\omega-\omega_c)t_o}M(\omega - \omega_c) + \frac{1}{2}e^{-j(\omega+\omega_c)t_o}M(\omega + \omega_c).$$

You should be able to confirm that this expression is the Fourier transform of the function $m(t - t_o)\cos(\omega_c t)$. Therefore, the nondispersive filter responds to the band-pass AM input $f(t) = m(t)\cos(\omega_c t)$ with the output $y(t) = m(t - t_o)\cos(\omega_c t)$, which is different from $f(t - t_o)$, which would be the response of a distortionless band-pass filter to the same input signal. Evidently, a nondispersive[1] filter delays only the envelope of an AM signal, while a distortionless band-pass filter delays the entire signal.

---

[1] The term *nondispersive* refers to the fact that the signal envelope is not distorted when the filter phase variation is a linear function of ω across the passband. Deviations from phase linearity generally cause

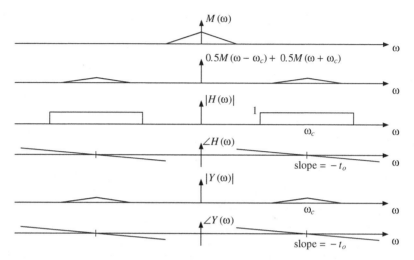

Figure 12.3 Frequency-domain representations of a low-pass signal $m(t) \leftrightarrow M(\omega)$, band-pass AM signal $m(t)\cos(\omega_c t) \leftrightarrow 0.5M(\omega - \omega_c) + 0.5M(\omega + \omega_c)$, a nondispersive bandpass filter $h(t) \leftrightarrow H(\omega)$, and the filter response $y(t) \leftrightarrow Y(\omega)$ with $y(t) = h(t) * m(t)\cos(\omega_c t) = m(t - t_0)\cos(\omega_c t)$ due to the AM input.

Since both types of filters with linear phase variation preserve the envelope integrity of the input, they both are compatible with the filter design goals stated above.

12.2 1st- and 2nd-Order Filters

Suppose we wish to design a filter having a frequency response approximating the ideal characteristics outlined in the previous section (flat response in the passband, steep drop-off in the stopband, and linear phase across the passband). In particular, suppose we wish to design a low-pass filter.

We might begin by contemplating the simple 1st-order RC circuit shown in Figure 12.4, which is a replica of Figure 5.1. Obviously, the amplitude and phase characteristics of $H(\omega)$, shown in Figures 12.4c and 12.4d, are far from being close approximations of the ideal features of distortionless low-pass filters. Although $H(\omega)$ may be "good enough" in certain situations, most applications would require an amplitude response that is more rectangular or a phase response that is more linear.

If a 1st-order circuit is not up to the job, then we must explore higher-order circuit options, i.e., more complex circuits with higher-order $\hat{H}(s)$. If $\hat{H}(s)$ is a high-order transfer function, then it can be realized in straightforward fashion by rewriting $\hat{H}(s)$

"spreading" of the envelope, which is known as dispersion. Also, the terms *phase delay* and *group delay* often are used to refer to the delay imposed by a filter on the carrier and envelope components, respectively, of AM signals. In distortionless filters, phase and group delays are equal; in nondispersive filters they are different.

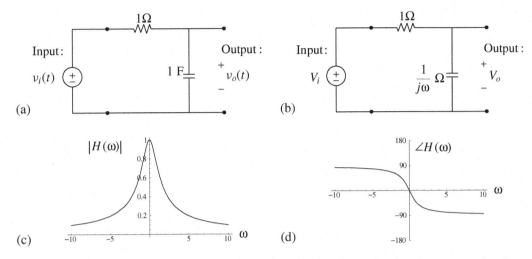

Figure 12.4 (a) A 1st-order low-pass filter circuit, (b) its phasor representation, (c) the amplitude response $|H(\omega)|$, and (d) phase response $\angle H(\omega)$ (in degrees).

as a product of second-order (and possibly first-order) sections and then implementing $\hat{H}(s)$ as a cascade of the corresponding low-order sections. The actual implementation is accomplished most easily using a cascade of *active* op-amp filter sections, because an op-amp draws essentially no current from the remainder of the circuit (the current is supplied by the op-amp power supply). Thus, each interconnected op-amp filter section continues to function as it would in isolation, so that the transfer function of the entire cascade system is simply the product of the individual transfer functions, which is $\hat{H}(s)$, as desired.

12.2.1 Active op-amp filters

Figures 12.5a and 12.5b show two low-pass op-amp filter circuits with 1st- and 2nd-order transfer functions

$$\hat{H}_a(s) = \frac{\frac{K}{RC}}{s + \frac{1}{RC}}$$

and

$$\hat{H}_b(s) = \frac{K\omega_o^2}{s^2 + 2\alpha s + \omega_o^2},$$

respectively, where (as shown in Exercise Problems 12.1 and 12.2)

$$K = 1 + \frac{R_1}{R_2}, \quad \omega_o = \frac{1}{\sqrt{R_3 R_4 C_1 C_2}},$$

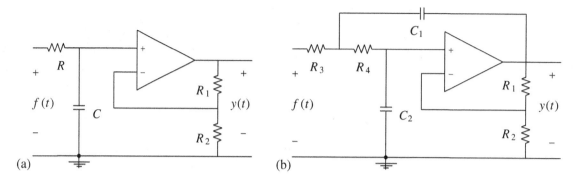

Figure 12.5 (a) 1st-order low-pass *active* filter circuit, and (b) 2nd-order low-pass active filter circuit known as the *Sallen-Key* circuit.

and

$$\alpha = \frac{1}{2R_3C_1} + \frac{1}{2R_4C_1} + \frac{1-K}{2R_4C_2}.$$

Both circuits are BIBO stable, with low-pass frequency responses

$$H_{a,b}(\omega) = \hat{H}_{a,b}(j\omega)$$

that can be controlled by capacitance and resistance values in the circuits. The op-amps in the circuits provide a DC amplitude gain $K \geq 1$ and the possibility, as discussed above, of cascading similar active filter circuits in order to assemble higher-order filter circuits having more ideal frequency response functions $H(\omega)$ than either $H_a(\omega)$ or $H_b(\omega)$.

Figure 12.6a illustrates the amplitude response curves $|H_Q(\omega)|$ of three different versions of the 2nd-order filter shown in Figure 12.5b, labeled by distinct values of $Q = \frac{\omega_o}{2\alpha}$. These responses are to be contrasted with $|H(\omega)|$ of Figure 12.6b describing the cascade combination of the same three systems. The phase response $\angle H(\omega)$ of the cascaded system also is shown in Figure 12.6b. Clearly, the cascade idea seems to provide a simple means of obtaining *practical* filter characteristics approaching those of ideal filters. In Section 12.3 we will learn a method for designing high-order transfer functions $\hat{H}(s)$ that can be implemented as a cascade of 2nd-order (and possibly first-order) sections to produce high-order op-amp filters having properties approximating the ideal. We close this section with a discussion of the Q parameter introduced above and its relation to pole locations and frequency responses of dissipative 2nd-order systems.

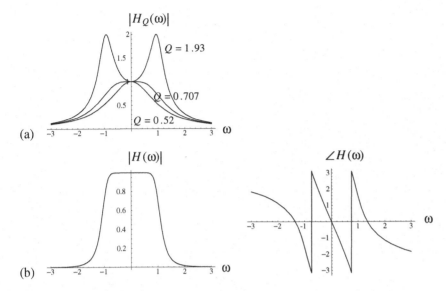

Figure 12.6 The amplitude response of (a) 2nd-order low-pass filters with $\omega_0 = 1\,\frac{rad}{s}$ and $Q \equiv \frac{\omega_0}{2\alpha} = 1.93, 0.707,$ and 0.52, and (b) a 6th-order filter obtained by cascading (multiplying) the filter curves shown in (a); phase response of the 6th-order filter also is shown in (b).

12.2.2   2nd-order systems and Q

Characteristic polynomials of stable 2nd-order LTIC systems (for which the realizable circuit of Figure 12.5b is an important example) can be expressed as

$$P(s) = (s - p_1)(s - p_2) = s^2 + 2\alpha s + \omega_o^2,$$

where p_1 and p_2 denote characteristic system poles confined to the LHP. The parameters $\omega_o = \sqrt{p_1 p_2}$ and $2\alpha \equiv -(p_1 + p_2)$ are real and positive coefficients having the ratio $\omega_o/2\alpha$ denoted as Q.

The zero-input response of such systems takes *underdamped*, *critically damped*, or *overdamped* forms illustrated in Table 12.1a, depending on the relative values of ω_o and α, known as the *undamped resonance frequency* and *damping coefficient*, respectively. Also, depending on the values of ω_o and α, the poles p_1 and p_2 either are both real valued (note the pole locations depicted in Table 12.1a) or form a complex conjugate pair (i.e., $p_2 = p_1^*$, specifically in underdamped systems). The parameter

Under-damped, critically damped, overdamped

$$Q \equiv \frac{\omega_o}{2\alpha}$$

is called the *quality factor* and it has a number of useful interpretations summarized in Table 12.1b, which are discussed below.

Quality factor Q

(a)

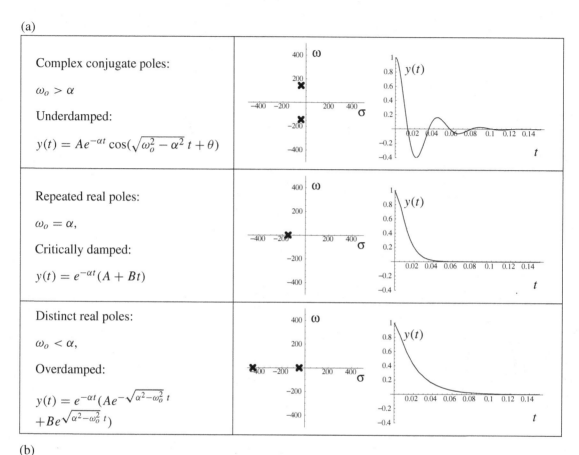

(b)

Quality factor $Q = \frac{\omega_o}{2\alpha}$	Poles at $s = -\alpha \pm \sqrt{\omega_o^2 - \alpha^2}$
$\frac{\text{center frequency}}{\text{3 dB bandwidth}}$	Band-pass filter $\hat{H}(j\omega)$
$\approx 2\pi \, \frac{\text{stored energy}}{\text{energy dissipated per period}}$	Energy interpretation
$\approx$ countable cycles of oscillations	Underdamped response

Table 12.1   (a) Zero-input response types and examples for dissipative 2nd-order systems with a characteristic polynomial $P(s) = s^2 + 2\alpha s + \omega_o^2$, where constants A, B, and θ depend on initial conditions and ×'s mark the locations of characteristic poles, and (b) quality factor with interpretations.

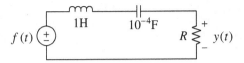

Figure 12.7   A 2nd-order series *RLC* band-pass filter.

When a 2nd-order system is used to implement a bandpass filter, such as the circuit shown in Figure 12.7 having the transfer function

$$\hat{H}(s) = \frac{\hat{Y}(s)}{\hat{F}(s)} = \frac{R}{s + \frac{1}{10^{-4}s} + R} = \frac{Rs}{s^2 + Rs + 10^4},$$

the quality factor Q turns out to be the ratio of *center-frequency* ω_o and 3-dB *bandwidth* 2α for the filter. To verify this, we first note that in the above transfer function $P(s) = s^2 + Rs + 10^4$, $2\alpha = R$, and $\omega_o = 10^2$, so that the corresponding filter frequency response is

$$H(\omega) = \hat{H}(j\omega) = \frac{2\alpha\omega}{2\alpha\omega + j(\omega^2 - \omega_o^2)}.$$

The peak amplitude response is then $|H(\omega_o)| = 1$ at a center frequency of $\omega = \omega_o$ **Center** as shown in Figure 12.8 for $Q = 5$ (underdamped), 0.5 (critically damped), and 0.3 **frequency ω_o** (overdamped). We can calculate the 3-dB bandwidth of the same filter as $\Omega = \omega_u - \omega_l$, where $\omega_{u,l}$ are 3-dB frequencies satisfying

$$|H(\omega_{u,l})|^2 = \frac{1}{2}.$$

For $\omega_u > \omega_o$ this requires $\omega_u^2 - \omega_o^2 = 2\alpha\omega_u$, implying

$$\omega_u = \alpha + \sqrt{\alpha^2 + \omega_o^2},$$

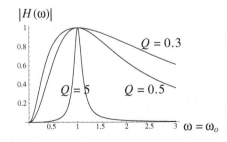

Figure 12.8   The amplitude response $|H(\omega)|$ of a 2nd-order band-pass filter with $Q = \frac{\omega_o}{2\alpha} = 5$, 0.5, and 0.3.

while for $0 < \omega_l < \omega_o$, we obtain

$$\omega_l = -\alpha + \sqrt{\alpha^2 + \omega_o^2}.$$

Thus, the 3-dB bandwidth of the filter is

$$\Omega = \omega_u - \omega_l = 2\alpha.$$

Therefore, as indicated in the first row of Table 12.1b, $Q = \frac{\omega_o}{2\alpha}$ is the center frequency to bandwidth ratio of the filter. Notice that ω_u and ω_l are not equidistant from ω_o except in the limit for very large Q.

The remaining interpretations of Q given in Table 12.1b have diagnostic value in the case of high Q (i.e., $Q \gg 1$), irrespective of system details (for electrical, mechanical, or any type of system modeled by a second-order transfer function). To understand the *energy interpretation* given in the second row of Table 12.1b, we first note that the envelope of the instantaneous power $|y(t)|^2$ for underdamped $y(t)$ (see Table 12.1a) decays with time-constant

$$\tau_d = \frac{1}{2\alpha},$$

while

$$T = \frac{2\pi}{\sqrt{\omega_o^2 - \alpha^2}} \approx \frac{2\pi}{\omega_o}$$

is the oscillation period for $y(t)$ for high Q. Consequently we have

$$Q = \frac{\omega_o}{2\alpha} \approx 2\pi \frac{\tau_d}{T} = 2\pi \frac{\tau_d \frac{|y(t)|_e^2}{2}}{T \frac{|y(t)|_e^2}{2}},$$

where $|y(t)|_e \equiv Ae^{-\alpha t}$ is the envelope function of $y(t)$, $T \frac{|y(t)|_e^2}{2}$ in the denominator is the same as $\int_t^{t+T} |y(\tau)|^2 d\tau$, the energy dissipated in one underdamped oscillation period T, and $\tau_d \frac{|y(t)|_e^2}{2}$ in the numerator is equivalent to $\int_t^{\infty} |y(\tau)|^2 d\tau$ (see Exercise Problem 12.5), the stored energy in the system at time t (can you see why?). Hence, the energy interpretation of Q follows as indicated in Table 12.1b.

Example 12.1

Suppose a 2nd-order system with an oscillation period of 1 s dissipates 1 J/s when its stored energy is 10 J. Approximate the damping coefficient α of the system.

Solution   Based on the energy interpretation of Q given above,

$$Q \approx 2\pi \frac{10\,\text{J}}{1\,\text{J}} = 20\pi \approx 63.$$

Clearly, this is a high Q system and therefore the oscillation frequency $\sqrt{\omega_o^2 - \alpha^2} \approx \omega_o = \frac{2\pi}{1}$. Hence,

$$\alpha = \frac{\omega_o}{2Q} = \frac{\frac{2\pi}{1}}{2(20\pi)} = \frac{1}{20} = 0.05 \text{ s}^{-1}.$$

Finally, for $Q \gg 1$, once again, we can write

$$Q = \frac{\omega_o}{2\alpha} \equiv \frac{\frac{2\pi}{T_o}}{2\alpha} \approx \frac{\frac{\pi}{\alpha}}{T},$$

where $T \approx T_o = \frac{2\pi}{\omega_o}$ is the period of underdamped oscillations in $y(t)$. Thus Q can be interpreted as the number of oscillation cycles of $y(t)$ occuring during a time-interval $\frac{\pi}{\alpha}$. Since the underdamped $y(t)$ has an envelope proportional to $e^{-\alpha t}$, and $y(\frac{\pi}{\alpha})/y(0) = e^{-\alpha \frac{\pi}{\alpha}} = e^{-\pi} \approx 4.3\%$, this duration $\frac{\pi}{\alpha}$ is just about the "lifetime" of the underdamped response (before the amplitude gets too small to notice). This justifies the *count interpretation* of Q given in the bottom row of Table 12.1b.

Example 12.2

A steel beam oscillates about 30 times before the oscillation amplitude is reduced to a few percent of the initial amplitude. What is the system Q and what is the damping rate if the oscillation period is 0.5 s?

Solution $Q \approx 30$, and therefore

$$\alpha = \frac{\omega_o}{2Q} \approx \frac{\frac{2\pi}{0.5}}{2 \cdot 30} \approx 0.2 \text{ s}^{-1}.$$

Before closing this section, it is worthwhile to point out that the limiting case with $Q \to \infty$ corresponds to a dissipation-free resonant circuit (a straightforward application of the energy interpretation of Q). Thus, high-Q systems are considered to be "near resonant." In the following section we will see that such near-resonant circuits are needed to build high-quality low-pass filters.

12.3 Low-Pass Butterworth Filter Design

Consider the low-pass amplitude response function

$$|H(\omega)| = \frac{1}{\sqrt{1 + \left(\frac{\omega}{\Omega}\right)^{2n}}},$$

where $\Omega > 0$ and $n \geq 1$. Plots of $|H(\omega)|^2$ for $n = 1, 2, 3,$ and 6 and $\Omega = 1 \frac{\text{rad}}{\text{s}}$ are shown in Figure 12.9. Clearly, $|H(\omega)|$ above describes a filter with a 3-dB bandwidth

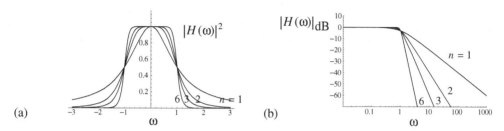

Figure 12.9　(a) The square of the amplitude response curves of 1st, 2nd, 3rd, and 6th-order Butterworth filters with a 3-dB bandwidth of $\Omega = 1\,\frac{rad}{s}$, and (b) decibel amplitude response plots of the same filters; labels n indicate the filter order.

Ω (which equals $1\,\frac{rad}{s}$ in Figure 12.9, where $|H(1)|^2 = \frac{1}{2} = -3\text{dB}$) and an amplitude response approaching the ideal with increasing n.

Stable low-pass filters with $|H(\omega)|$ as shown in Figure 12.9 are known as nth-order *Butterworth filters*. We next will describe how high-order Butterworth transfer functions $\hat{H}(s)$ can be designed having the desired Butterworth magnitude responses $|H(\omega)|$ described above. The high-order transfer function $\hat{H}(s)$ can be implemented by cascading 1st- and/or 2nd-order op-amp filter circuits as discussed earlier.

12.3.1　Finding the Butterworth $\hat{H}(s)$

The first step in designing and building Butterworth circuits is finding a stable and realizable transfer function $\hat{H}(s)$ that leads to the Butterworth amplitude response $|H(\omega)|$ given above. That in turn amounts to selecting appropriate pole locations, once the number of poles to be used is decided.

Given any stable LTIC system $\hat{H}(s)$ with a frequency response $H(\omega) = \hat{H}(j\omega)$,

$$|H(\omega)|^2 = H(\omega)H^*(\omega) = \hat{H}(j\omega)\hat{H}^*(j\omega) = \hat{H}(j\omega)\hat{H}(-j\omega).$$

Hence, for a Butterworth circuit with

$$|H(\omega)|^2 = \frac{1}{1 + \left(\frac{\omega}{\Omega}\right)^{2n}} = \frac{1}{1 + \left(\frac{j\omega}{j\Omega}\right)^{2n}} = \hat{H}(j\omega)\hat{H}(-j\omega),$$

the corresponding transfer function $\hat{H}(s)$ is constrained as

$$\hat{H}(s)\hat{H}(-s) = \frac{1}{1 + \left(\frac{s}{j\Omega}\right)^{2n}}.$$

This equation indicates that an nth-order Butterworth circuit must have a transfer function $\hat{H}(s)$ with n characteristic poles because the product $\hat{H}(s)\hat{H}(-s)$ has $2n$ poles corresponding to the $2n$ solutions of

$$1 + \left(\frac{s}{j\Omega}\right)^{2n} = 0.$$

After determining the poles of $\hat{H}(s)\hat{H}(-s)$ (by solving the equation above) we will assign the half of them from the LHP[2] to the Butterworth transfer function $\hat{H}(s)$. As we will see in Example 12.3 below, once the characteristic poles of the Butterworth $\hat{H}(s)$ are known, $\hat{H}(s)$ itself is easy to write down.

To determine the poles of $\hat{H}(s)\hat{H}(-s)$ we rearrange the equation above as

$$\left(\frac{s}{j\Omega}\right)^{2n} = -1 = e^{jm\pi},$$

where m is any odd integer $\pm 1, \pm 3, \pm 5 \cdots$. Taking the "$2n$th root" of both sides,

$$s = j\Omega e^{j\frac{m\pi}{2n}} = \Omega \angle (90° + \frac{\pi}{2}\frac{m}{n}) = \Omega \angle 90° (1 + \frac{m}{n}).$$

For m positive and odd, $m = 1, 3, \cdots < 2n$, this result gives the locations of n distinct LHP poles of $\hat{H}(s)\hat{H}(-s)$. The remaining n poles of $\hat{H}(s)\hat{H}(-s)$, obtained with m negative and odd, $m = -1, , -3, \cdots > -2n$, are all in the RHP. Furthermore, as we will see below in Example 12.3, the complex LHP poles of $\hat{H}(s)\hat{H}(-s)$ come in conjugate pairs, as needed for $\hat{H}(s)$ with real-valued coefficients.

Thus, here is the main result:

> The characteristic poles of a *stable* nth-order Butterworth filter with a 3-dB bandwidth $\Omega > 0$ are uniformly spaced around a semicircle in the LHP, having radius Ω, at locations
>
> $$s = \Omega \angle 90° (1 + \frac{m}{n}), \quad m = 1, 3, 5, \cdots < 2n.$$

Butterworth poles

The following examples illustrate the applications of this result.

Example 12.3
Determine the pole locations of a 3rd-order Butterworth filter with 3-dB bandwidth $\Omega = 10 \frac{\text{rad}}{\text{s}}$. Also, determine $\hat{H}(s)$ so that the system DC gain is 1.

---

[2]There are many ways of selecting n of the $2n$ poles of $\hat{H}(s)\hat{H}(-s)$ for $\hat{H}(s)$. However, for $\hat{H}(s)$ to have real-valued coefficients, the selection must be made so that $H(\omega) = \hat{H}(j\omega)$ is conjugate symmetric and, furthermore, the selection should include only LHP poles to assure stability.

Solution   Since $n = 3$ and $\Omega = 10$, the pole locations of the circuit transfer function are given by the formula

$$s = \Omega \angle 90° (1 + \frac{m}{n}) = 10 \angle 90° (1 + \frac{m}{3}), \quad m = 1, 3, 5 < 6.$$

Clearly, all three poles $p_1 = 10 \angle 120°$, $p_2 = 10 \angle 180°$, and $p_3 = 10 \angle 240°$ have magnitudes 10 and are positioned in the LHP around a semicircle as shown in Figure 12.10a. The angular separation between the neighboring poles on the semicircle is $\frac{180°}{n} = \frac{180°}{3} = 60°$.

The transfer function of the filter with DC gain of 1 can be written as

$$\hat{H}(s) = \frac{K}{(s - 10 \angle 120°)} \times \frac{1}{(s - 10 \angle 180°)} \times \frac{1}{(s - 10 \angle 240°)}$$

for some constant K to be determined. Since

$$\hat{H}(0) = -\frac{K}{10^3} \angle - (120 + 180 + 240)° = K \times 10^{-3},$$

it follows that

$$\hat{H}(0) = H(0) = 1$$

requires

$$K = 10^3 = \Omega^3.$$

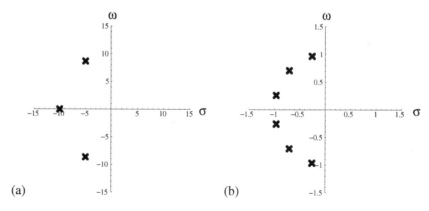

(a)   (b)

Figure 12.10   The pole locations of (a) a 3rd-order Butterworth filter with 3-dB frequency $\Omega = 10 \frac{rad}{s}$, and (b) 6th-order Butterworth filter with 3-dB frequency $\Omega = 1 \frac{rad}{s}$.

Example 12.4
How would the filter of Example 12.3 be implemented in a cascade configuration?

Solution   The transfer function determined in Example 12.3, with $K = 10^3$, can be expressed as

$$\hat{H}(s) = \frac{10}{(s - 10\angle 120°)} \times \frac{10}{s + 10} \times \frac{10}{(s - 10\angle - 120°)}$$

$$= \frac{10}{s + 10} \frac{10^2}{s^2 + 10s + 10^2}.$$

Hence, the system can be implemented by cascading a 1st-order system

$$\hat{H}_1(s) = \frac{10}{s + 10}$$

(e.g., the active filter shown in Figure 12.5a) with a 2nd-order system

$$\hat{H}_2(s) = \frac{10^2}{s^2 + 10s + 10^2}$$

(e.g., the active filter shown in Figure 12.5b).

Example 12.5
How would the filter of Example 12.3 be implemented in a parallel configuration?

Solution   Note that

$$\hat{H}(s) = \frac{10}{s + 10} \frac{10^2}{s^2 + 10s + 10^2} = \frac{10}{s + 10} + \frac{-10s}{s^2 + 10s + 10^2}.$$

Hence, the system can be implemented with a parallel connection of a
low-pass filter

$$\hat{H}_1(s) = \frac{10}{s + 10}$$

and a band-pass filter

$$\hat{H}_2(s) = \frac{-10s}{s^2 + 10s + 10^2}.$$

The pole locations of a 6th-order Butterworth low-pass filter with 3-dB band-
width $\Omega = 1 \frac{\text{rad}}{\text{s}}$ are shown Figure 12.10b. Note that the poles are positioned on a
LHP semicircle with radius 1 and with $30°$ angular separation between neighboring

poles, which once again agrees with angular spacing specified by the formula $\frac{180°}{n}$. Furthermore, the marked pole locations are the only possible locations in the LHP for 6 poles with $\frac{180°}{6} = 30°$ angular separations, magnitudes $\Omega = 1$, and complex conjugate pairings. The 6th-order filter can be implemented by cascading three 2nd-order low-pass filters. The frequency response magnitudes for the individual second-order sections are shown in Figure 12.11a.

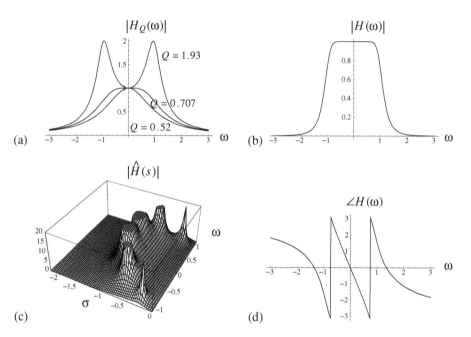

Figure 12.11 (a) The amplitude response $|H_Q(\omega)|$ of 2nd-order low-pass filters with a resonance frequency of $\omega_o = 1 \frac{\text{rad}}{\text{s}}$ and $Q = 1.93, 0.707,$ and 0.52, respectively, (b) the amplitude response $|H(\omega)|$ of a 6-pole Butterworth filter with a 3-dB bandwidth $\Omega = \omega_o = 1 \frac{\text{rad}}{\text{s}}$ obtained by cascading the three low-pass filters shown in (a), (c) a surface plot of the magnitude of the transfer function $\hat{H}(s)$ of the same 6-pole filter, and (d) the phase response $\angle H(\omega)$ of the same filter.

12.3.2   Cascade implementation of Butterworth filters

In general, the transfer function of an nth-order Butterworth filter with a 3-dB bandwidth Ω and unity DC gain can be expressed in cascade form as

$$\hat{H}(s) = \prod_{m=1,\text{odd}}^{n-1} \frac{\Omega^2}{s^2 + 2\Omega \sin(\frac{\pi}{2}\frac{m}{n})s + \Omega^2}$$

for n even or

$$\hat{H}(s) = \frac{\Omega}{s + \Omega} \prod_{m=1,\text{odd}}^{n-2} \frac{\Omega^2}{s^2 + 2\Omega \sin(\frac{\pi}{2}\frac{m}{n})s + \Omega^2}$$

for n odd. These expressions were obtained by noting that each pair of conjugate poles contribute to $\hat{H}(s)$ a 2nd-order component

$$\frac{\Omega^2}{(s - \Omega \angle \theta_m)(s - \Omega \angle - \theta_m)},$$

where $\theta_m = 90°(1 + \frac{m}{n})$ for odd $m < n - 1$. Multiplying out the denominator of this expression gives the more compact forms above (see Exercise Problem 12.8).

Note that the 2nd-order cascade components above correspond to underdamped systems with underdamped resonance frequency

$$\omega_o = \Omega,$$

damping coefficients

$$\alpha = \Omega \sin(\frac{\pi}{2}\frac{m}{n}) < \omega_o,$$

and quality factors

$$Q = \frac{\omega_o}{2\alpha} = \frac{1}{2\sin(\frac{\pi}{2}\frac{m}{n})} > \frac{1}{2}.$$

Figure 12.11a shows the amplitude response curves of three 2nd-order cascade components ($Q = 1.93$, 0.707, and 0.52) of a 6th-order Butterworth filter with a 3-dB bandwidth $\Omega = 1\,\frac{\text{rad}}{\text{s}}$. The amplitude and phase response of the 6th-order—or the "6-pole"—filter are shown in Figures 12.11b and 12.11d, respectively.

Notice that amplitude response curves of the three 2nd-order low-pass filters shown in Figure 12.11a are far from ideal (non-flat, rippled for higher Q, sluggish decay for low Q). However, their product—which is the amplitude response of the 6th-order Butterworth filter obtained by cascading the three filters—is flat almost all the way up to the 3-dB bandwidth frequency $1\,\frac{\text{rad}}{\text{s}}$ and drops reasonably fast beyond that. Also, the accompanying phase response curve shown in Figure 12.11d is reasonably linear up to the 3-dB frequency. Higher-order Butterworth filters will do even better in all these respects.

The Butterworth transfer function surface plot shown in Figure 12.11c clearly indicates the locations and distributions of the six LHP system poles of $\hat{H}(s)$ (same locations as in Figure 12.10b)—the poles are arranged around a semi-circle as expected. The pole-pair closest to the ω-axis is the contribution of the highest-Q subcomponent of the cascade; the same poles also are responsible for the peaks in the amplitude response curve of the highest-Q subsystem shown in Figure 12.11a. By contrast, the pole-pair furthest away from the ω-axis (and closest to the σ-axis) is the contribution of the lowest-Q subcomponent of the system with the narrowest frequency response curve shown in Figure 12.11a.

Example 12.6

A 5th-order (or a 5-pole) Butterworth filter with a 3-dB frequency of 5 kHz needs to be built using a cascade configuration. Choose the appropriate capacitor and resistance values C_1, C_2, R_1, R_2, R_3, and R_4 for the highest-Q subcomponent of the cascade assuming that the 2nd-order active filter shown in Figure 12.5b will be used.

Solution   A 5 kHz 3-dB frequency means that we have

$$\Omega = 2\pi(5000) = 10^4\pi \; \frac{\text{rad}}{\text{s}}.$$

The quality factors of the subcomponent circuits will be $Q = \frac{1}{2\sin(\frac{\pi}{2}\frac{m}{n})}$, with $n = 5$ and $m = 1, 3$. The higher Q, obtained with $m = 1$, is

$$Q = \frac{1}{2\sin(\frac{\pi}{2}\frac{1}{5})} = 1.618.$$

Hence, with $\omega_o = \Omega = 10^4\pi$ and $Q = 1.618$, the damping coefficient of the highest-Q subcomponent circuit must be

$$\alpha = \frac{\omega_o}{2Q} = \frac{10^4\pi}{3.236} \approx 9708.1.$$

Finally, the required DC amplification of the circuit is $K = 1$ (since a Butterworth filter has a DC gain of 1).

Therefore, the element values of the circuit in Figure 12.5b need to satisfy the constraints

$$K = 1 + \frac{R_1}{R_2} = 1,$$
$$\omega_o = \frac{1}{\sqrt{R_3 R_4 C_1 C_2}} = 10^4\pi,$$

and

$$\alpha = \frac{1}{2R_3C_1} + \frac{1}{2R_4C_1} + \frac{1-K}{2R_4C_2} = 9708.1.$$

Clearly, the first constraint allows for $R_1 = 0$ and $R_2 = \infty$ (i.e., the voltage-follower circuit) and the formula for α simplifies to

$$\frac{1}{2R_3C_1} + \frac{1}{2R_4C_1} = 9708.1.$$

Suppose we decide to use $R_3 = R_4 = 1\,\text{k}\Omega$. Then the second constraint equation can be solved for C_1 as

$$C_1 = \frac{1}{1000 \times 9708.1} = 1.03 \times 10^{-7}\,\text{F} = 103\,\text{pF}.$$

Substituting $R_3 = R_4 = 1\,\text{k}\Omega$ and $C_1 = 103\,\text{pF}$ into the first constraint equation, we find

$$C_2 = \frac{1}{10^8 \pi^2 \times 10^3 \times 10^3 \times 1.03 \times 10^{-7}} = 9.84 \times 10^{-9}\,\text{F} = 9.84\,\text{pF}.$$

In summary, $R_3 = R_4 = 1\,\text{k}\Omega, C_1 \approx 100\,\text{pF}, C_2 \approx 10\,\text{pF}, R_1 = 0$, and $R_2 = \infty$ give one possible solution of the filter design problem. Other values for C_1 and C_2 also can be chosen by starting with different values of R_3 and R_4.

12.3.3 Filters other than Butterworth low-pass

Clearly, in cascading several high- and low-Q filters to obtain improved filter characteristics, it is not necessary to follow the Butterworth design formula. A 6th-order Butterworth filter, for instance, is just one particular way of cascading three 2nd-order filters to obtain a nice 6th-order filter. Other cascade possibilities—corresponding to alternate pole locations in the LHP—exist, some of which may produce filter characteristics that are better suited for some applications.

One virtue of Butterworth filters is that they are "optimal" in the sense that, for a given filter order n, their amplitude response is *maximally flat*, meaning that the amplitude response curve has the maximum number of zero-valued derivatives at $\omega = 0$ and $\omega = \pm\infty$ among all possible amplitude response curves that can be obtained from rational-form transfer functions of a given order.

In other respects Butterworth filters are not optimal. For instance, for a given filter order n there are other low-pass filters (less flat and possibly "rippled" in the passband, in the stopband, or both) with faster amplitude decay from passband to stopband. In particular, "elliptic" filters offer the steepest decay from passband to stopband for a filter of a given order, but with a magnitude response that exhibits ripple (oscillatory behavior) in both the passband and stopband. The phase response of an elliptic filter tends to be somewhat less linear than that for a Butterworth filter. Another type of filter, called Chebyshev, is available in two types, one exhibiting ripple only in the passband and the other having ripple only in the stopband. In terms of decay rate and phase linearity, Chebyshev filters offer characteristics that are between those of Butterworth and elliptic filters.

Whereas the design of an nth-order Butterworth filter begins with

$$|H(\omega)| = \frac{1}{\sqrt{1 + \left(\frac{\omega}{\Omega}\right)^{2n}}},$$

the amplitude response of nth-order Chebyshev and elliptic filters replace

$$\left(\frac{\omega}{\Omega}\right)^{2n}$$

with nth-order Chebyshev polynomials and Jacobi elliptic functions, respectively. Tedious algebra is associated with the design of Chebyshev and elliptic filters, as well as with the design of high-order Butterworth filters, so filter design virtually always is performed using one of the standard computer software packages for signal procesing design and simulation, such as Matlab.

High-pass and band-pass filters generally are designed by first designing a low-pass transfer function and then transforming it to the transfer function of either a high-pass or band-pass filter. The transformation is accomplished by a simple change of variable. For example, if $\hat{H}(s)$ is the transfer function of a low-pass filter with passband cutoff frequency Ω, then a high-pass filter with passband cutoff frequency ω_h can be obtained by simply replacing s in $\hat{H}(s)$ with

$$\frac{\Omega\omega_h}{s}.$$

Similarly, a band-pass filter with lower passband cutoff frequency ω_l and upper passband cutoff frequency ω_u can be produced from a low-pass transfer function with cutoff Ω by replacing s in the low-pass transfer function with

$$\Omega\frac{s^2 + \omega_l\omega_u}{s(\omega_u - \omega_l)}$$

as illustrated in Figure 12.12. Because the high-pass and low-pass transformations replace the variable s with a nonlinear function of s, the mappings slightly distort the frequency response characteristic of the low-pass filter. This occurs through a nonuniform stretching or squashing of the frequency scale. So, for example, a band-pass response is not just a shifted version of the low-pass response. Nevertheless, the band-pass response takes on exactly the same set of values as the low-pass response – just at different frequencies. In general, the resulting distortion is small and the above transformations are very useful in practice.

Commonly used software packages permit the design of Butterworth, Chebyshev, elliptic and other filters using a single command. A single user instruction triggers both the design of a low-pass filter prototype $\hat{H}(s)$ and the transformation of the prototype to the desired type of filter.

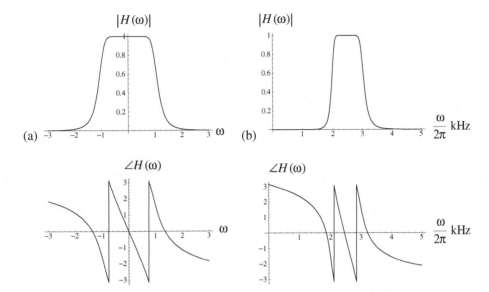

Figure 12.12 Amplitude and phase response curves for (a) a 6th-order Butterworth low-pass filter with $\Omega = 1$, and (b) a bandpass filter with $\omega_l/2\pi = 2$ kHz and $\omega_u/2\pi = 3$ kHz derived from the same low-pass filter.

EXERCISES

12.1   Derive the transfer function $\hat{H}_a(s)$ of the 1st-order active filter circuit depicted in Figure 12.5a.

12.2   Derive the transfer function $\hat{H}_b(s)$ of the Sallen-Key circuit depicted in Figure 12.5b.

12.3   Given the following transfer functions determine whether the system is underdamped, overdamped, or critically damped, and calculate the quality factor Q:

    (a) $\hat{H}_1(s) = \frac{s}{s^2+4s+400}$,

    (b) $\hat{H}_2(s) = \frac{s+5}{s^2+2000s+10^6}$,

    (c) $\hat{H}_3(s) = \frac{s^2+100}{s^2+20000s+10^6}$.

12.4   Identify each of the three systems defined in Problem 12.3 as low-pass, band-pass, or high-pass, and for the band-pass system determine the 3-dB bandwidth Ω.

12.5   Given that $y(t)$ is an underdamped zero-input response of the form given in the first row of Table 12.1a, show that

$$\int_{t}^{\infty} |y(\tau)|^2 d\tau \approx \tau_d \frac{|y(t)|_e^2}{2},$$

where $\tau_d = \frac{1}{2\alpha}$ and $|y(t)|_e = Ae^{-\alpha t}$ is the envelope of the underdamped $y(t)$, and explain why $\int_{t}^{\infty} |y(\tau)|^2 d\tau$ can be interpreted as the stored energy of the system at instant t. In performing the integral, make use of $\omega_o \gg \alpha$ to handle an integral with an oscillatory integrand.

12.6   The zero-input response of a 2nd-order band-pass filter is observed to oscillate about sixty cycles before the oscillation amplitude is reduced to a few percent of its initial amplitude. Furthermore, the oscillation period of the zero-input response is measured as 1 ms. Assuming that the maximum amplitude response of the system is 1, write an approximate expression for the frequency response $H(\omega) = \hat{H}(j\omega)$ of the system.

12.7   Determine the frequency response, 3-dB bandwidth, and quality factor Q of the following parallel RLC bandpass filter circuit, below, in terms of resistance R:

12.8   Given $\theta_m = 90°(1 + \frac{m}{n})$, verify that

$$(s - \Omega \angle \theta_m)(s - \Omega \angle - \theta_m)$$

multiplies out as

$$s^2 + 2\Omega \sin(\frac{\pi}{2} \frac{m}{n})s + \Omega^2.$$

12.9   Determine the pole locations of a 4th-order low-pass Butterworth filter with a 3-dB frequency of 1 kHz.

12.10   What is the transfer function of the highest-Q subcomponent of the 4th-order Butterworth filter described in Problem 12.9.

12.11   Assuming $R_3 = R_4 = 4\,k\Omega$, determine the capacitance values C_1 and C_2 for the 2nd-order circuit of Figure 12.5b to implement the transfer function of Problem 12.10.

12.12   Approximate the time delay of the filter described in Problem 12.10 by calculating the slope of the phase response curve of the filter at $\omega = 0$.

12.13 Determine the transfer function $\hat{H}(s)$ of a 3rd-order Butterworth low-pass filter having 3-dB cutoff frequency 15 kHz. Sketch the magnitude of the frequency response. Verify that $|\hat{H}(j\omega)|$ agrees with your sketch.

12.14 Design a 2nd-order Butterworth high-pass filter having cutoff frequency 50 Hz. Do so by first designing a Butterworth low-pass filter having cutoff frequency 1 Hz and then transforming it to a high-pass filter using the low-pass to high-pass transformation in Section 12.3.3. Sketch the magnitudes of the frequency responses for both the low-pass prototype and the high-pass filter.

12.15 Show that application of the low-pass to band-pass transformation in Section 12.3.3 results in a band-pass transfer function with frequency response satisfying

$$H_B(\omega_l) = H(-\Omega)$$

and

$$H_B(\omega_u) = H(\Omega)$$

where H_B is the band-pass frequency response with lower and upper cutoff frequencies ω_l and ω_u, respectively, and H is the frequency response of the low-pass filter with cutoff frequency Ω. Notice that these relations prove that the filter transformation moves the 3-dB cutoff frequencies of the low-pass filter to the desired frequencies for the band-pass filter.

A

Complex Numbers and Functions

A.1   Complex Numbers as Real Number Pairs

Complex arithmetic, an invention of the late sixteenth century,[1] is an arithmetic of *real number pairs*—such as $(2, 0)$, $(0, 1)$, $(\pi, 2.5)$, etc.—known as *complex numbers*.

According to the rules of complex arithmetic, complex numbers (a, b) and (c, d) are added and multiplied as

$$(a, b) + (c, d) = (a + c, b + d)$$

and

$$(a, b)(c, d) = (ac - bd, ad + bc),$$

respectively. Furthermore, the complex number $(a, 0)$ and real number a are *defined* to be the same; that is,

$$(a, 0) = a.$$

Two important consequences of these definitions are:

---

[1] The introduction of a consistent theory of complex numbers is credited to Italian mathematician Rafael Bombelli (1526–1572).

(1) Complex addition and multiplication are compatible with real addition and multiplication; for complex numbers $(a, 0)$ and $(c, 0)$, which also happen to be the reals a and c,

$$(a, 0) + (c, 0) = (a + c, 0) = a + c$$

and

$$(a, 0)(c, 0) = (ac - 0, 0 + 0) = (ac, 0) = ac,$$

as in real arithmetic. Thus, real arithmetic is embedded in complex arithmetic as a special case.

(2) The product of the complex number $(0, 1)$ with itself is

$$(0, 1)(0, 1) = (0 - 1, 0 + 0) = (-1, 0) = -1.$$

If we denote $(0, 1)$ as j, adopting the definition

$$j \equiv (0, 1),$$

we can express the preceding result more concisely as

$$j^2 = -1.$$

We refer to j as the *imaginary unit*,[2] since there exists no real number whose square equals -1. However, within the complex number system, $j = (0, 1)$ is no less real or no more imaginary than, say, $(1, 0) = 1$. Furthermore, because in the complex number system, $j^2 = (0, 1)(0, 1) = -1$ and $(-j)^2 = (0, -1)(0, -1) = -1$, the square root of -1 exists and equals $(0, \pm 1) = \pm j$. Because of this, we say $j = \sqrt{-1}$.

$$j^2 = -1$$
$$\sqrt{-1} = \pm j$$

Example A.1

Notice that, as a consequence of $j^2 = -1$, it follows that

$$\frac{1}{j} = \frac{1}{j}\frac{j}{j} = \frac{j}{-1} = -j.$$

---

[2]In math and physics books, the imaginary unit usually is denoted as i. We prefer j in electrical engineering, because i is often a current variable.

A.2 Rectangular Form

Given the real numbers a and b, and the imaginary number $j \equiv (0, 1)$,

$$a + jb = (a, 0) + (0, 1)(b, 0) = (a, 0) + (0 - 0, b + 0) = (a, 0) + (0, b) = (a, b).$$

Thus, $a + jb$ and (a, b) are simply two different ways of expressing the same complex number.

In the next section, we will explain how $a + jb = (a, b)$ can be plotted in the two-dimensional plane, where a is the horizontal coordinate and b is the vertical coordinate. Because of this ability to directly map $a + jb = (a, b)$ onto the Cartesian plane, both $a + jb$ and (a, b) are called *rectangular forms* of a complex number. To distinguish these two slightly different representations, however, we will refer to (a, b) as the *pair form* and reserve the term *rectangular form* for $a + jb$. The value a is called the *real part* of the complex number, whereas the value b is referred to as the *imaginary part*. This is dreadful terminology, because both a and b are real numbers. We are stuck with these descriptors, though, because they were adopted long ago and are in widespread use. Remember that the imaginary part of a complex number is real valued! The imaginary part of $a + jb$ is b, not jb.

When expressed in rectangular form, complex numbers add and multiply like algebraic expressions in the variable j. This is the main advantage in using the j notation. For example, the product of

$$X = (1, 1) = 1 + j$$

and

$$Y = (-1, 1) = -1 + j1$$

is, simply,

$$XY = (1 + j1)(-1 + j1) = -1 + j^2 + j - j = -2,$$

since $j^2 = -1$. The result is the same as that obtained with the original multiplication rule (which is hard to remember):

$$(1, 1)(-1, 1) = (-1 - 1, 1 - 1) = (-2, 0) = -2.$$

Also, notice that

$$X + Y = (1 + j1) + (-1 + j1) = 0 + j2 = j2,$$

in conformity with

$$(1, 1) + (-1, 1) = (0, 2).$$

All complex arithmetic, including subtraction and division, can be carried out by treating complex numbers as algebraic expressions in j. For example,

$$X - Y = (1 + j) - (-1 + j) = 2 + j0 = 2.$$

In complex division, we sometimes wish to express a given complex ratio, $\frac{X}{Y}$, in rectangular form. We can accomplish this by multiplying both X and Y by the complex conjugate of Y. For example,

$$\frac{X}{Y} = \frac{1 + j}{-1 + j} = \frac{(1 + j)(-1 - j)}{(-1 + j)(-1 - j)} = \frac{0 - j2}{2 + j0} = \frac{-j2}{2} = -j,$$

where, in the second line, we multiplied both the numerator and denominator by the complex conjugate of $-1 + j$, which is $-1 - j$. In general, the *complex conjugate*   **Complex-conjugate** of $a + jb$ is defined to be $a - jb$.

Example A.2
Given $P = (3, 6)$ and $Q = 2 + j$, determine $R = PQ$ and express the product in pair form.

Solution

$$R = PQ = (3, 6)(2 + j) = (3 + j6)(2 + j)$$
$$= 6 + 6j^2 + j3 + j12 = 0 + j15.$$

Hence, in pair form, $R = (0, 15)$.

Example A.3
Given $P = (3, 6)$, $Q = 2 + j$, and $R = PQ$, determine $\frac{P-Q}{R}$.

Solution   Since $R = j15$, as determined in Example A.2,

$$\frac{P - Q}{R} = \frac{(3 + j6) - (2 + j)}{j15} = \frac{1 + j5}{j15} = \frac{(1 + j5)j}{(j15)j}$$
$$= \frac{-5 + j}{-15} = \frac{1}{3} - j\frac{1}{15}.$$

Example A.4
Use the quadratic equation[3] to determine the roots of the polynomial

$$x^2 + x + 1.$$

---

[3]If $ax^2 + bx + c = 0$, then

$$x = \frac{-b \pm \sqrt{b^2 - 4ac}}{2a}.$$

Solution　　The roots of $x^2 + x + 1$ are the numbers x such that

$$x^2 + x + 1 = 0.$$

Using the quadratic equation, we find that the roots are given by

$$x = \frac{-1 \pm \sqrt{1 - 4}}{2} = \frac{-1 \pm \sqrt{-3}}{2}.$$

However, since $\sqrt{-3}$ is not a real number, $x^2 + x + 1$ has no real roots; in other words, there is no real number x for which the statement $x^2 + x + 1 = 0$ is true. That is, if we plot $f(x) = x^2 + x + 1$ versus x, where x is a customary real variable, then $f(x)$ does not cross through the x-axis.

But if we allow the variable x to take complex values, then $x^2 + x + 1 = 0$ is true for

$$x = -\frac{1}{2} \pm j\frac{\sqrt{3}}{2},$$

since two distinct roots of the number -3 are $\pm j\sqrt{3}$. Therefore, if x is a complex variable, the roots of the polynomial $x^2 + x + 1$ are $-\frac{1}{2} \pm j\frac{\sqrt{3}}{2}$. The notion of complex variables will be explored further in Section A.7.

While the pair form of complex numbers has conceptual importance, this form is cumbersome when we are performing calculations by hand. In practice, we usually will rely on rectangular form (with j notation) as well as the exponential and polar forms discussed next.

A.3　Complex Plane, Polar and Exponential Forms

A complex number

$$C = (a, b) = a + jb$$

can be envisioned as a point (a, b) on the 2-D Cartesian plane where a is the coordinate on the horizontal axis and b is the coordinate on the vertical axis. Since the reals a and b are called the *real* and *imaginary part*s of the complex number $C = (a, b) = a + jb$, we use the notation

$$a = \text{Re}\{C\}$$

and

$$b = \text{Im}\{C\}.$$

We label the horizontal and vertical axes of the 2D plane as Re and Im, respectively, as shown in Figure A.1a, where we have plotted several different complex numbers. When plotting complex numbers in this fashion, we call the 2-D plane the *complex plane*.

Clearly, (a, b) and $a + jb$ are equivalent ways of referencing the same point C on the complex plane. As illustrated in Figure A.1b, point C also can be *referenced* by another pair of numbers: its distance $|C|$ from the origin, which is the length of the dashed line connecting the origin to point C; and the angle $\angle C$ between the positive real axis and a straight line path from the origin to the point C. Using simple trigonometry, we have

Complex plane

Magnitude, $|C|$, and angle, $\angle C$, of $C = a + jb$

$$|C| = \sqrt{a^2 + b^2}$$

and

$$\angle C = \tan^{-1}(\frac{b}{a}),$$

which are said to be the *magnitude* $|C|$ and *angle* $\angle C$ of

$$C = (a, b) = a + jb,$$

respectively.

The formula given for $\angle C$ assumes $a \geq 0$; otherwise, $\pm 180°$ needs to be added to the value of $\tan^{-1}(\frac{b}{a})$ provided by your calculator. For instance, the angle of the

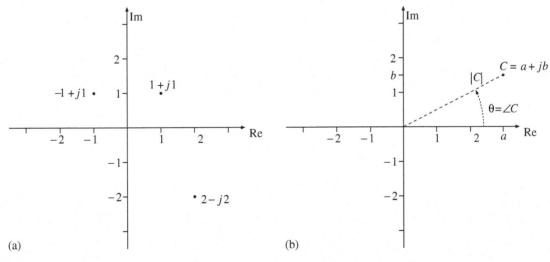

(a)   (b)

Figure A.1   The *complex plane* showing (a) the locations of complex numbers $1 + j1$, $-1 + j1$, and $2 - j2$, and (b) an arbitrary complex number $C = (a, b) = a + jb$ with magnitude, $|C|$, and angle, $\angle C = \theta$.

complex number

$$Y = -1 + j1$$

shown in Figure A.2 is not $-45°$, as $\tan^{-1}(\frac{b}{a})$ would indicate, but rather, $-45° \pm 180°$, as can be graphically confirmed. One version of $\angle(-1 + j)$,

$$-45° + 180° = 135°,$$

corresponds to counterclockwise rotation starting from the Re axis, while the second version,

$$-45° - 180° = -225°,$$

corresponds to clockwise rotation. Either is correct, because they correspond to the same complex number.

With the aid of Figure A.2, try to confirm that, for $X = 1 + j1$,

$$|X| = \sqrt{2} \quad \text{and} \quad \angle X = 45°;$$

while for $Z = 2 - j2$,

$$|Z| = 2\sqrt{2} \quad \text{and} \quad \angle Z = -45°.$$

Polar form
$|C|\angle C$

Because complex numbers can be represented by their magnitudes and angles in the two-dimensional plane, we call this *polar-form* representation. We write a complex number C in polar form as

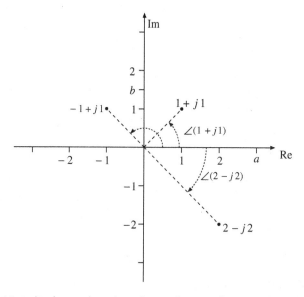

Figure A.2 Magnitudes and angles of complex numbers $1 + j1$, $-1 + j1$, and $2 - j2$ are indicated by dashed lines and arcs, respectively.

$$|C|\angle C.$$

So, X and Z from the preceding expression can be written as

$$X = \sqrt{2}\angle 45°$$

and

$$Z = 2\sqrt{2}\angle - 45°.$$

There is another version of polar-form representation that is much more convenient mathematically. To construct this new representation, we note that for any complex number $C = a + jb$, trigonometry allows us to recover a and b from the magnitude $|C|$ and angle $\theta \equiv \angle C$, as (see Figure A.1b)

$$a = \text{Re}\{C\} = |C|\cos\theta$$

and

$$b = \text{Im}\{C\} = |C|\sin\theta.$$

Hence,

$$C = a + jb = |C|(\cos\theta + j\sin\theta).$$

Now, as will be shown in Section A.5, there is a remarkably useful mathematical relation called *Euler's identity*, which states that

Euler's identity

$$e^{j\theta} = \cos\theta + j\sin\theta.$$

Applying Euler's identity to the previous expression for C, we have

$$C = |C|e^{j\theta},$$

which is called the *exponential form* of $C = a + jb$.

For example, exponential forms of

Exponential form $|C|e^{j\theta}$

$$X = 1 + j1,$$
$$Y = -1 + j1,$$
$$Z = 2 - j2$$

are (see Figure A.2)

$$X = \sqrt{2}e^{j45°},$$
$$Y = \sqrt{2}e^{j135°},$$
$$Z = 2\sqrt{2}e^{-j45°},$$

respectively.

Using the exponential form and ordinary algebraic rules, we have

$$XY = (\sqrt{2}e^{j45°})(\sqrt{2}e^{j135°}) = \sqrt{2} \times \sqrt{2} \times e^{j45°} \times e^{j135°}$$
$$= 2e^{(j45°+j135°)} = 2e^{j180°} = -2,$$

which is the same result as

$$XY = (1 + j1)(-1 + j1) = -1 + j^2 + j - j = -2.$$

Note that

$$2e^{j180°} = -2,$$

because

$$e^{j180°} = \cos 180° + j \sin 180° = -1.$$

(Alternatively, it is even easier to reach the same conclusion graphically on the complex plane.)

Likewise,

$$\frac{X}{Z} = \frac{\sqrt{2}e^{j45°}}{2\sqrt{2}e^{-j45°}} = \frac{1}{2}e^{j45°} \times e^{j45°} = \frac{1}{2}e^{j90°} = j\frac{1}{2}$$

and

$$XZ = (\sqrt{2}e^{j45°})(2\sqrt{2}e^{-j45°}) = \sqrt{2} \times 2\sqrt{2} \times e^{j(45°-45°)} = 4e^{j0} = 4.$$

As these examples illustrate, exponential form is convenient for complex multiplication and division.

Let us try a couple of examples where we convert from exponential to rectangular form, or vice versa.

Example A.5

Convert the following complex numbers from exponential to rectangular form:

$$C_1 = 7e^{j25°}, \quad C_2 = 7e^{-j25°},$$
$$C_3 = 2e^{j160°}, \quad C_4 = 2e^{-j160°}.$$

Solution   Using Euler's identity (and a calculator to evaluate the cosines and sines), we get

$$C_1 = 6.34 + j2.95, \quad C_2 = 6.34 - j2.95,$$
$$C_3 = -1.88 + j0.68, \quad C_4 = -1.88 - j0.68.$$

You should roughly sketch the locations of these points on the complex plane and check that these exponential-to-rectangular conversions seem correct.

Example A.6
Convert

$$C_1 = 3 + j4, \quad C_2 = -3 + j4,$$
$$C_3 = 3 - j4, \quad C_4 = -3 - j4$$

to exponential form.

Solution   To write $C = a + jb$ in the form $|C|e^{j\theta}$, use

$$|C| = \sqrt{a^2 + b^2}$$

and

$$\theta = \angle C = \tan^{-1} \frac{b}{a}.$$

We have

$$|C_1| = |C_2| = |C_3| = |C_4| = 5.$$

We find the angle for C_1 as

$$\angle C_1 = \tan^{-1} \frac{4}{3} = 53.13°.$$

For C_2 we seemingly have

$$\angle C_2 = \tan^{-1} \frac{-4}{3} = -53.13°.$$

However, the inverse tangent is multivalued, and your calculator gives only one of the two values. If $a > 0$, your calculator provides the correct value. As indicated earlier, if $a < 0$ then you must add or subtract $180°$ from the calculated value to give the correct angle. Thus,

$$\angle C_2 = -53.13° + 180° = 126.87°.$$

Similarly, we have $\angle C_3 = -53.13°$ and $\angle C_4 = -126.87°$. So finally,

$$C_1 = 5e^{j53.13°}, \quad C_2 = 5e^{j126.87°},$$
$$C_3 = 5e^{-j53.13°}, \quad C_4 = 5e^{-j126.87°}.$$

Visualization of complex numbers on the complex plane is an important skill. Remember that any complex number written in exponential form, $|C|e^{j\theta}$, is at a distance $|C|$ from the origin and at an angle θ from the positive real axis. Thus, $e^{j0°}$, $e^{j90°}$, $e^{j180°}$, and $e^{j270°}$ all lie on a circle of radius one and have values 1, j, -1, and $-j$, respectively. Similarly, $5e^{j0°}$, $5e^{j90°}$, $5e^{j180°}$, and $5e^{j270°}$ all lie on a circle of radius five and have values 5, $j5$, -5, and $-j5$, respectively. You should be able to picture these locations on the complex plane.

Radians versus degrees

Angles of complex numbers can be measured in units of degrees, as we have done so far, or in units of radians, with $360°$ and 2π radians signifying the same angle. By convention, $e^{j90°}$ and $e^{j\frac{\pi}{2}}$ are the same complex number because $90° = \frac{\pi}{2}$ radians. Likewise,

$$e^{-j\frac{\pi}{4}} = e^{-j45°} = \frac{1}{\sqrt{2}} - j\frac{1}{\sqrt{2}}.$$

The following equalities should be understood and *remembered*:

$$e^{\pm j\frac{\pi}{2}} = \pm j$$

$$e^{\pm j\pi} = -1$$

$$e^{\pm j2\pi n} = 1 \text{ for any integer } n.$$

Recall that exponential form was derived from polar form. So, for example, we can express $Y = -1 + j1 = \sqrt{2}e^{j\frac{3\pi}{4}}$ as $\sqrt{2}\angle 135°$. Since both exponential and polar forms express Y in terms of magnitude $|Y| = \sqrt{2}$ and angle $\angle Y = \frac{3\pi}{4}$ rad $= 135°$, their distinction is mainly cosmetic.

Example A.7
What is the magnitude of the product P of complex numbers $U = 1 + j2$ and $V = 3e^{j\frac{\pi}{2}}$?

Solution   Clearly,

$$P = UV = |U|e^{j\angle U}|V|e^{j\angle V} = |U||V|e^{j(\angle U + \angle V)}.$$

Therefore, the magnitude of P is

$$|P| = |U||V| = (\sqrt{1^2 + 2^2})3 = 3\sqrt{5}.$$

Example A.8
Given that $A = 2\angle 45°$ and $B = 3e^{-j\frac{\pi}{2}}$, determine AB and $\frac{A}{B}$.

Solution   Since $A = 2\angle 45° = 2e^{j\frac{\pi}{4}}$,

$$AB = (2e^{j\frac{\pi}{4}})(3e^{-j\frac{\pi}{2}}) = 6e^{j(\frac{\pi}{4} - \frac{\pi}{2})} = 6e^{-j\frac{\pi}{4}}.$$

Also,

$$\frac{A}{B} = \frac{2e^{j\frac{\pi}{4}}}{3e^{-j\frac{\pi}{2}}} = \frac{2}{3}e^{j(\frac{\pi}{4}+\frac{\pi}{2})} = \frac{2}{3}e^{j\frac{3\pi}{4}}.$$

Example A.9

Express AB and $\frac{A}{B}$ from Example A.8 in rectangular form.

Solution   Using Euler's identity, we first note that

$$e^{-j\frac{\pi}{4}} = e^{j(-\frac{\pi}{4})} = \cos(-\frac{\pi}{4}) + j\sin(-\frac{\pi}{4}) = \frac{1}{\sqrt{2}} - j\frac{1}{\sqrt{2}},$$

which also can be seen visually on the complex plane. Hence,

$$AB = 6e^{-j\frac{\pi}{4}} = 6(\frac{1}{\sqrt{2}} - j\frac{1}{\sqrt{2}}) = 3\sqrt{2} - j3\sqrt{2}.$$

Likewise,

$$e^{j\frac{3\pi}{4}} = \cos(\frac{3\pi}{4}) + j\sin(\frac{3\pi}{4}) = -\frac{\sqrt{2}}{2} + j\frac{\sqrt{2}}{2},$$

and so

$$\frac{A}{B} = \frac{2}{3}e^{j\frac{3\pi}{4}} = \frac{2}{3}(-\frac{\sqrt{2}}{2} + j\frac{\sqrt{2}}{2}) = -\frac{\sqrt{2}}{3} + j\frac{\sqrt{2}}{3}.$$

A.4   More on Complex Conjugate

For the four different forms of a complex number

$$C = (a, b) = a + jb = |C|e^{j\theta} = |C|\angle\theta,$$

the complex conjugate C^* is

$$C^* \equiv (a, -b) = a - jb = |C|e^{-j\theta} = |C|\angle -\theta.$$

The last two equalities can be verified by using Euler's formula. Try it!

In the rectangular and exponential forms, we obtain the conjugate by changing j to $-j$, whereas in polar form the algebraic sign of the angle is reversed. Thus, for $X = 1 - j1$, we have

$$X^* = 1 - (-j)1 = 1 + j1;$$

whereas, for $Y = e^{j\frac{\pi}{6}}$,

$$Y^* = e^{-j\frac{\pi}{6}};$$

and for $Z = 2\angle - \frac{\pi}{4}$,

$$Z^* = 2\angle\frac{\pi}{4}.$$

These same changes in sign work even for more complicated expressions, such as the pair

$$Q = \frac{j3e^{-j\frac{\pi}{4}}}{(1+j2)2^{(1+j)}}5\angle 30°$$

and

$$Q^* = \frac{-j3e^{j\frac{\pi}{4}}}{(1-j2)2^{(1-j)}}5\angle -30°.$$

Graphically, the conjugate C^* is the point on the complex plane opposite C on the "other side" of the Re axis. For example, $j^* = -j$. Also, notice that the complex conjugate of a complex conjugate gives the original number; that is, $(C^*)^* = C$.

The product of

$$C = a + jb$$

and

$$C^* = a - jb$$

is

$$CC^* = |C|^2,$$

since, using the exponential form, we have $CC^* = (|C|e^{j\theta})(|C|e^{-j\theta}) = |C|^2$ (which, in turn, equals $a^2 + b^2$). Thus, CC^* is always real.

The absolute value $|-2|$ of the real number -2 is 2. The absolute value $|1 + j1|$ of the complex number $1 + j1 = \sqrt{2}e^{j\frac{\pi}{4}}$ is its magnitude $\sqrt{2}$, that is, its distance from the origin of the complex plane. The absolute values of $-1 - j1$ and $1 - j1$ also are $\sqrt{2}$. Since, for an arbitrary C, $CC^* = |C|^2$, it follows that $|C| = \sqrt{CC^*}$ (the positive root, only).

Taking the sum of $C = a + jb$ and $C^* = a - jb$, we get

$$C + C^* = 2a = 2\text{Re}\{C\},$$

yielding

$$\text{Re}\{C\} = \frac{C + C^*}{2}.$$

The difference gives

$$C - C^* = j2b = j2\text{Im}\{C\},$$

implying that

$$\text{Im}\{C\} = \frac{C - C^*}{j2}.$$

Thus, for example,

$$\frac{1 - j2}{j2} + \frac{1 + j2}{-j2} = 2\text{Re}\{\frac{1 - j2}{j2}\} = \text{Re}\{-j(1 - j2)\} = \text{Re}\{-2 - j\} = -2.$$

A.5   Euler's Identity

The function e^x of real variable x has an infinite series[4] expansion

$$e^x = 1 + x + \frac{x^2}{2} + \frac{x^3}{3!} + \frac{x^4}{4!} + \cdots + \frac{x^n}{n!} + \cdots.$$

The complex function e^C of complex variable C is *defined* as

$$e^C \equiv 1 + C + \frac{C^2}{2} + \frac{C^3}{3!} + \frac{C^4}{4!} + \cdots + \frac{C^n}{n!} + \cdots,$$

so that e^x is a special case of e^C (corresponding to $C = x$, obviously).

For $C = j\phi$, where ϕ is real, e^C evaluates to

$$e^{j\phi} = (1 - \frac{\phi^2}{2} + \frac{\phi^4}{4!} - \cdots) + j(\phi - \frac{\phi^3}{3!} + \frac{\phi^5}{5!} - \cdots) = \cos\phi + j\sin\phi,$$

because $j^2 = -1$, $j^3 = -j$, $j^4 = 1$, etc., and

$$1 - \frac{\phi^2}{2} + \frac{\phi^4}{4!} - \cdots$$

---

[4]The Taylor series of e^x, about $x = 0$, is obtained as $\sum_{n=0}^{\infty} \frac{d^n}{dx^n} e^x |_{x=0} \frac{x^n}{n!}$. Note that $\frac{d^n}{dx^n} e^x = e^x$ for any n, leading to the series quoted above. The series converges to e^x for all x.

and

$$\phi - \frac{\phi^3}{3!} + \frac{\phi^5}{5!} - \cdots$$

are the series expansions of $\cos \phi$ and $\sin \phi$, respectively. Therefore, for real ϕ,

$$\text{Re}\{e^{j\phi}\} = \cos \phi, \quad \text{Im}\{e^{j\phi}\} = \sin \phi,$$

and

$$e^{j\phi} = \cos \phi + j \sin \phi.$$

**Euler's
identity**

The last statement is Euler's identity, which we introduced without proof in Section A.3. Throughout this textbook, we make use of both Euler's identity and its conjugate,

$$e^{-j\phi} = \cos \phi - j \sin \phi.$$

Euler's identity and its corollary

$$\text{Re}\{e^{j\phi}\} = \cos \phi$$

are essential for understanding the phasor technique discussed in Chapter 4.

Euler's identity and its conjugate imply that

$$\cos \phi = \frac{e^{j\phi} + e^{-j\phi}}{2}$$

and

$$\sin \phi = \frac{e^{j\phi} - e^{-j\phi}}{j2}.$$

These formulas will be used often; so they, along with Euler's formula, should be committed to memory.

Example A.10
Given that

$$4(e^{j3} + e^{-j3}) = A \cos(\chi),$$

determine A and χ.

Solution   Using the identity

$$\cos \phi = \frac{e^{j\phi} + e^{-j\phi}}{2},$$

we have

$$4(e^{j3} + e^{-j3}) = \frac{A}{2}\left(e^{jX} + e^{-jX}\right).$$

Therefore, $A = 8$ and $\chi = 3$.

Example A.11

Express the function $\frac{e^{j5t} - e^{-j5t}}{2}$ in terms of a sine function.

Solution   Using the identity

$$\sin\phi = \frac{e^{j\phi} - e^{-j\phi}}{j2},$$

we find

$$\frac{e^{j5t} - e^{-j5t}}{2} = j\frac{e^{j5t} - e^{-j5t}}{j2} = j\sin(5t).$$

A.6   Complex-Valued Functions[5]

A function $f(t)$ is said to be real valued if, at each instant t, its numerical value is a real number. For example,

$$f(t) = \cos(2\pi t)$$

is real valued. By contrast, complex-valued functions take on values that are complex numbers. For example,

$$f(t) = e^{j2\pi t} = \cos(2\pi t) + j\sin(2\pi t) = (\cos(2\pi t), \sin(2\pi t))$$

is complex valued, since, for instance, $f(\frac{1}{4}) = e^{j\frac{\pi}{2}} = j$ is a complex number. A complex-valued function can be expressed in exponential, rectangular, and pair forms, as just illustrated.

Although voltage and current signals generated and measured in electrical circuits are always real-valued functions, there are at least two reasons why complex-valued functions are useful in the analysis of real-world signal processing systems. First, real-valued signals can be expressed in terms of complex-valued functions, as, for instance,

$$\cos(\omega t) = \text{Re}\{e^{j\omega t}\}$$

---

[5]This section can be studied after Chapter 4, in preparation for Chapter 5.

and

$$\sin(\omega t) = \text{Im}\{e^{j\omega t}\}.$$

In the case of linear time-invariant circuits, it is mathematically simple to calculate output signals when input signals are complex exponentials. If the true input is a cosine instead, then the output is simply the real part of the calculated complex output. This is the phasor method of Chapter 4.

A second instance where complex-valued signals are useful is in the modeling of certain types of communication or other signal processing systems where a pair of channels use sinusoidal signals that are $90°$ out of phase. The sinusoids can be concisely represented as

$$e^{j\omega t} = (\cos(\omega t),\ \sin(\omega t)).$$

The mathematical representation and manipulation for such systems is tremendously simplified through the use of complex signals.

As a simple example of complex signal representation, consider the derivative operation $\frac{d}{dt}$, which converts the complex function $e^{j\omega t}$ into $j\omega e^{j\omega t}$ (i.e., $j\omega$ times the function itself). This multiplicative scaling by $j\omega$ is simpler to remember and express than are the conversions of $\cos(\omega t)$ into $-\omega \sin(\omega t)$ and $\sin(\omega t)$ into $\omega \cos(\omega t)$. Pictorially, a *differentiator*, which can be implemented by a linear circuit, can be described in terms of the input–output rule

$$e^{j\omega t} \longrightarrow \boxed{\text{Diff}} \longmapsto j\omega e^{j\omega t},$$

where the function on the left is the system input and the function on the right is the system output. This is a concise representation of a much more complicated real-world situation, which can be understood by expressing the input and output functions in pair form (with the help of Euler's identity), given as

$$(\cos(\omega t),\ \sin(\omega t)) \longrightarrow \boxed{\text{Diff}} \longmapsto \omega(-\sin(\omega t),\ \cos(\omega t)).$$

Physically, this corresponds to a *pair* of differentiators, one of which converts a cosine input $\cos(\omega t)$ into $-\omega \sin(\omega t)$ and the second converting a sine input $\sin(\omega t)$ into $\omega \cos(\omega t)$.

Likewise, the pair form of

$$e^{j\omega t} \longrightarrow \boxed{\text{Lowpass}} \longmapsto \frac{1}{1+j\omega} e^{j\omega t}$$

specifies how a *low-pass filter* circuit—another linear system—converts its cosine and sine inputs into outputs (as discussed in detail in Chapter 5).

In summary, even though real-world systems and circuits (such as differentiators or filter circuits) process (i.e., act upon) only real-valued signals, the models of such systems used for analysis and design purposes can be constructed in terms of complex-valued functions whenever it is advantageous to do so. The advantages are made abundantly clear in Chapter 4 and later chapters.

Example A.12
Given a linear filter circuit described by

$$e^{j\omega t} \longrightarrow \boxed{\text{Filter}} \longrightarrow \frac{1}{1+j\omega}e^{j\omega t},$$

what is the filter output $y(t)$ if the input is $f(t) = \cos(2t)$?

Solution   According to the stated input–output relation,

$$e^{j2t} \longrightarrow \boxed{\text{Filter}} \longrightarrow \frac{1}{1+j2}e^{j2t}.$$

Since

$$e^{j2t} = (\cos(2t), \ \sin(2t))$$

and

$$\frac{e^{j2t}}{1+j2} = \frac{e^{j2t}}{\sqrt{5}e^{j\tan^{-1}2}} = \frac{e^{j(2t-\tan^{-1}2)}}{\sqrt{5}}$$

$$= \frac{1}{\sqrt{5}}(\cos(2t - \tan^{-1}2), \sin(2t - \tan^{-1}2)),$$

the input–output relation implies that

$$\cos(2t) \longrightarrow \boxed{\text{Filter}} \longrightarrow \frac{\cos(2t - \tan^{-1}2)}{\sqrt{5}}.$$

Therefore, the input

$$f(t) = \cos(2t)$$

produces the output

$$y(t) = \frac{1}{\sqrt{5}}\cos(2t - \tan^{-1}2).$$

A.7 Functions of Complex Variables[6]

Let us return to Example A.3 from Section A.2. In that example, we were asked to determine the roots of the polynomial $x^2 + x + 1$. That is, we were asked to find the values of x for which

$$x^2 + x + 1 = 0.$$

Figure A.3a shows a plot of

$$f(x) = x^2 + x + 1.$$

Clearly, there is no value of x for which this function passes through zero. But, isn't this function, which is a second-order polynomial, required to have two roots? The answer is no, not if x is a real variable!

An nth-order polynomial is guaranteed to have n roots only if the polynomial is a function of a complex variable.[7] A *complex variable* x is an ordered pair of *real* variables—say, x_r and x_i—much like a complex number C is an ordered pair of real numbers (a, b). We can define

$$x \equiv (x_r, x_i)$$

where x_r and x_i are called the real and *imaginary* parts of x, respectively. Alternatively, we can write $x = x_r + jx_i$.

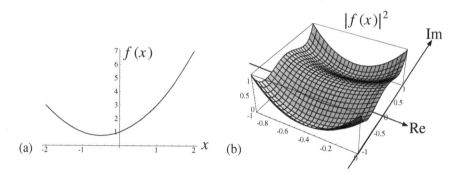

Figure A.3   (a) Plot of function $f(x) = x^2 + x + 1$ of the real variable x, and (b) surface plot of squared magnitude of function $f(x)$ of the complex variable $x = (x_r, x_i) = x_r + jx_i$ over the complex plane.

---

[6]This section can be studied after Chapter 10, in preparation for Chapter 11.

[7]In high school you probably worked with functions of only a real variable. However, in cases where polynomials were factored and discovered to have complex roots, it was implicitly assumed that the variable was complex (whether you were told that or not!).

If x is a complex variable, then the function $f(x)$ can be expanded as

$$f(x) = f((x_r, x_i)) = f(x_r + jx_i) = (x_r + jx_i)^2 + x_r + jx_i + 1.$$

Because this function depends on two real-valued variables, x_r and x_i, we can contemplate plotting it in 3-D, as a surface over the x_r–x_i plane. There is one catch, though: $f(x)$ itself is complex valued. That is, the value of $f(x)$ for a given x is generally a complex number. For example, if $x_r = 0$ and $x_i = 1$, then $f(x) = (j)^2 + j + 1 = j$. How do we plot $f(x)$ if it is complex valued?

The answer is that we must make *two* plots. Our first plot, or sketch, can be the real part of $f(x)$ as a 3-D surface over the x_r–x_i plane, which also happens to be the complex plane. The second sketch can be the surface plot of the imaginary part of $f(x)$. These two sketches together would fully describe $f(x)$. Alternatively, we could sketch the magnitude of $f(x)$ and the angle of $f(x)$, both as surfaces over the x_r–x_i plane, which also would fully describe $f(x)$. Let's go ahead and calculate, and then plot, the *square* of the magnitude of $f(x)$, and check whether there are any values of x for which the squared magnitude hits zero (in which case $f(x)$ itself must, of course, be zero).

The squared magnitude of $f(x)$ is the square of the real part of $f(x)$ plus the square of the imaginary part of $f(x)$. We have

$$f(x) = (x_r + jx_i)^2 + x_r + jx_i + 1 = x_r^2 + j2x_rx_i - x_i^2 + x_r + jx_i + 1.$$

So,

$$\text{Re}\{f(x)\} = x_r^2 - x_i^2 + x_r + 1$$

and

$$\text{Im}\{f(x)\} = 2x_rx_i + x_i = x_i(2x_r + 1),$$

giving

$$|f(x)|^2 = \text{Re}^2\{f(x)\} + \text{Im}^2\{f(x)\} = (x_r^2 - x_i^2 + x_r + 1)^2 + x_i^2(2x_r + 1)^2.$$

A 3-D surface plot of the squared magnitude $|f(x)|^2$ is shown in Figure A.3b. It appears that this plot may hit zero at two locations, but it is difficult see precisely. An examination of the preceding expression for $|f(x)|^2$ can help. This quantity can be zero only if both terms are zero. The second term, $x_i^2(2x_r + 1)^2$, can be zero only if either $x_i = 0$ or $x_r = -1/2$. With the first choice, it is impossible for $x_r^2 - x_i^2 + x_r + 1$ to be zero (remember, x_r is a real variable). However, if

$$x_r = -1/2,$$

then

$$x_r^2 - x_i^2 + x_r + 1 = 0$$

when

$$x_i = \pm\sqrt{3}/2.$$

Thus, the squared magnitude of $f(x)$, which is the plot shown in Figure A.3b, hits zero when

$$x = (x_r, x_i) = (-1/2, \pm\sqrt{3}/2).$$

This agrees with the result we obtained in Section A.2 by using the quadratic formula.

Plots such as the one in Figure A.3b help us visualize functions of complex variables. The key is that a complex variable is a pair of real variables, and so plots of this sort always must be made over the 2-D complex plane. In Chapter 10 we introduce the Laplace transform, which is a function of a complex variable. You may have seen the Laplace transform used as an algebraic tool to assist in solving differential equations. In this course, we will need to have a deeper understanding of the functional behavior of the Laplace transform and we will sometimes want to plot either the magnitude or squared magnitude, using ideas similar to those expressed here. Doing so will give us insight into the frequency response and stability of signal processing circuits.

B

Labs

Lab 1: *RC*-Circuits

Lab 2: Op-Amps

Lab 3: Frequency Response and Fourier Series

Lab 4: Fourier Transform and AM Radio

Lab 5: Sampling, Reconstruction, and Software Radio

Lab 1: RC-Circuits

Over the course of five laboratory sessions you will build a working AM radio receiver that operates on the same principles as commercially available systems. The receiver will consist of relatively simple subsystems examined and discussed in class. We will build the receiver up slowly from its component subsystems, mastering each as it is added.

In Lab 1 you will begin your AM receiver project with a study of RC circuits. Although RC circuits, consisting of resistors R and capacitors C, are simple, they can perform many functions within a receiver circuit. They are often used as audio filters (e.g., the circuitry behind the "bass" and "treble" knobs), and as you will see later in this lab, envelope detectors, with the inclusion of diodes. Lab 1 starts with an exercise that will familiarize you with sources and measuring instruments to be used in the lab, and continues with a study of characteristics of capacitors and steady-state and transient behavior in RC circuits. Then you convert an RC filter circuit into an envelope detector and test it using a synthetic AM signal.

1 Prelab

Prelab exercises are meant to alert you to topics to be covered in each lab session. Make sure to complete them before coming to lab since their solutions often will be essential for understanding/explaining the results of your lab measurements.

(1) For the circuit of Figure 1, calculate the following:

    (a) The RC time constant.

    (b) The voltage $v(t)$ across the capacitor 1 ms after the switch is closed, assuming the capacitor is initially uncharged—express in terms of V_s.

    (c) The initial current $i(0^+)$ that will flow in the circuit.

(2) Suppose you are given an unknown capacitor. Describe an experimental technique that you could use to determine its value.

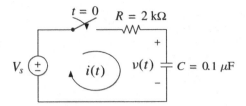

Figure 1 *RC* circuit for prelab exercise

2   Laboratory Exercise

- Equipment: Function generator, oscilloscope, protoboard, RG-58 cables with BNC connectors, Y-cables, and wires.
- Components: 50 Ω resistor, 2 kΩ resistor, 0.1 μF capacitor, and 1N54 diode.

2.1   Generating and measuring waveforms

In the lab you will use a *function generator*, HP 33120A shown in Figure 2a, to produce electrical signal waveforms—sinusoids, square waves, triangle waves, and many others over a wide range of amplitudes and frequencies—applied across various circuit components. The components, e.g., resistors, diodes, and so on, will be interconnected to one another on a *protoboard* (see Figure 2b) and you will use an *oscilloscope*, HP 54645D shown in Figure 2c, to display and measure signal waveforms. For connections you will use coax cables with BNC and/or Y-endings and wires of various lengths (see Figure 2d) as needed. User's manuals for the HP 33120A and HP 54645D can be downloaded from the class web site. The following sequence of exercises should familiarize you with the lab equipment:

Figure 2 Your (a) function generator, (b) protoboard, (c) oscilloscope, (d) cables and wires, and (e) resistor on a protoboard with source and probe connections to (a) and (c).

(1) Place a 50 Ω resistor on your protoboard and use Y-cables to connect the signal generator (output port) and oscilloscope (input port) across the resistor terminals—see Figure 2e or ask your TA for help if you are confused about this step.

(2) Press the power buttons of the scope and the function generator to turn both instruments on.

(3) **(a)** Set the function generator to produce a co-sinusoid output with 5 kHz frequency and 4 V peak-to-peak amplitude:

- Press "Freq," "Enter Number," "5," "kHz."
- Press "Ampl," "Enter Number," "4," "Vpp."

(b) Press "Auto-scale" on the scope and adjust, if needed, vertical and horizontal scales further so that the scope display exhibits the expected co-sinusoid waveform produced by the generator.

(c) Sketch what you see on the scope, labeling carefully horizontal and vertical axes of your graph in terms of appropriate time and voltage markers.

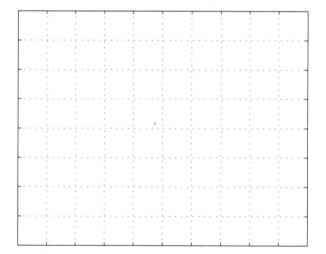

(4) The *default* setting of your function generator is to produce the specified voltage waveform (e.g., 4 V peak-to-peak co-sinusoid as in above measurement) for a 50 Ω resistive load. An alternate setting, known as "High Z", allows you to specify the generator output in terms of *open circuit* voltage. To enter High Z mode you can use the following steps (needed every time after turning on the generator):

- Press "shift" and "enter" to enter the "MENU" mode.
- Press ">" three times until "D sys Menu" is highlighted.
- Press "V" twice until "50 Ohm" is highlighted.

- Press ">" once to select "High Z."
- Press "Enter."

Without modifying the protoboard connections used in step 3, switch the function generator to High Z mode as described above, and then reset the output amplitude of the function generator to 4 V peak-to-peak once again. Observe and make a sketch of the modified scope output.

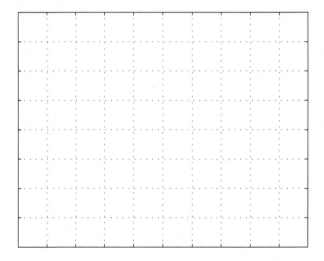

(5) Remove the 50 Ω resistor from the protoboard without modifying the remaining connections to the function generator and scope. Observe and make a sketch of the modified scope output once again.

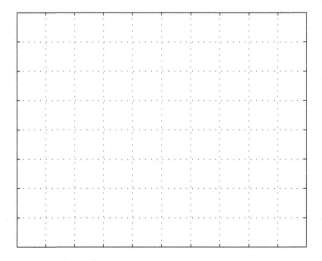

(6) Based on observations from steps 3, 4, and 5, and the explanation for High Z mode provided in step 4, determine the output resistance (Thevenin) of the function generator and the input (load) resistance of the scope. Explain your reasoning.

(7) Measure the period of the co-sinusoidal signal displayed in step 5 by using the time cursors of the oscilloscope and compare the measured period with the inverse of the input frequency.

(8) Repeat step 7 using a square wave with a 50% duty cycle (i.e., the waveform is positive 50% of the time and negative also 50% of the time) instead of the co-sinusoid.

When you power up your function generator it will come up by default in 50 Ohm mode. Remember to switch it to High Z mode if you want to specify open-circuit voltage outputs (as in the next section, and in most experiments).

2.2 Capacitor characteristics

In this section, you will study the characteristics of the capacitor in the network of Figure 3. Once you have built the network on the protoboard, complete these steps:

(1) Apply a 20 kHz co-sinusoidal signal $v_{in}(t)$ measuring 10 V peak-to-peak to the RC circuit as shown in Figure 3. Display voltages $v_{in}(t)$ and $v_C(t)$ using Channels 1 and 2 of the oscilloscope, respectively.

   (a) What is the phase difference in degrees between $v_{in}(t)$ and $v_C(t)$? This can be approximated from inspection, but obtain an accurate measurement using the time cursors on the oscilloscope. Record your measurement.

   (b) Repeat step 1 with a 1 kHz co-sinusoidal wave input.

   (c) Is $v_C(t)$ leading or lagging $v_{in}(t)$? Hint: If one waveform is leading another, "it happens" earlier in time.

(2) Make sure that you change the input frequency back to 20 kHz for this question. Determine the current $i(t)$ in the circuit by first measuring the voltage $v_R(t)$ across the 2 kΩ resistor. You can measure $v_R(t)$ without moving your oscilloscope test probes—instead, invert Channel 2 and add the two channels. By KVL, $v_{in} = v_R + v_C$, so $v_R = v_{in} - v_C$. This sum can be displayed on the oscilloscope using the +/- function button. What is the phase shift of the current relative to $v_{in}(t)$?

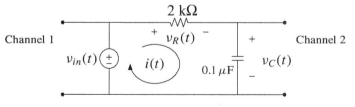

Figure 3 *RC* circuit for laboratory exercises.

2.3 *RC* time constants

To investigate the transient response of your *RC* circuit, change the function generator output to a 100 Hz square wave measuring 1 V peak-to-peak with a 0.5 V DC offset. As in the previous section, Channel 1 should display $v_{in}(t)$, and Channel 2 should display $v_C(t)$.

Across its positive half-period, the square wave approximates a unit step and the capacitor is charged. During the negative half-period, the input voltage is zero and the capacitor discharges through the resistor starting from a peak voltage value v_{max} and with a time constant RC. Thus, the voltage across the capacitor over the negative one-half period is

$$v_C(t) = v_{max} e^{-\frac{t}{RC}},$$

and, when $t = RC$,

$$v_C = v_{max} e^{-1} \approx 0.368 v_{max}.$$

(1) Measure the time constant by determining the amount of time required for the capacitor voltage to decay to 37% of v_{max} on the scope. Use the measurement cursors and record the result, and sketch what you see on the oscilloscope.

(2) Compare this value to your theoretical value for RC (give percent error).

2.4 Frequency response

RC circuits have the very useful property that the co-sinusoidal voltage measured across the capacitor or resistor changes in response to a change in the frequency of the input co-sinusoid. This property provides the basis of the filters (and envelope detectors) commonly found in AM radio receivers. You will investigate and graphically display in this section the frequency dependence of capacitor voltage $v_C(t)$ when the amplitude of $v_{in}(t)$ is held constant but its frequency is varied.

(1) Change the input waveform $v_{in}(t)$ back to a sine wave measuring 10 V peak-to-peak with a zero DC offset. Adjust the frequency f of the co-sinusoidal $v_{in}(t)$ to each of the following, and record the amplitude of the capacitor voltage $v_C(t)$ for each.

Frequency f	v_c amplitude (V)
100 Hz	
500 Hz	
1 kHz	
5 kHz	
10 kHz	
50 kHz	

(2) Plot the amplitude versus frequency data collected above in the box below—
note that the axes are labeled using a logarithmic scale.

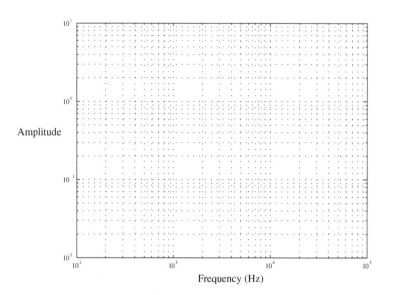

2.5 AM signal and signal envelope

An AM radio signal consists of a high-frequency sinusoid whose amplitude is *modulated*, or scaled, by a message signal. The message signal is the speech or music recorded at the radio station, and the objective of the AM radio receiver is to recover this message signal to present it through a loudspeaker to the listener.

Figure 4 illustrates an AM signal with a co-sinusoidal amplitude modulation. The horizontal axis represents time, and the vertical axis is signal voltage. You can see that the amplitude of the high-frequency sinusoid changes over time. In this case, the message signal is a low-frequency sinusoid. You can imagine the message signal as a low-frequency sinusoid connecting the peaks of the high-frequency sinusoid. Because the message signal connects the peaks of the high-frequency sinusoid, it is also called the AM signal's *envelope*, and a circuit that detects this envelope, producing the message signal, is called an *envelope detector*.

The function generator can create an AM signal for you as follows:

(1) Connect the output of the function generator directly to Channel 1 of the oscilloscope.
(2) Set the function generator to create an AM signal with $f_c = 13$ kHz and 4 V peak-to-peak amplitude modulated by an 880 Hz sine-wave with 80% modulation depth:

  • Create a 13 kHz sine wave measuring 4 V peak-to-peak, no DC offset.

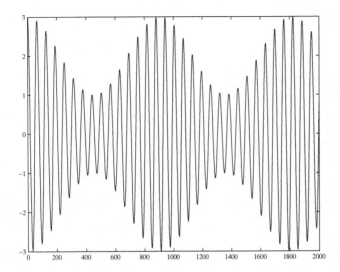

Figure 4   Illustration of AM radio signal.

- Press "Shift," then "AM" to enable amplitude modulation.
- Press "Shift," then "Freq" to set the message signal frequency to 880 Hz.
- Press "Shift," then "Level" to set the modulation depth to 80%. This adjusts the height of the envelope.

(3) Adjust the oscilloscope until the display resembles Figure 4.

2.6   Envelope detector circuit

With a simple modification to your *RC* circuit, you can create an envelope detector that will recover the message signal contained in the function generator's AM signal. Change your circuit to match Figure 5b. Make sure you reconnect the function generator to the circuit, as indicated.

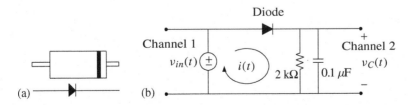

Figure 5   (a) Circuit symbol and physical diagram for a diode, and (b) an "envelope detector" circuit using a diode.

(1) View the voltage across the capacitor on the oscilloscope. What do you see? Use the measurement cursors to determine the frequency of the output. Is this the message signal?

(2) Can you explain how the circuit works? *Hints:* The diode in the circuit will be conducting part of each cycle and non-conducting in the remainder—figure out how the capacitor voltage will vary when the diode is conducting and non-conducting. When is the *RC* time constant in the circuit relevant—when the diode is conducting or not?

Also consider these questions:

    (a) Is the envelope detector circuit linear or nonlinear? Explain.

    (b) Recall that a capacitor is a charge-storage device. When you turn on your radio, the capacitor will store an unknown charge. Will it be necessary to account for this in the design of your receiver circuit? Why or why not?

Important!   Leave your envelope detector assembled on your protoboard! You will need it in future lab sessions.

THE NEXT STEP

Already you have come far: Your circuit can detect the message-containing envelope of an AM signal. In the next lab session, you will explore amplifier circuits. You will build an audio amplifier and connect your envelope detector to it so that you can listen to the message signal you recovered in this session.

Lab 2: Op-Amps

Labs 2 and 3 study operational amplifiers, or "op-amps." Op-amps were originally developed to implement mathematical operations in analog computers, hence their name. Today, op-amps are commonly used to build amplifiers, active filters, and buffers. You will work with many of these circuits in lab. The following table contrasts ideal versus practical characteristics of typical op-amps.

Op-Amp Property	Ideal	in practice
Gain A	∞	very large, $\sim 10^6$ and constrained by supply voltage
R_i	∞	very large, $\sim 10^6 \, \Omega$
R_o	0	very small, $\sim 15 \, \Omega$
Frequency response	flat	gain depends on frequency

The objective of this experiment is to gain experience in the design and construction of operational amplifier circuits. You also will examine some nonideal op-amp behavior.

By the end of Lab 2, you will have designed and built two amplifiers for your radio circuit. Once you have verified that they work, you will connect them to your envelope detector from Lab 1 and listen to synthetic AM signals.

1 Prelab

In the prelab exercises, you will review the analysis of op-amp circuits and design amplifiers for your radio circuit.

(1) Assuming ideal op-amps, derive an expression for the output voltage v_o in the circuit of Figure 1(a).

(2) Write two KCL equations in terms of v_x that relate the output voltage to input voltage for the circuit in Figure 1(b) (you do not need to solve the KCL equations). Simplify your expressions as much as possible.

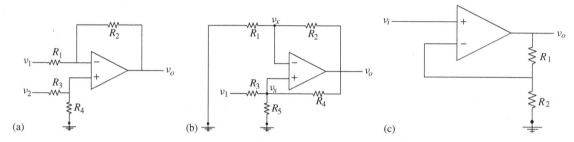

(a) (b) (c)

Figure 1 Circuits for prelab analysis.

(3) Again assuming an ideal op-amp, derive an expression for the output voltage v_o in the circuit in Figure 1(c).

  (a) For a gain $\frac{v_o}{v_i}$ of 2, how must R_1 and R_2 be related?

  (b) For a gain $\frac{v_o}{v_i}$ of 11, how must R_1 and R_2 be related?

  (c) How do you build two amplifiers, one with a gain of two and one with a gain of 11, given four resistors: one 2 kΩ, two 10 kΩ, and one 20 kΩ?

  (d) Using the pin-out diagram in Figure 2 as a reference, draw how you will wire the amplifier with a gain of 11. Draw in resistors and wires as needed.

2 Laboratory Exercise

- Equipment: Function generator, oscilloscope, protoboard, loudspeaker, audio jack, RG-58 cables, Y-cables, and wires.
- Components: two 741 op-amps, one 0.1 μF capacitor, one 33 μF capacitor, one 2 kΩ resistor, two 10 kΩ resistors, one 20 kΩ resistor, and one 100 kΩ resistor.

2.1 Using the 741 op-amp

The 741 op-amp contains a complex circuit, composed of 17 bipolar-junction transistors, 5 diodes, 11 resistors, and a capacitor that provides internal compensation to prevent oscillation. When wiring your circuits, however, you only have to make the usual connections: the inverting and noninverting inputs, the output, and the positive and negative DC supplies. Figure 2 labels and describes the pins.

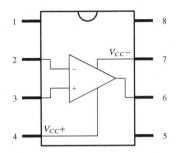

Pins 1,5	offset null correction—not used
2	inverting input $(-)$
3	noninverting input $(+)$
4	$V_{CC}-$ supply set to -12 V (DC)
7	$V_{CC}+$ supply set to -12 V (DC)
6	output voltage v_o
8	not connected

Figure 2 Pin-out diagram for the 741 op-amp.

- The DC supplies must be set carefully to power the op-amp without destroying it.
- The DC sources should be set to $+12$ V for $V_{CC}+$ supply and -12 V for $V_{CC}-$ supply.
- The $V_{CC}+$ and $V_{CC}-$ supplies always stay at the same pins throughout the experiment. Do not switch them or you will destroy the op amp!
- When you are ready to turn the power on, always *turn on the DC supplies first*, then the AC supply.
- When you turn the circuit off, always *turn off the AC first* and then the DC.

2.2 An op-amp amplifier

(1) Configure the op-amp amplifier circuit shown in Figure 3 on your protoboard using component values of $R_1 = 10$ kΩ, $R_2 = 10$ kΩ, $R_f = 100$ kΩ, and $C = 0.1$ μF. *In this and all future circuits using op-amps, connect the circuit ground to supply ground (the black terminal labeled "N" between + and −)—your TA can show you how to do this.* We recognize this circuit as an op-amp integrator in the limit as $R_f \to \infty$, so we will call the circuit with $R_f = 100$ kΩ an integrating amplifier.

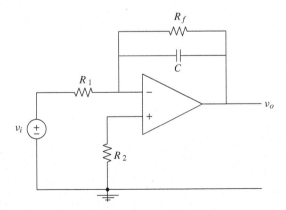

Figure 3 Integrating amplifier.

(2) Apply a 500 Hz, 6 V peak-to-peak square wave to the input (with 50% duty cycle). Sketch the input and output waveforms. Explain the shape of the output waveform, and how it confirms that the circuit acts as an effective integrator.

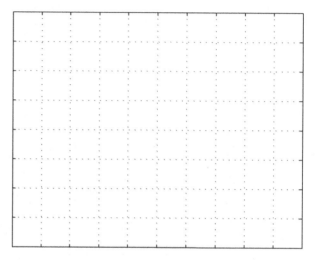

(3) Switch the input waveform from square to co-sinusoid and decrease the frequency to 100 Hz. Sketch the input and output waveforms again.

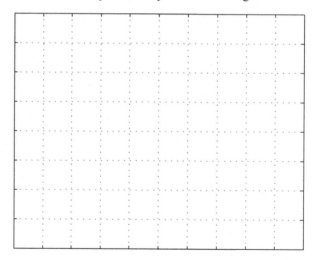

(4) Now slowly increase the input amplitude to 16 V peak-to-peak and describe what happens to the shape of the output waveform. How do you explain what you see? What is the peak-to-peak amplitude of the output waveform when the input amplitude is 16 V peak-to-peak? If you were to increase the amplitude of the input further, would that increase the amplitude of the output? Why or why not?

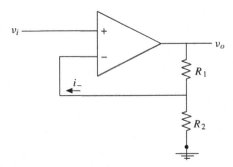

Figure 4   Noninverting amplifier, showing the signal i_-.

2.3   Noninverting amplifiers

Now you will build the two amplifiers you designed in the prelab and connect them to your envelope detector.

(1) First, build the noninverting amplifier with a gain of 11 that you designed in the prelab. Make sure you connect the DC supplies and the bench ground to your circuit before you apply any voltage to the op-amp inputs. Also, place the circuit near the center of your protoboard, to alllow for other circuits to be built on both sides.

(2) Apply a 100 Hz sine wave input v_i with 1 V peak-to-peak amplitude. Measure v_i and v_o with the oscilloscope. Calculate the voltage gain $\frac{v_o}{v_i}$. How does this compare with the theoretical value? What could account for the difference?

(3) Turn off the AC input, and then turn off the DC supplies. Without modifying the noninverting amplifier you just built, build the noninverting amplifier with a gain of two that you designed in the prelab. Place your new circuit to the right side of your protoboard, allowing for some extra space in between the two amplifiers.

(4) Turn on the DC supplies and connect the AC input (same as in step 2) to the new amplifier (gain 2). Observe again the voltage gain $\frac{v_o}{v_i}$. Does it agree with the design value?

(5) Remove the AC input. Place your envelope detector circuit (from Lab 1) in between the two amplifier circuits as shown in Figure 5. Note that a 33 μF capacitor is included in the envelope detector circuit to remove the DC component from the input signal of the last amplifier (recall that a capacitor acts as an open circuit at DC). Have your TA check your circuit before you continue.

The three-stage circuit that you just built will be part of your radio receiver in Lab 4. As you saw in Lab 1, the envelope detector stage will recover the message from an AM signal. In order for it to work, however, the voltage of input signal must be sufficiently high to turn the diode on and off—hence the first amplifier. The second amplifier, which follows the envelope detector, increases the voltage of the recovered message signal in order to drive a loudspeaker. It also provides a buffer between the

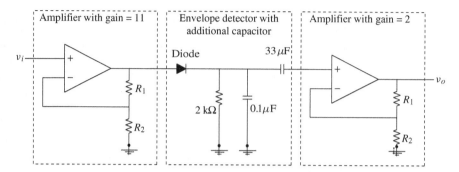

Figure 5 Three-stage circuit for radio.

envelope detector and the loudspeaker, preventing the loudspeaker from changing the time constant of the tuned envelope detector.

Until now, we have been displaying signals on our oscilloscopes. However, signals in the audio frequency range also can be "displayed" acoustically using loudspeakers. We next will use a corner of our protoboard, an audio jack, and a pair of speakers to *listen* to a number of signal waveforms:

(1) Connect the function generator output to the speaker input (via the protoboard and audio jack) and generate and listen to an 880 Hz signal. Repeat for a 13 kHz signal. Describe what you hear in each case—in what ways do the 880 Hz and 13 kHz audio signals sound different to your ear?

(2) To test your three-stage circuit, create an AM signal with the function generator (almost like in Lab 1):

- Set the function generator to create a 13 kHz sine wave measuring 0.2 V peak-to-peak, no DC offset.
- Press "Shift," then "AM" to enable amplitude modulation.
- Press "Shift," then "Freq" to set the message-signal frequency to 880 Hz.
- Press "Shift," then "Level" to set the modulation depth to 80%. This adjusts the height of the envelope.
- Listen to the AM signal on the loudspeaker.

Turn on the DC supplies, and connect the function generator to the input of your three-stage circuit from step 5 above. Connect the output of the same circuit to the oscilloscope. Sketch what you see on the oscilloscope display. How does it compare to the waveform you obtained in the last part of Lab 1? On the function generator, press "Shift," then "Freq" and sweep the message-signal frequency from 100 Hz and 2000 Hz. Describe what you see.

(3) Disconnect the oscilloscope and replace it with a loudspeaker. What do you hear? Sweep the message signal frequency from 100 Hz to 2000 Hz again. Describe how the sound changes as the frequency is swept, do you recognize the 880 Hz sound from step 1 above?

Important! Leave your circuit on the protoboard but return the audiojack to your TA.

The next step

In the next laboratory session, you will continue working with op-amps to build an active filter that will remove noise from your radio signal. You also will use the active filter to study the Fourier series.

Lab 3: Frequency Response and Fourier Series

In this lab you will build an active bandpass filter circuit with two capacitors and an op-amp, and examine the response of the circuit to periodic inputs over a range of frequencies. The same circuit will be used in Lab 4 in your AM radio receiver system as an intermediate frequency (IF) filter, but in this current lab our main focus will be on the frequency response $H(\omega)$ of the filter circuit and the Fourier series of its periodic input and output signals. In particular we want to examine and gain experience with the response of linear time-invariant circuits to periodic inputs.

1 Prelab

(1) Determine the compact-form trigonometric Fourier series of the square wave signal, $f(t)$, with a period T and amplitude A shown in Figure 1. That is, find c_n and θ_n such that

$$f(t) = \frac{c_0}{2} + \sum_{n=1}^{\infty} c_n \cos(n\omega_o t + \theta_n),$$

where $\omega_o = \frac{2\pi}{T}$. Notice $\frac{c_0}{2} = 0$. How could you have determined that without any calculation?

(2) Consider the circuit in Figure 2 where $v_i(t)$ is a co-sinusoidal input with some radian frequency ω.

   (a) What is the phasor gain $\frac{V_o}{V_i}$ in the circuit as $\omega \to 0$? (Hint: How does one model a capacitor at DC—open or short?)

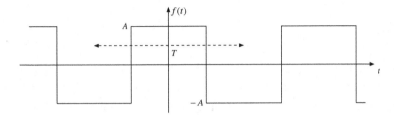

Figure 1 Square wave signal for prelab.

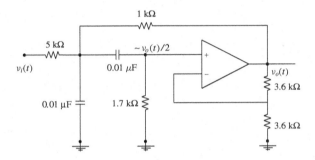

Figure 2 Circuit for analysis in prelab and lab.

(b) What is the gain $\frac{V_o}{V_i}$ as $\omega \to \infty$? (Hint: Think of capacitor behavior in the limit as $\omega \to \infty$.)

(c) In view of the answers to parts (a) and (b), and the fact that the circuit is second-order (it contains two energy storage elements), try to guess what kind of filter the system frequency response $H(\omega) \equiv \frac{V_o}{V_i}$ implements— low-pass, high-pass, or band-pass? The amplitude response $|H(\omega)|$ of the circuit will be measured in the lab.

2 Laboratory Exercise

- Equipment: Function generator, oscilloscope, protoboard, and wires.
- Components: 741 op-amp, two 0.01 μF capacitors, one 1 kΩ resistor, one 1.7 kΩ resistor, two 3.6 kΩ resistors, and one 5 kΩ resistor.

2.1 Frequency response H(ω)

The *frequency response* $H(\omega)$ of a linear and dissipative time-invariant circuit contains all of the key information about the circuit that is needed to predict the circuit response to arbitrary inputs. Its magnitude $|H(\omega)|$ is known as the *amplitude response* and $\angle H(\omega)$ usually is referred to as the *phase response*. In this section, you will construct

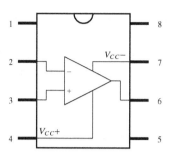

Pins 1,5	offset null correction—not used
2	inverting input (−)
3	noninverting input (+)
4	$V_{CC}-$ supply set to −12 V (DC)
7	$V_{CC}+$ supply set to −12 V (DC)
6	output voltage v_o
8	not connected

Figure 3   Pin-out diagram for the 741 op-amp.

an active bandpass filter circuit and measure its amplitude response over the frequency range 1–20 kHz. Do the following:

(1) Construct the circuit shown in Figure 2 on your protoboard. For now, do not connect it to the three-stage circuit from Lab 2. Remember the rules for wiring the 741 op-amp, which are repeated in Figure 3.

(2) Turn on the DC supplies, then connect a 1 kHz sine wave with amplitude 1 V peak-to-peak as the AC input $v_i(t)$. Display $v_i(t)$ and $v_o(t)$ on different channels of the oscilloscope and verify the waveforms.

(3) Increase the function generator frequency from 1 kHz to 20 kHz in 1 kHz increments. At each frequency enter the magnitude of the phasor voltage gain $\frac{V_o}{V_i}$ in the graph shown below—$\frac{V_o}{V_i}$ is the system frequency response $H(\omega)$ and its magnitude $\frac{|V_o|}{|V_i|}$ is the system amplitude response $|H(\omega)|$.

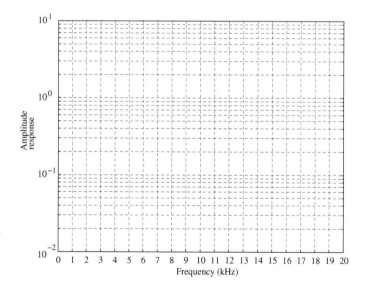

(4) The *center frequency* $\omega_o = 2\pi f_o$ of a band-pass response $H(\omega)$ is defined as the frequency at which the amplitude response $|H(\omega)|$ is maximized. What is the center frequency f_o in kHz units, and what is the maximum amplitude response $|H(\omega_o)|$ of the circuit? Estimate f_o and $|H(\omega_o)|$ from your graph as accurately as you can.

(5) The *3 dB cutoff frequencies* $\omega_u = 2\pi f_u$ and $\omega_l = 2\pi f_l$ are the frequencies above and below $\omega_o = 2\pi f_o$ at which the amplitude response $|H(\omega)|$ is $\frac{1}{\sqrt{2}} \approx 0.707$ times its maximum value $|H(\omega_o)|$. The same frequencies also are known as *half-power* cutoff frequencies since at frequencies ω_u and ω_l the output signal power is one half the value at ω_o, assuming equal input powers at all three frequencies. Mark these cutoff frequencies as f_u and f_l on the horizontal axis of your amplitude response plot shown above.

(6) Determine the *3 dB bandwidth* $B \equiv f_u - f_l$ of the bandpass filter in kHz units and calculate the *quality factor* of the circuit defined as $Q \equiv \frac{\omega_o}{2\pi B} = \frac{f_o}{B}$.

2.2 Displaying Fourier coefficients

In order to display the Fourier coefficients of a periodic signal on the oscilloscope, we can use the built in FFT function.[1] On the "FFT screen" of your scope the horizontal axis will represent frequency ω (like in the frequency response plot of the last section) normalized by 2π, and you will see narrow spikes positioned at values $f = \frac{\omega}{2\pi}$ equal to harmonic frequencies $n\frac{\omega_o}{2\pi}$ of the periodic input signal; spike amplitudes will be proportional to compact Fourier coefficients $c_n = 2|F_n|$ in dB. The next set of instructions tells you how to view the single Fourier coefficient (namely c_1) of a co-sinusoidal signal:

(1) Connect Channel 1 of your oscilloscope to the input terminal of your circuit from the previous section. Connect the output of the circuit to Channel 2.

(2) Set the function generator to a 1 kHz sinusoid with amplitude 500 mV peak-to-peak.

(3) Set the oscilloscope to compute the Fourier transform:

    (a) Press "$+/-$."

    (b) Turn on Function 2.

    (c) Press "Menu."

    (d) Press "Operation" until FFT appears.

    (e) Set the Operand to Channel 1.

    (f) Press "FFT Menu" (default window setting "Hanning" should be used).

    (g) Adjust the time/div knob to set the frequency span to 24.4 kHz and the center frequency to 12.2 kHz.

---

[1] *FFT* stands for *fast Fourier transform* and it is a method for calculating Fourier transforms with sampled signal data—see Example 9.26 in Section 9.3 of Chapter 9 to understand the relation of windowed Fourier transforms to Fourier coefficients.

(4) Observe the output signal's Fourier coefficient by setting the Operand to Channel 2 under the function menu. How does the FFT display change as you sweep the frequency of the input from 1 kHz to 20 kHz? Describe what you see and briefly explain what is happening.

2.3   Fourier coefficients of a square wave

Now you will introduce a periodic signal with a more interesting set of Fourier coefficients—a square-wave:

(1) Change the function generator setting to create a 15 kHz square wave with amplitude 0.5 V peak-to peak as the input to your circuit.

(2) Display the FFT of the square wave at the filter input by setting the Operand to Channel 1. Keeping in mind your result for Problem 1 of the Prelab, describe what you see on the screen.

(3) Display both the input and output in the time domain on the scope. Explain the output taking into account the frequency response shape of your filter circuit.

(4) Repeat for a 10 kHz square wave and a 5 kHz square wave, explaining the outputs.

(5) Still using the 5 kHz square wave, set the oscilloscope to display the FFT of the output. What do you see on the scope? How does the Fourier domain representation confirm your explanation of the output in the time domain?

(6) Repeat for a 10 kHz square wave and a 15 kHz square wave.

Important!   Leave your active filter assembled on your protoboard! You will need it in the next lab session.

The next step

The active filter is the last component you will build for the AM radio receiver. In Lab 4, you will combine your components from Labs 1 through 3 to create a working AM radio receiver. The frequency-domain techniques you learned this week will be essential to following the AM signal through each stage of the receiver system.

Lab 4: Fourier Transform and AM Radio

In Lab 4, you finally will connect all of your receiver components and tune in an AM radio broadcast. You will follow the radio signal through the entire system, from antenna to loudspeaker, in both the time domain and the frequency domain.

1 Prelab

You should prepare for this lab by reviewing Sections 8.3 and 8.4 of the text on AM detection and superhetrodyne receivers, familiarizing yourself with your own receiver design shown in Figure 1, and answering the following questions:

(1) Suppose you want to tune your AM receiver in the lab to WDWS, an AM station broadcasting from Champaign-Urbana with a carrier frequency $f_c = \frac{\omega_c}{2\pi} = 1400$ kHz. Given that the IF (Intermediate Frequency) of your receiver is $f_{IF} = \frac{\omega_{IF}}{2\pi} = 13$ kHz, to which LO (Local Oscillator) frequency $f_{LO} = \frac{\omega_{LO}}{2\pi}$ should you set the function generator input of the mixer in your receiver (see Figure 1) to be able to detect WDWS? (Hint: There are two possible answers; give both of them.)

(2) Repeat 1, supposing you wish to listen to WILL, an AM broadcast at $f_c = 580$ kHz.

(3) Sketch the amplitude response curve $|H_{IF}(\omega)|$ of an *ideal* IF filter designed for an IF of $f_{IF} = \frac{\omega_{IF}}{2\pi} = 13$ kHz and a filter bandwidth of 10 kHz. Label the axes of your plot carefully using appropriate tick marks and units.

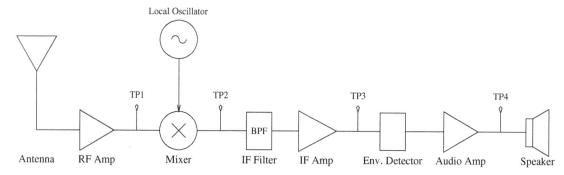

Figure 1 Superheterodyne AM receiver.

2   Laboratory Exercise

- Equipment: Function generator, oscilloscope, protoboard, and wires.
- Components: Three-stage circuit from Lab 2, band-pass filter from Lab 3, RF amplifier, and mixer.

2.1   Fourier transform

Your scope is capable of displaying the Fourier transform of its input signal. We have already used this feature in Lab 3 in "observing" the Fourier coefficients of periodic signal inputs. In this section we will learn how to examine nonperiodic inputs in the frequency domain.

(1) No circuit is used for this part of the laboratory. Connect the function generator's output to Channel 1 of the oscilloscope.

(2) Set the function generator to create a 1 kHz square wave with amplitude 1 V peak-to-peak. Turn on burst mode by pressing "Shift" then "Burst." In this mode the generator outputs a single rectangular pulse, (i.e., $rect(t/\tau)$), which is repeated at a 100 Hz rate. Sketch the burst signal output that you see on the scope display. Confirm that pulses are generated with at a 100 Hz rate (that is, 100 pulses per second, or a pulse every 10 ms).

(3) Set the oscilloscope to display the magnitude of the Fourier transform of a segment of the input signal (containing a single rectangle):

- Press "+/−."
- Turn on Function 2.
- Press "Menu."
- Press "Operation" until FFT appears.
- Set the Operand to Channel 1.
- Press "FFT Menu."

- Set the window to "Hanning."[1]
- Set the units/div to 10 dB and the ref level to 0 dBV.
- Set the frequency span to 24.4 kHz and the center frequency to 12.2 kHz.
- The oscilloscope now displays $|F(\omega)|$ in dB units, defined as $20\log |F(\omega)|$, where $F(\omega)$ is the Fourier transform of a windowed (see the footnote regarding Hanning window) segment of the oscilloscope input $f(t)$. Since $20\log|F(\omega)| = 10\log|F(\omega)|^2$, the scope display is also related to the *energy spectrum* $|F(\omega)|^2$ of the segment of $f(t)$. We will refer to the display as the *frequency spectrum* of the input. Note that the spectrum is shown only over positive frequencies $f = \frac{\omega}{2\pi}$ within the frequency band specified in the last step above.

(4) Sketch the frequency spectrum $|F(\omega)|$ (in dB) of the oscilloscope input and compare with theory. (What is the Fourier Transform of a *rect* function?)

(5) Change the input to a sinc pulse, and again sketch the frequency spectrum and compare with theory. To do this, press "Shift," "Arb," select "sinc," and press "enter." Set the time/div to 500 μs and the frequency span to 48.8 kHz.

2.2   AM signal in frequency domain

Amplitude Modulation (AM) is a communications scheme that allows many different message signals to be transmitted in adjacent band-pass channels. Before the message signal is multiplied by the high-frequency sinusoidal carrier, a DC component is added so that the voltage of the message signal always is positive. This makes it easy to recover the message signal from the envelope of the carrier. In Lab 1, you synthesized an AM signal with the function generator and then displayed it on the oscilloscope in the time domain. Let's see how the same AM signal looks in the frequency domain. (Hint: Use the modulation property to interpret what you will see on scope display!)

(1) No circuit is used for this part of the laboratory. Connect the function generator's output to Channel 1 on the oscilloscope.

(2) Set the function generator to create an AM signal with $f_c = 13$ kHz and 1 V peak-to-peak amplitude modulated by an 880 Hz sine-wave with 80% modulation depth.

- Create a 13 kHz sine wave measuring 1 V peak-to-peak, with no DC offset.
- Press "Shift," then "AM" to enable amplitude modulation.
- Press "Shift," then "<," then "∨," to select the shape of the message signal. You can select a sine, square, triangle, or arbitrary waveform by pressing ">." For this part of the lab, select a sine waveform.

---

[1] As a result of this choice, the incoming signal $f(t)$ is effectively multiplied with $w_H(t) \equiv$ rect$(\frac{t}{T})\cos^2(\frac{\pi t}{T})$ prior to Fourier transformation. This procedure limits the length of the Fourier transform segment to duration T of the window function $w_H(t)$. Alternative window functions such as $w_B(t) = $ rect$(\frac{t}{T})$ give rise to frequency spectra $|F(\omega)|$ with different resolution and sidelobe details.

- Press "Enter" to save the change and turn off the menu.
- Press "Shift," then "Freq" to set the message signal frequency to 880 Hz.
- Press "Shift," then "Level" to set the modulation depth to 80%. This adjusts the DC component added to the message signal before modulation.

(3) Set your oscilloscope to display the frequency spectrum of the input:

- Press "+/−."
- Turn on Function 2.
- Press "Menu," then "Operation" until FFT appears.
- Set the time/div to 2 ms.
- Set the units/div to 10 dB and the ref level to 0 dBV.
- Set the center frequency to 12.2 kHz.
- Set the frequency span to 24.4 kHz.

(4) Sketch the AM signal and its frequency spectrum using the oscilloscope display and explain what you see in terms of the modulation property of the Fourier transform.

(5) Change the shape of the message signal in the modulation menu from SINE to SQUARE. Do not change the shape or frequency of the 13 kHz carrier signal. Explain what you see. (Hint: See Example 9.26 in Section 9.3 of Chapter 9.)

2.3 AM radio receiver

The most popular AM communications receiver is the superheterodyne receiver, which was developed for greater sensitivity and selectivity. A block diagram for the superheterodyne receiver is shown in Figure 1.

The antenna, RF amplifier, and frequency mixer all rely on electrical components *not covered* in this textbook, but their effects on the incoming signal should be familiar. You built the remaining components of the circuit in Labs 1 through 3. In this section, you will combine all of the components to tune in an AM radio broadcast and follow the signal from the antenna to the loudspeaker.

Perform the following steps:

(1) Connect the RF amplifier and frequency mixer modules provided. Make sure the labels on the boxes are right side up and facing you. Connect a Y-cable to the input stage of the RF amplifier—this will serve as a crude AM antenna. Use the DC source to power the modules by connecting +12 V to the blue terminals, −12 V to the purple terminals, and the bench ground ("N") to the black terminals.

(2) Connect the output of the frequency mixer to the input of your band-pass filter from Lab 3 (which will take the place of the ideal IF filter discussed in Problem 3 in the Prelab). Connect the supply ground ("N") to your circuit's signal ground, and connect the DC supplies to your op-amp.

(3) Connect the output of the band-pass filter to the input of your three-stage circuit from Lab 2. Make sure all of the signal grounds are connected, and connect the DC supplies to your op-amps.

(4) Turn on the DC supplies. Then connect the function generator to the local-oscillator input on the frequency mixer. Tune in AM 1400 by selecting an appropriate mixing frequency for the local oscillator as described below. (Hint: Look back to Problem 1 in the Prelab.)

(5) Now you will follow the processing of the received RF signal into an audible signal by displaying the time and frequency domain of the four test-point signals on the oscilloscope. The test points are described below.

When displaying the signals from each of the test points, it will be your task to select an appropriate time scale for the time-domain waveform and center frequency and frequency span for its Fourier spectrum. If you select inappropriate numbers, all you will see on the oscilloscope is noise. You may ask the TA for hints, but give your choice some thought and discuss it with your partner before displaying the signal.

Test point 1: RF amplifier   The antenna picks up the AM radio broadcast we are trying to tune, along with many other unwanted signals. The antenna signal is then passed through an *LC* resonator called the preselector, which acts as a crude band-pass filter. The preselector removes some unwanted frequencies from the antenna signal while preserving the frequencies associated with the desired AM radio broadcast. The output of the preselector is amplified by the RF amplifier stage.

Connect an oscilloscope probe to the RF amplifier output (TP1). Sketch the waveform in both the time and frequency domains. Set the oscilloscope for the FFT as before, but set the center frequency to 1.2 MHz and the frequency span to 2.4 MHz. You should see at least one AM signal in the frequency domain, but nothing discernible in the time domain. Touching your finger to one of the antenna terminals will amplify the signal considerably.

Test point 2: frequency mixer   The mixer multiplies the RF signal with the signal coming from the local oscillator, which is tuned so that all subsequent processing is independent of the radio broadcast's carrier frequency. In most commercial receivers, the mixer is tuned to produce an Intermediate Frequency (IF) of 455 kHz. For our purposes, an IF of 13 kHz will suffice. The band-pass filter and envelope detector you built are suited for an IF of 13 kHz.

The function generator will be our local oscillator, abbreviated as LO. To produce an IF of 13 kHz, the LO must be set at a frequency 13 kHz above or below the carrier signal of the station we are trying to tune. We will be tuning AM 1400, so set the LO to be a 1413 kHz or a 1387 kHz sine wave with an amplitude of 400 mV peak-to-peak. Make sure that the function generator's AM feature is turned off.

You may have to adjust LO frequency slightly to better tune the station. Once the station is tuned, probe TP2 and sketch the output in both the time domain and the frequency domain. At TP2, the signal should be very noisy, just as it was at TP1.

Test point 3: IF filter and amplifier   The IF filter is used to select the signal centered on the IF frequency and to reject all other signals. Receivers employing higher IFs typically include ceramic IF filters that operate on a piezoelectric principle. Although small and inexpensive, ceramic filters can have very sharply tuned responses, which are needed with large IF compared to AM bandwidth. With lower IF, such as 13 kHz, a sharply tuned response is not necessary, and thus even the low-Q op-amp-based band-pass filter from Lab 3 that we are using is more than adequate.

Depending on the AM signal strength from the antenna, and the noise level, you may find it necessary to add gain to the IF amplifier. Feel free to experiment with different gain values (remember the design equation from Lab 2) to get an output that can be demodulated by the envelope detector. An IF gain of about 30 is not uncommon.

Probe the signal at TP3. Sketch the time waveform and the frequency spectrum.

Test point 4: envelope detector and audio amplifier   The envelope detector then recovers the message signal from the IF signal. Probe TP4 and sketch the time-domain waveform and frequency spectrum.

At this point, if all stages of the AM radio behave as expected, hook up the output of the audio amplifier to the speaker. Do you hear what you expect to hear?

The next step

Now that your AM radio receiver is complete, you will turn your attention to a "software radio" implementation of the same design using digital signal processing in the next lab.

Lab 5: Sampling, Reconstruction, and Software Radio

Until this point, your study of signals and systems has concerned only the continuous-time case,[1] which dominated the early history of signal processing. About 50 years ago, however, the development of the modern computer generated research interest in digital signal processing (DSP), a type of discrete-time signal processing. Although hardware limitations made most real-time DSP impractical at the time, the continuing maturation of the computer has been matched with a continuing expansion of DSP. Much of that expansion has been into areas previously dominated by continuous-time systems: our telephone network, medical imaging, music recordings, wireless communications, and many more.

You do not need to worry whether the time and effort you have invested in studying continuous-time systems will be wasted because of the growth of DSP— digital systems are practically always hybrids of analog and digital sub-systems. Furthermore, many DSP systems are linear and time-invariant, meaning that the same analysis techniques apply, although with some modifications. In this lab, you will explore some of the parallels between continuous-time systems and DSP with a "software radio" designed to the same specifications as the receiver circuit you developed on your protoboard.

---

[1] The term "continuous time" is used generically to refer to signals that are functions of a continuous independent variable. Often that variable represents time, but it may instead represent distance, etc. "Discrete time" is used in the same way.

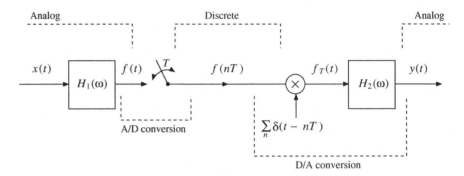

Figure 1 A conceptual system that samples a bandlimited continuous-time signal $f(t)$ and reconstructs a continuous-time output $y(t)$. An A/D converter generates the samples $f(nT)$ of its analog input $f(t)$. An ideal D/A converter generates a signal $f_T(t) = \sum_n f(nT)\delta(t - nT)$ from samples $f(nT)$ and low-pass filters $f_T(t)$ with $H_2(\omega)$ to produce an analog $y(t)$. In a practical D/A converter the impulse train $\sum_n \delta(t - nT)$ is replaced by a practical pulse train $\sum_n p(t - nT)$ such that $p(t) * h_2(t)$ is a close approximation of a delayed $\operatorname{sinc}(\frac{\pi}{T}t)$.

1 Prelab

Our software radio is typical of many DSP systems in that both the available input and required output are continuous-time signals. The conversion of a continuous-time input signal to a discrete-time signal is called *sampling* (or A/D conversion), and the conversion of a discrete-time signal to a continuous-time output signal is called *reconstruction* (or D/A conversion). As discussed in class, *samples* $f(nT)$ of a *bandlimited* analog signal $f(t)$ can be used to reconstruct $f(t)$ exactly when the *sampling interval* T and *signal bandwidth* $\Omega = 2\pi B$ satisfy the *Nyquist criterion* $T < \frac{1}{2B}$.

This is illustrated by the hypothetical system shown in Figure 1, where the analog signal $f(t)$ defined at the output stage of a low-pass filter $H_1(\omega)$ has a bandwidth $\Omega = 2\pi B$ limited by the bandwidth $\Omega_1 = 2\pi B_1$ of the filter. The A/D converter extracts the samples $f(nT)$ from $f(t)$ with a sampling interval of T. D/A conversion of samples $f(nT)$ into an analog signal $y(t)$ can be envisioned as low-pass filtering of a hypothetical signal $f_T(t) = \sum_n f(nT)\delta(t - nT)$ using the filter $H_2(\omega)$. With an appropriate choice of $H_2(\omega)$, the system output $y(t)$ will be identical to $f(t)$ so long as $T < \frac{1}{2B_1}$. The reason for that easily can be appreciated after comparing the Fourier transforms $F(\omega)$ and $F_T(\omega)$ of signals $f(t)$ and $f_T(t)$ with the help of Figure 2.

The following prelab exercises concern the system shown in Figure 1. Assume that $T = \frac{1}{44100}$ s (i.e., the sampling frequency is $T^{-1} = 44100$ Hz) and signal $x(t)$ has a Fourier transform $X(\omega)$ shown in Figure 3.

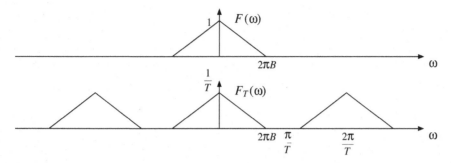

Figure 2 An example comparing the Fourier transforms of signals $f(t)$ and $f_T(t)$ defined in Figure 1. Since for $|\omega| < \frac{\pi}{T}$ the two Fourier transforms have the same shape, low-pass filtering of $f_T(t)$ yields the original analog signal $f(t)$. $F_T(\omega)$ is constructed as a superposition of replicas of $\frac{F(\omega)}{T}$ shifted in ω by all integer multiples of $\frac{2\pi}{T}$ (see item 25 in Table 7.2 in the text).

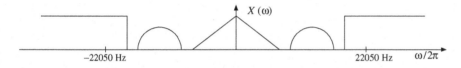

Figure 3 Fourier transform of $x(t)$ for prelab questions.

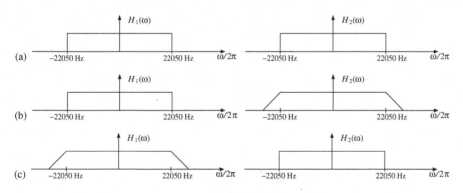

Figure 4 Filter frequency responses for the prelab problems.

(1) For frequency responses $H_1(\omega)$ and $H_2(\omega)$ in Figure 4a, sketch the Fourier transforms of $f(t)$, $f_T(t) = \sum_n f(nT)\delta(t - nT)$, and $y(t)$. Is $y(t)$ a perfect reconstruction of $x(t)$? Is $y(t)$ a perfect reconstruction of $f(t)$?

(2) Now, consider an ideal $H_1(\omega)$ but a nonideal $H_2(\omega)$ given by Figure 4b. The signals $f(t)$ and $f_T(t)$ are unchanged, but sketch the Fourier transform of the new $y(t)$. Is $y(t)$ a perfect reconstruction of $f(t)$?

(3) Now suppose a nonideal $H_1(\omega)$ and an ideal $H_2(\omega)$ given by Figure 4c. Sketch the Fourier transform of $f(t)$, $f_T(t)$, and $y(t)$. Is $y(t)$ a perfect reconstruction of $f(t)$?

(4) Discuss the role of filters $H_1(\omega)$ and $H_2(\omega)$ in the system examined above. In what ways do they impact the system output?

2 Laboratory Exercise

- Equipment: Function generator, PC with a sound card, and wires.
- Components: antenna, RF amplifier, mixer, stereo jack, and 33 μF capacitor.
- Software: MATLAB (R2006b) and softRx.m.

2.1 Sampling and reconstruction

In this section, you will observe a real system much like the one you studied in the prelab exercises. The lab also will illustrate the phenomenon of aliasing, which occurs when the analog input is undersampled.

(1) Connect the function generator's output to the "mic in" jack at the back of the computer at your lab station. Use BNC "Y" cables to bring the signal from the lab equipment to the protoboard. Use stereo jacks and stereo cables to run the signal from the protoboard to the computer. Ask your TA if you need help.

(2) In Windows, go to Start, Control Panel, and click Sound, Speech, and Audio Devices. Next click on Adjust the system volume. Under the volume tab turn the Device Volume all the way up. Then in the Audio tab click on Volume... under Sound Recording. Make sure the Microphone box is selected and turn the Microphone level all the way down. Also make sure the "wave" setting is all the way up. Leave this window open, as it may be necessary to change the gain if the signal is too strong or too weak during the lab exercises.

(3) In Windows, go to Start, Programs, MATLAB, R2006b and click MATLAB R2006b (MATLAB versions are continously updated. If you do not see MATLAB R2006b, start the latest version installed on the computer). At MATLAB command prompt, type "softRx." This will launch the graphical user interface (GUI) for the software AM receiver shown in Figure 5.

(4) Select "Output = Unprocessed Input" from the pull-down options menu near the top left corner of the softRx GUI, in which case the analog "mic in" signal is sampled and reconstructed as an analog signal as depicted in Figure 6 and explained in the caption. Note the resemblance of the diagram to the system studied in the prelab. The low-pass filters shown in Figure 6 are part of the sound card and serve the same role as those in the prelab. *Important:* You should use the zoom buttons at the top left corner of the GUI to zoom in on the waveform if it looks like a solid line on the screen.

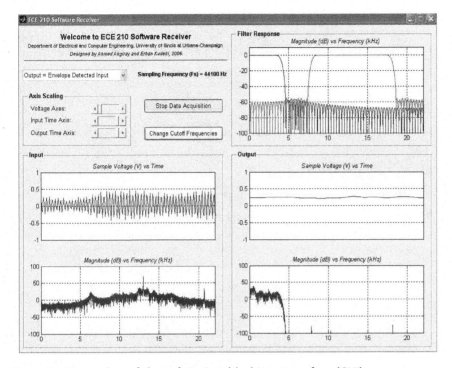

Figure 5   Screenshot of the softRx Graphical User Interface (GUI).

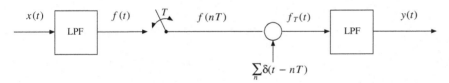

Figure 6   Block diagram of the system implemented by softRx when the "Output = Unprocessed Input" option is selected. Note that no signal processing is applied to samples $f(nT)$ before D/A conversion.

(5) Set the function generator to produce a 5 kHz sinusoid with amplitude 500 mV peak-to-peak. Connect the 33 μF capacitor in parallel with the stereo jack (adding this capacitor will prevent saturation of the input port; you can remove it to see its effect). Click the "Start Data Acquisition" button in the GUI to begin sampling and reconstruction. Describe what you see in the time and frequency plots of the output in the GUI.

(6) Slowly sweep the input frequency up to 19 kHz. Does the output look like a perfect reconstruction of the input signal?

(7) Now, slowly sweep the input frequency from 19 to 25 kHz, passing through 22.05 kHz. What is the significance of the frequency 22.05 kHz? What happens in the time and frequency domain? Does this look like a perfect reconstruction or do we have an *aliased* component at the output? Which component(s) in the system could be improved to reduce the aliasing effect? Note that the answer to the last question is not the sampling frequency of the sound card or the sound card itself.

(8) Finally, sweep the input frequency from 25 to 30 kHz. What do you observe?

2.2   Digital filtering

In this section, you will examine the digital filter option of softRx.

(1) Select the "Output = IF Filtered Input" setting in the GUI. The samples $f(nT)$ will be processed by the digital IF filter shown in Figure 7 to generate samples $g(nT)$. You will be asked to enter two cutoff frequencies for the filter. Accept the default values for this step.

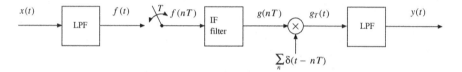

Figure 7   Block diagram for the Software Receiver when the "Output = IF Filtered Input" option is selected.

(2) Note that the filter frequency-response depicted in the GUI depends on the cutoff frequency inputs. Change them, after clicking "Change Cutoff Frequencies," from $f_{cl} = 8$ and $f_{cu} = 18$ kHz to 10 and 16 kHz, respectively, and observe the new filter response.

(3) Switch the input waveform from the sinusoid to a square wave with $f = 1$ kHz and describe what you see at the filter output.

2.3   Receiving synthetic AM

Now you will generate a synthetic AM signal and process it with softRx.

(1) Set the function generator to create an AM signal:

- Set the function generator to create a 13 kHz sine wave measuring 500 mV peak-to-peak, no DC offset.
- Press "Shift," then "AM" to enable amplitude modulation.
- Press "Shift," then "<," then ∨ to select the shape of the message signal. You can select a sine, square, triangle, or arbitrary waveform by pressing ">." For this part of the lab, select a sine waveform.

- Press "Enter" to save the change and turn off the menu.
- Press "Shift," then "Freq" to set the message signal frequency to 880 Hz.
- Press "Shift," then "Level" to set the modulation depth to 80%. This adjusts the DC component added to the message signal before modulation.

(2) Select the "Output = Unprocessed Input" option. Sketch and describe what you see on the GUI output.

(3) Select the "Output = IF Filtered Input" option. Sketch and describe what you see.

(4) Select the "Output = Envelope Detected Input" option. This introduces an envelope detector to the system as shown in Figure 8. You will be asked to enter three cut-off frequencies, two for the bandpass filter and one for the lowpass filter of the envelope detector. Sketch and describe what you see on the output panel. Overall, what does this system accomplish?

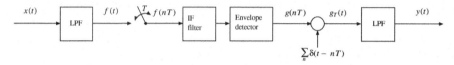

Figure 8 Block diagram for the Software Receiver when "Output = Envelope Detected Input" is selected.

2.4 Receiving broadcast AM

In this last section, you will mix a broadcast AM signal to an IF of 13 kHz, as in Lab 4. You then will process the IF signal with softRx and eventually listen to its detected version.

(1) Connect the antenna, RF amplifier, and frequency mixer modules provided (as in Lab 4). Make sure the labels on the boxes are right side up and facing you. Use the DC source to power the modules by connecting $+12$ V to the blue terminals, -12 V to the purple terminals, and the bench ground ("N") to the black terminals. Subsequently turn on the DC supply.

(2) Use the function generator as your local oscillator (LO). Generate a sinusoid with 400 mV peak-to-peak amplitude (do not forget to turn off the AM feature of the function generator). What frequencies could be used as the LO frequency to tune to AM 1400 kHz via a 13 kHz IF? Select one of those frequencies and connect the function generator to the LO input of the frequency mixer.

(3) Connect the output of the frequency mixer to "mic in" on the sound card (the 33 μF capacitor should be taken out). Select the "Output = Unprocessed Input" option in the GUI. Sketch and describe what you see on the output panel.

(4) Select the "Output = IF Filtered Input" option. Sketch and describe the output that you see.

(5) Repeat 4 with the "Output = Envelope Detected Input" option. How do the signals in each of these settings resemble those you observed in the continuous-time case during Lab 4?

(6) Connect a pair of loudspeakers to "speaker out" to listen to AM 1400. Is your software radio working as expected? Note: You might need to move the antenna or change how you are holding the antenna in order to increase signal quality.

(7) Explore sound quality changes as you vary the following parameters:

   (a) Increase the IF frequency to 16 kHz by varying the LO frequency.

   (b) Set $f_{cl} = 14$ kHz and $f_{cu} = 18$ kHz.

   (c) Set the lowpass filter cutoff frequency, f_c, to 2 kHz.

The End

Congratulations on completing the ECE 210 labs! Over these five labs you have learned and applied the most important principles of continuous-time signals and systems and explored their parallels in discrete-time signals and systems. Advanced coursework in ECE will require you to apply these principles again and again. You are well prepared!

C

Further Reading

1) D. M. Bressoud, *A Radical Approach to Real Analysis*, 2nd ed. Washington, D. C: The Mathematical Association of America, 2006.
2) J. W. Brown and R. V. Churchill, *Complex Variables and Applications*, 7th ed. New York: McGraw Hill, 2003.
3) B. P. Lathi, *Linear Systems and Signals*, 2nd ed. Oxford, England: Oxford Univ. Press, 2004.
4) M. J. Lighthill, *Introduction to Fourier Analysis and Generalized Functions*. Cambridge, England: Cambridge University Press, 1964.
5) J. H. McClellan, R. W. Schafer, and M. A. Yoder, *Signal Processing First*. Upper Saddle River, NJ: Prentice Hall, 2003.
6) P. J. Nahin, *Science of Radio*, 2nd ed. New York: Springer-Verlag, 2001.
7) J. W. Nilsson and S. A. Riedel, *Electric Circuits*, 8th ed. Upper Saddle River, NJ: Prentice-Hall, 2008.
8) A. Papoulis, *The Fourier Integral and its Applications*. New York: McGraw Hill, 1962.
9) J. G. Proakis and D. G. Manolakis, *Digital Signal Processing: Principles, Algorithms, and Applications,* 4th ed. Upper Saddle River, NJ: Prentice Hall, 2006.
10) A. V. Oppenheim, R. W. Schafer, and J. R. Buck, *Discrete-Time Signal Processing*, 2nd ed. Upper Saddle River, NJ: Prentice Hall, 1999.
11) D. Rutledge, *The Electronics of Radio*. Cambridge, England: Cambridge University Press, 1999.
12) A. S. Sedra and K. C. Smith, *Microelectronic Circuits*, 5th ed. Oxford, England: Oxford University Press, 2003.

Index